Analytical Techniques of Celestial Mechanics

Victor A. Brumberg

Analytical Techniques of Celestial Mechanics

Springer

Professor Dr. Victor A. Brumberg
Russian Academy of Sciences, Institute of Applied Astronomy
197042 St. Petersburg, Russia

With 5 Figures

ISBN-13: 978-3-642-79456-8 e-ISBN-13: 978-3-642-79454-4
DOI: 10.1007/978-3-642-79454-4

CIP data applied for

Softcover reprint of the hardcover 1st edition 1995

SPIN: 10480927 55/3144 - 5 4 3 2 1 0 - Printed on acid-free paper

Preface

The aim of this book is to describe contemporary analytical and semi-analytical techniques for solving typical celestial-mechanics problems. The word "techniques" is used here as a term intermediate between "methods" and "recipes". One often conceives some method of solution of a problem as a general mathematical tool, while not taking much care with its computational realization. On the other hand, the word "recipes" may nowadays be understood in the sense of the well-known book *Numerical Recipes* (Press et al., 1992), where it means both algorithms and their specific program realization in Fortran, C or Pascal. Analytical recipes imply the use of some general or specialized computer algebra system (CAS). The number of different CAS currently employed in celestial mechanics is too large to specify just a few of the most preferable systems. Besides, it seems reasonable not to mix the essence of any algorithm with its particular program implementation. For these reasons, the analytical techniques of this book are to be regarded as algorithms to be implemented in different ways depending on the hardware and software available.

The book was preceded by *Analytical Algorithms of Celestial Mechanics* by the same author, published in Russian in 1980. In spite of there being much common between these books, the present one is in fact a new monograph. The main recipe of the 1980 book was to use a specialized (Poisson-series) processor for celestial-mechanics problems to represent the solution by very long Poisson series. The main recipe of this book is to use a universal CAS (Maple, Mathematica, etc.) to develop specialized celestial-mechanics software and to represent the solution in a more compact form, based, for instance, on elliptic-function theory. Now it becomes obvious that the increase of the number of terms in Poisson series is by no means an efficient tool for improving the accuracy of analytical theories of celestial mechanics, and one should try to construct theories in a compact form.

The book deals with algorithms of typical problems without detailed application to the motion of specific celestial objects. These algorithms embrace the Keplerian processor in various forms, the expansion of the disturbing function in the cases of planets and satellites, methods of separation of slow and fast variables in elements and coordinates, the construction of general planetary theory, etc. The book differs from existing textbooks in its concentration on computational aspects of analytical celestial mechanics and may be considered an introductory treatise on a new branch of celestial mechanics known as the algorithmization of celestial mechanics. The author hopes that the book will be useful for all those who deal with the practical application of celestial-mechanics techniques, not necessarily in just the field of celestial mechanics.

This book was begun during the author's stay as visiting professor at the National Astronomical Observatory (NAO), Tokyo in 1992–93 and was completed during his stay as a Humboldt-award researcher at the University of Tübingen in 1993–94. The author is sincerely grateful to T. Fukushima, H. Kinoshita and Y. Kozai (NAO), H. Ruder and M. Soffel (University of Tübingen) and the Humboldt Foundation for these extremely favourable possibilities for scientific research. Particular thanks are due to Dmitry Brumberg as the author's tutor and main assistant in TEX. The author is indebted to Springer-Verlag for invaluable help in producing this book, and most particularly to Mr. Mark Seymour for improving the English text.

Contents

Introduction

This book deals with a domain of celestial mechanics which one might call the computation of motion. The subject of this domain is to calculate analytically or numerically the coordinates and velocities of celestial bodies based on some physical model of motion and qualitative mathematical picture of motion. To achieve the greatest efficiency one therefore combines general methods of applied mathematics with specialized celestial-mechanics techniques. In the past, celestial mechanics was a stimulus for developing and testing various tools of applied analysis. This interrelation of celestial mechanics and applied mathematics still exists nowadays, although not so explicitly as in the past.

The book is mainly concerned with analytical and semi-analytical techniques of celestial mechanics. There was and still is much controversy surrounding the question what kind of technique — numerical or analytical — should be preferred. In fact, this controversy means very little. One should combine both kinds of technique, according to a specific problem and the aim of the research (a particular or general solution, the interval of validity of the solution, accuracy considerations, etc.). If we apply numerical techniques to celestial mechanics, they have no important peculiarities compared with the general techniques of applied mathematics. It is only in the domain of analytical techniques that celestial mechanics has created and continues to create its own state of the art. The arsenal of analytical techniques of celestial mechanics is in a process of permanent improvement. This becomes particularly evident when we consider the excellent treatise by Brown and Shook (1933), which presents the computational techniques of classical celestial mechanics.

Compared with classical methods modern techniques have become logically simpler and at the same time more universal. The tendency to economize the extent of the calculation necessarily underlying all classical methods involved many transformations of variables that were not caused by the physical character of the problem under consideration. Thanks to computer facilities it has become possible nowadays to remove such transformations and make clearer the essence of the techniques themselves. Moreover, now there is no need to take into account all the particular features of a problem; instead it might be reasonable to develop techniques adapted to solving a broader range of problems. Even the form of published papers in celestial mechanics has been drastically changed. Pages with the endless formulae of classical papers have been replaced by a short description of the technique employed, followed by an indication of the possible use of a relevant program. Lengthy ephemerides of the major planets and the Moon published in astronomical almanacs have given way to a more compact presentation with the aid of series in Chebyshev polynomials.

Along with traditional expansions in series in powers of small parameters, more and more frequently now one applies iteration techniques. Of course, iteration techniques are rather time-consuming and demand much computer memory to store the intermediate results. In addition, all the operations should also be performed with a higher accuracy, as needed in techniques based on series expansions. Nevertheless, these difficulties atone for the compact formulation of iteration techniques and their programming convenience. Along with linear, Picard-type iterations in contemporary celestial mechanics one often meets quadratic, Newton-type iterations that enable one to attain the necessary accuracy for a smaller number of iterations (but with more complicated calculations for each step of the iteration process).

One further characteristic feature of modern techniques is the use of universal and specialized computer algebra systems (CAS). The most cumbersome operations in constructing analytical and semi-analytical theories of motion are done now with the aid of advanced celestial-mechanics software, including, in particular, Keplerian processors and generators of special functions employed in celestial mechanics. Contemporary analytical celestial mechanics mainly deals not with "hand-made" formulae but the computer derivation of formulae and the computer implementation of specific techniques.

Lastly, one should note the tendency towards the more compact representation of analytical theories of motion. One way to achieve this is to construct in analytical form the corrections to some nominal numerical solution corresponding to some specific initial data and values of parameters. This form retains the advantages of a purely analytical theory with a much smaller number of terms in the resulting series. Another way is to try to construct closed-form theories. This is usually achieved under some simplifications of a given problem. Finally, one can try to use some special functions (for example, elliptic functions) which may advantageously replace the traditional trigonometric functions.

All these tendencies are reflected in the techniques considered in the present book.

So that readers can see what they may gain by reading this book we describe below its key features.

The book starts in Chapter 1 with a description of Poisson series and computer operations on them. In fact, Poisson series (PS), i.e. multi-trigonometric series with multi-power series coefficients, occur not only in celestial mechanics but in many other domains of physics and mechanics. The peculiarity of such series in celestial mechanics is that they may consist of many thousands of terms involving many difficulties in their manipulation. After characterizing the most widespread forms of Poisson series as well as general purpose and specialized computer algebra systems for manipulating them (PS processors) (Section 1.1), we discuss typical analytical operations with Poisson series, also including a very important aspect of their numerical evaluation (Section 1.2). Then we consider a particular problem of packing indices by means of their numeration and give a new numeration formula for trigonometric indices. This technique may be of importance when in the permanent compromise between saving computer time or computer memory one is more interested in the latter.

Chapter 2 might be the most important for practical applications. It contains the solution of the two-body problem, which serves as a basis for solving much more complicated problems. But this is not merely a collection of formulae, as in most textbooks on this subject, but rather a collection of recipes (the Keplerian processor) for producing such formulae and their different generalizations by computer (Sections 2.1, 2.2 and 2.4). This enables one to use them immediately in perturbation theory (Section 2.6). Along with this we give some new generalizations of classical expansions of elliptic motion obtained by "hand" manipulation (Section 2.3). We have also developed a new technique based on elliptic-function expansions and presented up until now only in the journal literature (Section 2.5).

Hence, the content of the first two chapters gives an insight into the software facilities needed for contemporary celestial mechanics. No doubt in the future such facilities for the PS and Keplerian processors will be implemented in a specialized software package. But nowadays it seems premature to standardize such a programming package.

The next two chapters are devoted to polynomial and quasi-polynomial systems of differential equations in celestial mechanics. Examples of such systems are furnished by N-planet problems (Section 3.1) and perturbed two-body problems in Kustaanheimo–Stiefel variables (Section 3.2) or in Hansen coordinates by replacing the trigonometric functions of Euler angles by Euler parameters (Section 3.3). We describe a general algorithm to solve any polynomial system in the form of Taylor expansions in powers of time (Section 4.1). To deal with the secular polynomial system occurring, for example, in the N-planet problem one may use the solution in a purely trigonometric form based on Birkhoff normalization (Section 4.2).

In Chapter 5, which deals with the satellite disturbing function, after consideration of the well-known algorithm for the recurrence determination of terms in the spherical-function expansion (Sections 5.1 and 5.2) we give a new expression for the most general expansion of the satellite disturbing function due to the non-sphericity of the central body, taking into account the inclinations of the equator of the figure and the equator of rotation with respect to an arbitrary reference plane (Section 5.3). To overcome difficulties related to highly eccentric orbits one may replace the mean anomaly and orbital eccentricity of a satellite by its elliptic anomaly and Jacobi nome of elliptic functions introduced in Section 2.5. In addition, we describe a technique based on contact elements instead of osculating ones to take into account the effects of rotation of the reference system (Section 5.4).

Chapter 6, which is devoted to the expansion of the planetary disturbing function, demonstrates the progress achieved in modern celestial mechanics. Without computer technology it seemed infeasible to derive and employ the expansions for an arbitrary reference plane. Now such expansions have become routine, and we describe several algorithms for constructing them, even embracing the case of nearly intersecting orbits (Sections 6.1 and 6.2). For the problem of the motion of the major planets we prefer not to fix any definite form of expansion but to develop a rather general expansion which may be easily transformed into a desirable form by the facilities of the PS and Keplerian processors. To overcome difficulties related to the large ratios of the semi-major axes one may again introduce elliptic-function expansions (Section 6.3).

It might be possible to standardize some of the expansions of Chapters 5 and 6 and to give their program implementation. But we prefer not to impose such standards. Based on the techniques of Chapters 5 and 6 it is easy to produce by means of computer algebra any expansion which seems to be the most satisfactory for a specific problem.

We begin Chapter 7 by reproducing the well-known iteration versions of classical methods for determing perturbations in rectangular coordinates in combination with some useful techniques for taking the relevant quadratures (Section 7.1). As preliminary steps to a general planetary theory (GPT), the main celestial mechanics problem considered in this book, we apply iteration techniques to construct the intermediate orbits in the N-planet-problem (Section 7.2) and the two-body-problem expansions in powers of eccentricity and inclination variables (Section 7.3). The latter expansions may be regarded as one more (polynomial) form of the Keplerian processor.

Chapter 8 has a compilative character, dealing with the perturbation-in-elements methods of Krylov and Bogoljubov (Section 8.1), von Zeipel (Sections 8.2 and 8.3) and Kolmogorov and Arnold (Section 8.3). The main idea is to show that the facilities of PS processors enable one to generate these methods by means of computer algebra without any difficulties. One may ask the obvious question why the most popular methods today, those of Hori and Deprit, have been ignored. The evident answer is that simply owing to their widespread use there was no need to include them here. The more sophisticated answer is that we present the perturbation-in-elements methods only to show how they realize the principle of separation of fast and slow angular variables. The von Zeipel method, for example, is perfectly adapted to this purpose, relying on nothing but this principle. Frequent objections about the complexity of this method owing to the requirement of series inversion are not too serious because this inversion is now a standard operation of PS processors. In this regard we should like to note once again that there are many alternatives in celestial mechanics such as numerical or analytical investigations, orbital elements or rectangular coordinates, explicitly given right-hand members of differential equations or a Hamiltonian of canonical systems, etc. It is not possible, and in fact it is not necessary, to choose a definite alternative once and for ever. In this book preference is given to analytical techniques in rectangular coordinates, while the use of other tools is not avoided when they seem to be more satisfactory.

Chapter 9 deals just with the separation of fast and slow variables in rectangular coordinates. In classical methods of celestial mechanics one uses the general solution of the equations of variations for the two-body problem involving secular terms. Instead, one may not solve the equations of variations but only transform them into Jordan form (Section 9.1) with immediate application to the perturbed two-body problem (Section 9.2). A new transformation will exclude the fast variables and result in a secular system describing the evolution of slow variables (Section 9.3).

This technique is developed in detail in Chapter 10 applied to GPT, the general planetary theory without secular terms. The construction of GPT represents yet another problem of celestial mechanics regarded not so long ago as absolutely intractable. Now it becomes evident that all the technical difficulties associated with GPT may be overcome. The first-order theory with respect to the planetary masses might even be presented in closed form using elliptic-type quadratures. The application of elliptic-function expansions reduces the length of the resulting GPT series, making them quite manageable.

Chapter 11, the last one, presents outlines of the application of GPT techniques to construct the GPT-compatible theories of motion of the minor planets and the Moon. This is evidently not a very urgent problem in these pragmatic times, but in every branch of science one should always have in mind problems for future development.

The book may be read without any preliminary knowledge of celestial mechanics, but for better comprehension of the domain of applicability of the proposed techniques it is desirable for the reader to be familiar with some standard textbooks, such as that by Brouwer and Clemence (1961).

In some chapters one may meet different designations for the same quantities or, on the other hand, the same symbol may designate different quantities. This is due to the convenience of some formulations and cannot lead to misunderstanding for an attentive reader. Note the designation for the Kronecker symbol:

$$\delta_{ij} = \begin{cases} 0, & i \neq j, \\ 1, & i = j, \end{cases}$$

and the use of the generalized factorial (Pochhammer symbol):

$$(\alpha)_k = \alpha(\alpha+1)\dots(\alpha+k-1), \qquad (\alpha)_0 = 1.$$

E(z) denotes the integral part of the real number z and

$$\sum_{j=1}^{N}{}^{(i)}$$

means summation over all integer values of j from 1 to N excluding $j = i$. Besides the frequent use of Jacobi elliptic functions the following special functions of general mathematics will be met in this book:

$F(\alpha, \beta, \gamma, z)$,	hypergeometric function (series or polynomial);
$J_k(z)$,	Bessel coefficient;
$L_k^\lambda(z)$,	Laguerre polynomial;
$\Phi(\alpha, \gamma, z)$,	confluent hypergeometric function;
$C_k^\lambda(z)$,	Gegenbauer polynomial;
$P_k(z)$,	Legendre polynomial
$P_{kj}(z)$,	Ferrers' associated Legendre function;
$Q_\lambda^\mu(z)$,	associated Legendre function of the second kind;
$\Gamma(z)$,	gamma-function;
$F(\varphi, k)$,	incomplete elliptic integral of the first kind;
$K(k)$,	complete elliptic integral of the first kind;
$E(\varphi, k)$,	incomplete elliptic integral of the second kind;
$E(k)$,	complete elliptic integral of the second kind;
$Z(u)$,	Jacobi zeta function.

The following special functions of celestial mechanics will also be encountered:

$F_{kjl}(i)$,	Kaula inclination function;
$A_{kjl}(i)$,	generalized inclination function;
$X_k^{n,m}(e)$,	Hansen coefficient;
$B_k^{n,m}(q)$,	elliptic Hansen coefficient;
$P_k^j(n, m)$,	Newcomb operator;
$b_n^{(k)}(\alpha)$,	Laplace coefficient;
$c_n^{(k)}(a, a')$,	symmetrical Laplace coefficient.

The definitions of these functions are given in the text.

1 The Poisson-Series Processor

1.1 Universal CAS and Poisson-Series Processors

The main mathematical objects to be dealt with in celestial mechanics are so-called Poisson series. They are used to construct analytical theories of the motion of celestial bodies or at least to find corrections in a literal form to some approximate basic solution corresponding to specific initial numerical values and numerical values of the parameters. A series of power and trigonometric terms of the form

$$S = \sum S_{i_1 \ldots i_m}^{j_1 \ldots j_n} x_1^{i_1} \ldots x_m^{i_m} \begin{matrix} \cos \\ \sin \end{matrix} (j_1 y_1 + \ldots + j_n y_n) , \tag{1.1.1}$$

with $x = (x_1, \ldots, x_m)$ and $y = (y_1, \ldots, y_n)$ being power and trigonometric variables, respectively, is called an $(m,\, n)$ Poisson series. The summation in (1.1.1) is performed over all integer (positive and negative) values of indices $i = (i_1, \ldots, i_m)$ and $j = (j_1, \ldots, j_n)$. Sometimes it is appropriate to consider the exponential form of the Poisson series,

$$S = \sum S_{i_1 \ldots i_m}^{j_1 \ldots j_n} x_1^{i_1} \ldots x_m^{i_m} \exp \mathrm{i}(j_1 y_1 + \ldots + j_n y_n) . \tag{1.1.2}$$

The summation, subtraction and multiplication of two Poisson series as well as the differentiation and integration of a Poisson series with respect to any variable will again result in a Poisson series (the only exception is integration with respect to a trigonometric variable that does not occur in every term of the series). This makes Poisson series convenient for formal manipulation in a literal form.

The coefficients S_i^j of series (1.1.1) are represented by rational or floating-point numbers. For series (1.1.2) these two types of arithmetics should be combined with the complex-number form, although in many celestial-mechanics problems the coefficients of (1.1.2) are real numbers (or purely imaginary numbers).

Along with Poisson series (1.1.1) one often has to deal with series where the trigonometric variables are linear functions of an argument t and the coefficients are polynomials in t,

$$y_k = \omega_k t + y_k^{(0)} , \qquad S_i^j = S_i^j(t) . \tag{1.1.3}$$

The summation, subtraction and multiplication of these series are analogous to the corresponding operations with Poisson series. But the differentiation and integration of these series with respect to t involve the appearance of a quantity

$$(j\omega) = j_1\omega_1 + \ldots + j_n\omega_n$$

as a factor or a divisor. If the frequencies $\omega = (\omega_1, \ldots, \omega_n)$ are given numerically then the form of the series will be preserved. But if it is necessary to retain the quantity $(j\omega)$ analytically then the initial form of the series will be changed, resulting in so-called echeloned Poisson series. Each term of such a series is characterized not only by the power monomial $(x)^i$ and the trigonometric argument (jy) but by the associated set of quantities $(k\omega)$.

In the more complicated case the frequencies ω depend on the power variables x. If the power variables x may be regarded as small parameters and the frequencies ω represent series in powers of x then the factors and divisors $(j\omega)$ are combined with the monomials $(x)^i$, retaining form (1.1.1) of the Poisson series. But if the reciprocals of the divisors $(j\omega)$ cannot be expanded in powers of x then one meets more complicated series compared with (1.1.1), namely a trigonometric series with n angular variables y, the coefficients being rational functions of action variables x. Such series occur, for instance, in the analytical theory of motion of the Moon, where the expansions of the reciprocals of the divisors $(j\omega)$ in powers of x are not very convenient owing to their slow convergence (Bec et al., 1973).

In solving celestial-mechanics problems with the aid of elliptic functions it may be appropriate to replace these functions by their trigonometric expansions, the coefficients being the rational functions of the Jacobi nomes (and, possibly, of some other quantities). In this case we again meet the trigonometric series in y with rational coefficients of x. Jacobi nomes are usually small enough to enable one to expand the rational coefficients in powers of the nomes, restoring the original form (1.1.1) of the Poisson series. It is also possible to replace Jacobi elliptic functions by the ratios of theta functions. However, since these involve ratios of two Poisson series this is not used in analytical calculations.

Yet the trigonometric series in y with rational coefficients of x are not so widely applied in celestial mechanics as the Poisson series. In what follows we shall deal mainly with Poisson series in form (1.1.1) or (1.1.2) with possible modification (1.1.3) provided that this modification does not violate the form of the Poisson series.

Specialized celestial-mechanics software packages dealing with Poisson series are called PS processors. There are many PS processors that have demonstrated their efficiency in solving actual celestial-mechanics problems. For illustration, it is sufficient to indicate such PS processors as MAO (Rom, 1970; Deprit and Miller, 1989), ESP (Rom, 1971), TRIGMAN (Jefferys, 1970, 1972), CAMAL (Barton et al., 1970a,b), SPASM (Cherniack, 1973),

Broucke's processor (Broucke and Carthwaite, 1969; Broucke, 1989), Dasenbrock's processor (Dasenbrock, 1973), PARSEC (Richardson, 1989), UPP (Brumberg et al., 1989) and many others. Some of these enable one to manipulate a wider class of objects than just Poisson series. For example, the ESP processor deals with echeloned series, the CAMAL and SPASM processors may operate with rational functions as well, etc. A survey of PS processors may be found in the paper by Henrard (1989). A lot of celestial-mechanics problems have been solved with the aid of PS processors. Many examples of the application of PS processors were presented at IAU Colloquium No. 109 "Applications of Computer Technology to Dynamical Astronomy" (the proceedings of this colloquium were published in *Celestial Mechanics*, vol. 45, Nos. 1–3, 1988/89). Amongst the most impressive applications of PS processors to practical celestial-mechanics problems note the theory of motion of Enceladus and Dione (Jefferys and Ries, 1975), the theory of the Galilean satellites (Lieske, 1977), the Earth-satellite theory (Alfriend et al., 1977a,b) and the solution of Hill's problem in lunar theory (Henrard, 1978). The currently most advanced semi-analytical theories of the major planets VSOP87 (Bretagnon and Francou, 1988) and the Moon ELP2000-85 (Chapront-Touzé and Chapront, 1988) have been constructed using the specialized software package dealing with Poisson series developed at the Bureau des Longitudes.

PS processors are always being improved in efficiency and universality. In particular, their universality means their independence (so far as possible) of a specific computer, the type of Poisson series (forms (1.1.1) and (1.1.2) with possible modifications) and their internal representation. The numbers of power and trigonometric variables, the ranges of the associated indices, the degree of the polynomials $S_i^j(t)$ in modified form (1.1.3) and the arithmetic structure of the coefficients should remain unfixed, to be prescribed by the user for any specific problem. Such universality of PS processors is ensured by their hierarchical architecture. On the lowest level there are subroutines to deal with the coefficients of the terms and their indices. On the next level there are subroutines to manipulate separately the terms of the series. The typical operations with Poisson series such as sum, multiplication, differentiation, integration, etc. form the third level. Finally, on the highest level there are subroutines to perform quite complicated analytical operations described in the next section. Just these subroutines make PS processors an effective tool for solving the equations of celestial mechanics in the literal form.

One may ask if it is really necessary to have specialized celestial-mechanics software systems when a lot of general CAS (computer algebra systems) are available. Indeed, many of the problems of celestial mechanics have already been solved using general mathematics software tools. Note, in particular, the derivation of two-body Taylor expansions with FORMAC (Sconzo et al., 1965; LeSchack and Sconzo, 1968), the determination of short-period luni-solar perturbations in Earth satellite theory with APL (Fischer, 1972), the construction of the inclination functions with SYMBAL (Campbell, 1972), the application of MACSYMA for Poisson-series manipulation (Fateman, 1974), the development of MALISIAS (MATHEMATICA Lie simplifications in artificial-satellite theory) with MATHEMATICA (Palacián, 1992), etc. Brief comparative analysis of contemporary CAS such as REDUCE, MACSYMA, SCRATCHPAD, MATHEMATICA and MAPLE from the point of view of celestial-mechanics applications is given by Laskar (1990). However, simply as a result of the general mathematical character of such CAS their use for solving large celestial-mechanics problems involving very long Poisson series turns out to be less efficient compared to the application of PS processors. The situation may change in the future because the growth of the number of terms in Poisson series is no longer regarded as an effective tool for improving celestial-mechanics theories of motion, and one tries to construct solutions in a more compact form applying simplification techniques (Deprit and Ferrer, 1989) or the new types of expansions stemming, for instance, from the theory of elliptic functions. General CAS are of great use for such more compact expansions (Klioner, 1992). It seems that the most effective tool is to develop specialized celestial-mechanics systems for symbolic manipulation with a specific class of objects like Poisson series on the basis of some general CAS.

1.2 Operations with Poisson Series

We describe here the most important operations of the highest level together with their conventional names. Realization of these operations is of course different in various PS processors.

1. The binomial expansion. If A denotes a Poisson series it may often be necessary to find the binomial expansion $(1+A)^r$ retaining the terms from A^k to A^l inclusive. As a result we obtain the Poisson series

$$B = \sum_{s=k}^{l} \frac{(-1)^s(-r)_s}{(1)_s} A^s \tag{1.2.1}$$

denoted by

$$B = \mathsf{BINOME}(A, r, k, l)\,.$$

The complete binomial expansion evidently corresponds to $k = 0$ and $l = \infty$. With $l = \infty$ the summation in (1.2.1) is performed up to the first negligible term.

All series operations in PS processors have a formal character. The smallness of a term of a series is defined as a rule by the analytical order of the term. This order is calculated by means of the formula

$$N = \sum_{r=1}^{m} p_r i_r \,, \tag{1.2.2}$$

where p_r are the weight functions of the power variables x_r, and i_r are the corresponding indices. Even in the case of purely trigonometric series it may be useful to introduce a fictitious power variable to characterize the analytical order of smallness of the terms of the series. Besides this analytical characteristic, the smallness of a term of the series may be determined by the numerical value of its coefficients. Sometimes it occurs that the coefficients of the series terms do not decrease although the numerical estimation of the x variables permits one to draw a conclusion about the smallness of these terms. In such a case it is appropriate to perform the normalization

$$x_r = \varepsilon_r x'_r \,, \tag{1.2.3}$$

where ε_r are numerical coefficients characterizing the order of the smallness of the corresponding variables x_r, and x'_r are new variables of order 1. The former coefficient is now multiplied by the factor $\varepsilon_1^{i_1} \dots \varepsilon_m^{i_m}$ and becomes small.

Such smallness criteria determine, in particular, the practical summation limit in (1.2.1) with $l = \infty$.

2. Elementary functions. Given Poisson series A it is easy to define the elementary functions of A in the form of a Poisson series:

$$B = \mathsf{EXP}(A) = \sum_{s=0}^{\infty} \frac{A^s}{(1)_s} \,, \tag{1.2.4}$$

$$B = \mathsf{SIN}(A) = \sum_{s=0}^{\infty} \frac{(-1)^s}{(1)_{2s+1}} A^{2s+1} \,, \tag{1.2.5}$$

$$B = \mathsf{COS}(A) = \sum_{s=0}^{\infty} \frac{(-1)^s}{(1)_{2s}} A^{2s} \,, \tag{1.2.6}$$

etc. All these definitions are self-evident.

3. The Taylor expansion. Let $A(z)$ be a Poisson series with z denoting all its power and trigonometric variables x and y. Let δz_r $(r = 1, 2, \dots, q;\ q = m+n)$ be Poisson series in variables z. Considering δz_r as small variations of z_r one may define a Taylor expansion $A(z+\delta z)$. Retaining in this expansion the terms with a summation order from k to l inclusive with respect to the variations δz_r one gets a new Poisson series in z

$$B = \sum_{s=k}^{l} \sum_{r_1, \dots, r_q} \frac{1}{(1)_{r_1} \dots (1)_{r_q}} \frac{\partial^s A(z)}{\partial z_1^{r_1} \dots \partial z_q^{r_q}} (\delta z_1)^{r_1} \dots (\delta z_q)^{r_q} \tag{1.2.7}$$

denoted

$$B = \mathsf{TAYLOR}(A, \delta z, k, l) .$$

The internal summation is performed over all non-negative values of $r_1, \ldots, r_q$ satisfying the relation

$$r_1 + \ldots + r_q = s .$$

If $l = \infty$ the external summation is taken as in (1.2.1) up to the first term with vanishing contribution into the resulting expression.

4. Inversion of series. Many analytical methods of celestial mechanics involve the problem of finding the inverse transformation for

$$u = v + P(v) , \tag{1.2.8}$$

u, v and P being q-dimensional vectors. The components of $P(v)$ are Poisson series which are regarded as small quantities. This is the case when the series $P(v)$ are proportional to some small parameter ε. In practical problems one also often meets the case where the components of v are small power variables and the series $P(v)$ begin with at least second-degree terms with respect to these variables. In such cases the inversion of (1.2.8) may be easily performed by means of Picard iterations. Indeed, the inverse transformation has the form

$$v = u + Q(u) \tag{1.2.9}$$

with Poisson series $Q(u)$ satisfying the equation

$$Q(u) = -P(u + Q) . \tag{1.2.10}$$

Implicit equation (1.2.10) is solved by iterations

$$Q^{(n+1)}(u) = -P(u + Q^{(n)}) , \qquad n = 1, 2, \ldots \tag{1.2.11}$$

with the initial value

$$Q^{(1)}(u) = -P(u) . \tag{1.2.12}$$

In symbolic form one may write

$$Q = \mathsf{INVERSE}(P, N) ,$$

N being the number of iterations to be performed. Each iteration generally increases the power of ε or the degree of the power variables by one. The expansion of the right-hand member of (1.2.11) is performed by the preceding subroutine.

The problem of the inversion of (1.2.8) may be regarded as a particular case of the Lagrange implicit-function theorem. A convenient algorithm for the recurrent construction of the resulting series has been proposed by Feagin and Gottlieb (1971). Introducing a small parameter ε one may rewrite (1.2.8) in the form

$$u = v + \varepsilon P(v) .$$

Representing solution (1.2.9) of the inverse problem by means of the series

$$Q(u) = \sum_{n=1}^{\infty} \varepsilon^n Q^{(n)}(u)$$

one easily finds from the results of the cited paper that

$$Q^{(n+1)} = \frac{1}{n} \sum_{k=1}^{n} \frac{\partial Q^{(n+1-k)}}{\partial u} Q^{(k)} , \quad n = 1, 2, \ldots \tag{1.2.13}$$

with the initial value (1.2.12). If, as above, q denotes the dimension of the vectors u, v, P and Q then in component form one has

$$Q_i^{(n+1)} = \frac{1}{n} \sum_{k=1}^{n} \sum_{j=1}^{q} \frac{\partial Q_i^{(n+1-k)}}{\partial u_j} Q_j^{(k)} , \quad i = 1, 2, \ldots, q .$$

Another algorithm to solve the Lagrange implicit equation has been elaborated by Deprit (1979) based on the Lie transformation. In the original form of this algorithm u and v are scalars and $P(v)$ in (1.2.9) is of the form

$$P(v) = -\sum_{n=1}^{\infty} \frac{\varepsilon^n}{(1)_n} \varphi_n(v) .$$

If $F = F(v)$ is a given function of v developable into the series

$$F(v) = \sum_{n=0}^{\infty} \frac{\varepsilon^n}{(1)_n} F_{n,0}(v) \tag{1.2.14}$$

then this algorithm enables one to find the function

$$F^*(u) \equiv F\big(v(u)\big) = \sum_{n=0}^{\infty} \frac{\varepsilon^n}{(1)_n} F_{0,n}(u) , \tag{1.2.15}$$

where $v = v(u)$ is the inversion of (1.2.8). Having determined the sequence of the auxiliary functions $W_n = W_n(v)$,

$$W_1 = \varphi_1 ,$$

$$W_{n+1} = \varphi_{n+1} + \sum_{m=0}^{n-1} \frac{(-1)^m (-n)_m}{(1)_m} W_{m+1} \frac{d}{dv} \varphi_{n-m} , \quad n \geq 1 ,$$

one finds coefficients $F_{0,n}$ from the recurrences

$$F_{n-k,k} = F_{n-k+1,k-1} + \sum_{m=0}^{n-k} \frac{(-1)^m(-n+k)_m}{(1)_m} W_{m+1} \frac{d}{dv} F_{n-k-m,k-1}\,,$$

$$k = 1, 2, \ldots, n\,, \quad n \geq 1\,. \tag{1.2.16}$$

The inversion $v = v(u)$ of (1.2.8) is obtained here as a particular case for $F_{0,0}(v) = v$, $F_{n,0} = 0$ $(n > 0)$. The extension for the multidimensional case is evident. A comparative analysis of the Deprit algorithm and the straightforward iteration technique (1.2.9)–(1.2.12) is given by Klioner (1992).

In any realization the operation INVERSE is one of the key operations of the highest level with Poisson series. Some methods of perturbation theory were considered inefficient if they involved the inversion of the series. Implementing the INVERSE operation within the PS processors removes this technical difficulty.

5. Raising to a negative fractional power. In real applications the raising of Poisson series to a negative fractional power is one of the most frequently used operations. Denoted symbolically

$$B = \mathsf{RAISE}(A, -q)$$

this operation enables one to find

$$B = A^{-q}\,, \tag{1.2.17}$$

A being a given Poisson series and q a positive real number. It is assumed therewith that one has an approximate initial value C for the resulting Poisson series B. The number q may be given either as

$$q = \frac{1}{M} \tag{1.2.18}$$

or as

$$q = \frac{M}{2}\,, \tag{1.2.19}$$

M being a natural number.

As in the case of the operation INVERSE the operation RAISE is usually realized with the use of the binomial expansion or by means of the iterations

$$B_{n+1} = B_n + \delta B_n\,, \qquad B_1 = C\,, \quad n = 1, 2, \ldots\,. \tag{1.2.20}$$

Problem (1.2.18) has been considered by Broucke (1971a). Let R be the series

$$R = 1 - C^M A\,, \tag{1.2.21}$$

which is small provided that the initial approximation C is good. Then

$$B = C(1-R)^{-1/M}. \tag{1.2.22}$$

Hence, the resulting series B may be obtained using the binomial expansion. In the iterative solution by substituting (1.2.20) into (1.2.17) and retaining only first-degree terms in δB_n one has

$$(B_n^M + MB_n^{M-1}\delta B_n)A = 1$$

and within the same order of accuracy

$$\delta B_n = \frac{1}{M}B_n(1 - B_n^M A). \tag{1.2.23}$$

Equations (1.2.20) and (1.2.23) describe Newtonian-type iterations. This process may also be used in combination with the binomial expansion.

In the case of problem (1.2.19) one should distinguish between odd and even values of M. The binomial-expansion equivalent to (1.2.21) and (1.2.22) results in

$$R = 1 - C^2A^M, \qquad B = C(1-R)^{-1/2}, \quad M = 1,3,5,\ldots \tag{1.2.24}$$

for odd values of M and

$$R = 1 - CA^{M/2}, \qquad B = C(1-R)^{-1}, \quad M = 2,4,6,\ldots \tag{1.2.25}$$

for even values of M. In the iterative solution the substitution of (1.2.20) into (1.2.17) gives

$$(B_n^2 + 2B_n\delta B_n)A^M = 1.$$

From this it follows that

$$\delta B_n = \tfrac{1}{2}B_n(1 - B_n^2A^M), \quad M = 1,3,5,\ldots \tag{1.2.26}$$

or

$$\delta B_n = B_n(1 - B_nA^{M/2}), \quad M = 2,4,6,\ldots \tag{1.2.27}$$

Here it is also possible to combine the binomial expansion and iterative process. For example, to obtain the series

$$D = (1-R)^{-1} \tag{1.2.28}$$

occurring in (1.2.25) one may replace the binomial expansion by the quadratic iterations following from (1.2.27) and recommended by Jefferys (1970):

$$D_{n+1} = D_n(1 + R^{2^n}), \qquad D_1 = 1 + R, \quad n = 1,2,\ldots. \tag{1.2.29}$$

6. Substitution. Modern PS processors provide facilities to substitute analytical expressions for the power or trigonometric variables of Poisson series. One typical operation is the linear changing of the polynomial and trigonometric variables

$$x_i = \sum_{r=1}^{m} A_{ir} u_r \,, \qquad y_j = \sum_{s=1}^{n} B_{js} v_s \tag{1.2.30}$$

with constant matrices $A = \|A_{ir}\|$, $B = \|B_{js}\|$, the elements of B being integer numbers. This operation, based on term-by-term transformation (1.2.30), leads to a new Poisson series with polynomial variables u_i $(i = 1, 2, \ldots, m)$ and trigonometric variables v_j $(j = 1, 2, \ldots, n)$. Another typical operation is the substituting of Poisson series in place of power variables or trigonometric functions of any trigonometric variable. This subroutine (SUBST) may be realized by applying BINOME and TAYLOR operations.

7. Evaluation. Fast evaluation of Poisson series in PS processors may be regarded as a part of the general problem of the numerical–analytical interface (Brumberg et al., 1989). Its solution involves highly efficient algorithms (EVALUATE) for substituting numerical values into symbolic expressions as well as special programming facilities for evaluating expressions outside CAS. Usually the CAS internal facilities for substituting numerical values into symbolic expressions are not too effective since the objective of the CAS internal representation is to perform symbolic (not numerical!) manipulation. The evaluation problem in celestial mechanics is of particular importance owing to the very large number of terms in Poisson series. The symbolic expressions derived by CAS are often used as the right-hand members of the ordinary differential equations to be integrated numerically. In any discussion of observations it is necessary to evaluate the analytical theories of motion of celestial bodies for the numerous moments of time. Therefore, the evaluation problem has been always treated in PS processors with great care. The method of Shelus and Jefferys (1975) reduces the problem to the evaluation of power series by converting the sines and cosines of multiple arguments to the sum of products of powers of the sine and cosine of each variable separately. The technique developed by Coffey and Deprit (1980) reduces the evaluation of Poisson series to abundant use of trigonometric addition theorems provided that some tables of sines and cosines of specific linear combinations of trigonometric variables are computed in advance. Chapront (1982, 1984) succeeded in evaluating Poisson series by Fourier–Chebyshev approximations preserving the fast variables under the sign of the trigonometric functions and using Chebyshev polynomial approximations for the slowly changing variables. Constructing this approximation may be facilitated by using specialized CAS including manipulation with Chebyshev polynomial expansions.

Perhaps the best tool to produce an efficient interface between symbolic manipulation and numerical calculation is provided by the generation of evaluation programs. Such generation tools are available in many universal CAS but when dealing with extended symbolic expressions they turn out not to be so efficient. The best efficiency may be achieved in specialized evaluation systems manipulating a specific class of expressions like polynomials or Poisson series. One such system, called SPRINT, has been developed by N. N. Vasiliev (see Brumberg et al., 1989). We describe below the basic principles underlying this system.

Considering that any Poisson series of general form may be reduced to a power-series representation, one may use the algorithm to evaluate a polynomial of many variables. Such polynomials often turn out to be very sparse so that the usual techniques like the Horner method are not of much use here. SPRINT deals with the binary-tree representation of such sparse polynomials and transforms this representation into the evaluation program.

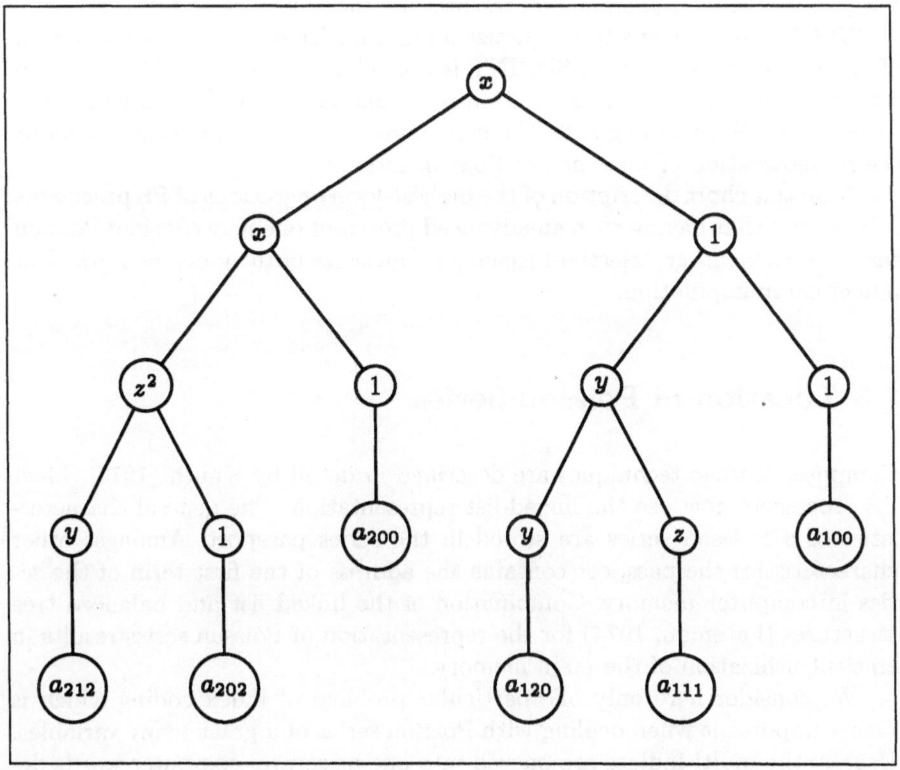

Fig. 1. Evaluation of the polynomial $P = a_{212}x^2yz^2 + a_{202}x^2z^2 + a_{200}x^2 + a_{120}xy^2 + a_{111}xyz + a_{100}x$ in the SPRINT system

Polynomial P (see Fig. 1 for illustration) is represented by a recursive procedure such as a binary tree with marked nodes. All the end nodes correspond to the coefficients of P. All the non-end nodes are related to some multi-indices. These multi-indices are not, in general, the multi-indices of terms of P. The recursive procedure begins by finding the common subindex of all terms of P. The root of the constructing tree is marked with this multi-index i. This subindex is removed from all terms of P, i.e. all multi-indices should be diminished by i. P is split into two polynomials P_1 and P_2, P_1 being the initial maximal segment of terms of P having the non-trivial common subindex and P_2 being the remainder of P. P_1 and P_2 are taken as the left and right subtrees, respectively. The same procedure is now applied to P_1 and P_2, and so on. If any polynomial reduces to a single free term then it corresponds to the end node.

In this manner the evaluation of P consists of (1) simultaneous calculation of all monomials X_i of the non-end nodes of the tree and (2) tree traversal with multiplication at every node of the relevant monomial X_i by the sum of the values of the polynomials represented by the appropriate subtrees.

SPRINT enables one to synthesize a program for simultaneous evaluation of any set of Poisson series. SPRINT is not related to any specific CAS and may be used to synthesize evaluation programs in combination with a great variety of CAS. In calling SPRINT it remains only to give information about the representation of a particular Poisson series.

From this short description of the highest-level operations of PS processors it is evident that having such an advanced processor one may consider Poisson series as elementary celestial-mechanics functions both under analytical or numerical manipulation.

1.3 Location of Poisson Series

Computer storage techniques are described in detail by Knuth (1973). Most PS processors now use the linked-list representation. The general characteristics of a Poisson series are stored in the series passport. Amongst other characteristics the passport contains the address of the first term of the series in computer memory. Combination of the linked list and balanced tree structures (Fateman, 1974) for the representation of Poisson series results in efficient utilization of the main memory.

We consider here only one particular problem of index coding which is rather important when dealing with Poisson series of a great many variables. Usually the multi-indices are packed into one or several computer words dependent on the number of indices, the interval over which they change and the length of the computer word. For a large number of indices their packing may turn out to be not so convenient. An alternative is their coding by means of numerating all admissible index combinations.

Consider first the case of m polynomial indices $i_1, \ldots, i_m$ when each index may take only non-negative values from 0 to some number K. In many celestial-mechanics problems the polynomial variables act as small parameters and one may neglect the terms if their order exceeds K. Therefore, all admissible non-negative values of the indices satisfy the equations

$$i_1 + i_2 + \ldots + i_m = k\,, \quad k = 0, 1, \ldots, K\,. \tag{1.3.1}$$

The number of solutions of one equation (1.3.1) with fixed k is

$$L(m,k) = \frac{(m)_k}{(1)_k} = \frac{(k+1)_{m-1}}{(1)_{m-1}} \tag{1.3.2}$$

and the total number of solutions of all equations (1.3.1) will be

$$M(m,K) = \sum_{k=0}^{K} L(m,k) = L(m+1,K) = \frac{(K+1)_m}{(1)_m}\,. \tag{1.3.3}$$

In the usual packing each index i_r $(r = 1, 2, \ldots, m)$ is assumed to run through values from 0 to K. With the numeration technique we take into account that only the sum of all indices i_r may attain the value K. Let us numerate all admissible index combinations in increasing order relative to k. Within the group for one and the same value of k the combinations are numerated with increasing integer number $i_1 i_2 \ldots i_m$, each i_r being regarded as the corresponding decimal. One-to-one correspondence between the ordinary number of an index combination and the actual index values is determined by the formula

$$N(i_1, \ldots, i_m) = L\Big(m+1, \sum_{r=1}^{m} i_r\Big) - \sum_{q=1}^{m-1} L\Big(m+1-q, \sum_{r=q+1}^{m} i_r - 1\Big)\,. \tag{1.3.4}$$

This number N, which takes values from 1 to $M(m,K)$, replaces the usual index packing.

To facilitate calculations made using this formula the necessary values of $L(n,k)$, $n = 2, 3, \ldots, m+1$, $k = -1, 0, \ldots, K$ may be found in advance and stored. The elements of the matrix $L(n,k)$ are determined most easily by the recurrence relation

$$L(n,k) = L(n,k-1) + L(n-1,k)\,, \tag{1.3.5}$$

$$n = 3, \ldots, m+1, \quad k = 0, \ldots, K\,,$$

with initial values

$$L(n,-1) = 0, \quad n = 2, \ldots, m+1$$

and

$$L(2,k) = k+1, \quad k = -1, 0, \ldots, K\,.$$

The calculation of N by (1.3.4) for a given set $i_1, \ldots, i_m$ presents no difficulties. The inverse problem of finding the set $i_1, \ldots, i_m$ for a given N is much more complicated. This problem may be solved by the consecutive dichotomic searching for N amongst the elements of every row of the matrix $L(n, k)$ starting with the last row. At the first step of this algorithm one finds for a given N a number k such that

$$L(m+1, k-1) < N < L(m+1, k). \tag{1.3.6}$$

Then

$$\sum_{r=1}^{m} i_r = k.$$

If it turns out that

$$N = L(m+1, k),$$

the process is over and

$$i_1 = k, \qquad i_2 = \ldots = i_m = 0.$$

In the general case one puts

$$N_1 = L(m+1, k) - N$$

and finds a number k_1 such that

$$L(m, k_1 - 1) < N_1 < L(m, k_1). \tag{1.3.7}$$

Then

$$\sum_{r=2}^{m} i_r = k_1.$$

If it turns out that

$$N_1 = L(m, k_1)$$

the process is over with

$$i_1 = k - k_1 - 1, \qquad i_2 = 1 + k_1, \qquad i_3 = \ldots = i_m = 0.$$

Proceeding further one finds for a given

$$N_s = N_{s-1} - L(m+2-s, k_{s-1} - 1)$$

a number k_s such that

$$L(m+1-s, k_s - 1) < N_s < L(m+1-s, k_s). \tag{1.3.8}$$

Then

$$\sum_{r=s+1}^{m} i_r = k_s \,.$$

If for some s

$$N_s = L(m+1-s, k_s)$$

the process is over, resulting in the values

$$i_{s+1} = 1 + k_s \,, \qquad i_{s+2} = \ldots = i_m = 0 \,,$$

$$i_s = k_{s-1} - k_s - 1 \,,$$

$$i_{s-1} = k_{s-2} - k_{s-1} \,, \qquad \ldots \,, \qquad i_1 = k - k_1 \,.$$

This process is repeated for $s = 2, 3, \ldots, m-1$ and at most for $s = m-1$ it comes to an end. Each searching (1.3.6)–(1.3.8) is performed by dichotomy.

In the case of trigonometric indices their numeration is realized in a somewhat more complicated way. Let each trigonometric index j_r $(r=1,2,\ldots,n)$ be changing from $-J$ to $+J$. In celestial-mechanics problems one often deals with series where the number J may be large but the sum

$$j_1 + j_2 + \ldots + j_n = k \tag{1.3.9}$$

represents a fairly small integer by its absolute value. One may assume that k runs over the values

$$k = 0, \pm 1, \pm 2, \ldots, \pm K \,, \quad 0 < K < J \,. \tag{1.3.10}$$

Therefore, the total number of admissible combinations will be much smaller compared with the case when each index changes independently the others.

Let $L(n,k,J)$ be the number of possible index combinations having a fixed sum k and satisfying the restriction

$$-J \le j_r \le J \,. \tag{1.3.11}$$

It is evident that

$$L(n, -k, J) = L(n, k, J) \,.$$

The total number of solutions of all equations (1.3.9) with conditions (1.3.10) and (1.3.11) will be

$$M(n, K, J) = \sum_{k=0}^{K} (2 - \delta_{k0}) L(n, k, J) \,. \tag{1.3.12}$$

The numbers $L(n,k,J)$ are calculated by the recurrence relation

$$L(n, k, J) = \sum_{r=1}^{2J+1} L(n-1, k+J+1-r, J) \tag{1.3.13}$$

with the initial values

$$L(2,k,J) = \big|2J - |k| + 1\big| . \tag{1.3.14}$$

The sets of indices $j_1, j_2, \ldots, j_n$ are numerated in groups with the same value of k in the order

$$k = -K, -K+1, \ldots, -1, 0, 1, \ldots, K-1, K .$$

Within each group they are numerated in increasing order of the natural integers $j_1^* j_2^* \ldots j_n^*$ with $j_r^* = j_r + J$ so that $0 \le j_r^* \le 2J$. If for a given k the integers m and l are determined from the equation

$$k + nJ = 2Jm + l , \quad 0 \le l < 2J \tag{1.3.15}$$

then the first set in a group with a given k is

$$\begin{aligned} j_1 &= \ldots = j_{n-m-1} = -J, \\ j_{n-m} &= l - J, \\ j_{n-m+1} &= \ldots = j_n = J \end{aligned} \tag{1.3.16}$$

and the last set in this group is

$$\begin{aligned} j_1 &= \ldots = j_m = J, \\ j_{m+1} &= l - J, \\ j_{m+2} &= \ldots = j_n = -J . \end{aligned} \tag{1.3.17}$$

The numeration formula generalizing (1.3.4) for the case of trigonometric indices takes the form

$$\begin{aligned} N(j_1, \ldots, j_n) = \sum_{r=-K}^{k} L(n,r,J) - \sum_{q=1}^{n-2} \sum_{r=k_{q-1}-J}^{k_q - 1} L(n-q,r,J) - \\ - L\big(2, J + \max\{0, j_{n+1} + j_n\} + 1 - j_n, J\big) \end{aligned} \tag{1.3.18}$$

with

$$k_s = \sum_{r=s+1}^{n} j_r , \quad s = 0, 1, \ldots, n-2 . \tag{1.3.19}$$

The inverse problem of finding the set $j_1, \ldots, j_n$ for a given value of N is solved by at most $n-1$ searches through the elements of $L(n-q,r,J)$, $q = 0, 1, \ldots, n-2$. The first step is to find the number k_0 such that

$$\sum_{r=-K}^{k_0-1} L(n,r,J) < N < \sum_{r=-K}^{k_0} L(n,r,J) . \tag{1.3.20}$$

Then

$$\sum_{r=1}^{n} j_r = k_0 = k\,.$$

If it turns out that

$$N = \sum_{r=-K}^{k_0} L(n, r, J)$$

then the search is over, resulting in the last term (1.3.17) of the group with the value k. In the general case one puts

$$N_1 = \sum_{r=-K}^{k_0} L(n, r, J) - N$$

and continues the process. For a given N_s $(s = 1, 2, \ldots, n-2)$ one finds k_s such that

$$\sum_{r=k_{s-1}-J}^{k_s - 1} L(n-s, r, J) < N_s < \sum_{r=k_{s-1}-J}^{k_s} L(n-s, r, J)\,. \tag{1.3.21}$$

In such a way one obtains sum (1.3.19) and the value $j_s = k_{s-1} - k_s$. If it turns out that

$$N_s = \sum_{r=k_{s-1}-J}^{k_s} L(n-s, r, J)$$

then the process is over. One determines m_s and l_s from the equation

$$k_s + 1 + (n-s)J = 2Jm_s + l_s\,, \quad 0 \le l_s < 2J\,,$$

resulting in

$$j_{s+1} = \ldots = j_{m_s+s} = J,$$

$$j_{m_s+s+1} = l_s - J,$$

$$j_{m_s+s+2} = \ldots = j_n = -J.$$

In the general case one puts

$$N_{s+1} = N_s - \sum_{r=k_{s-1}-J}^{k_s - 1} L(n-s, r, J)$$

and continues the process.

If the process did not stop at any value $s = 1, 2, \ldots, n-2$ then having determined N_{n-1} one obtains

$$j_n = N_{n-1} - J + \max\{0, k_{n-2}\}$$

and

$$j_{n-1} = k_{n-2} - j_n \,.$$

Each searching (1.3.20) and (1.3.21) is performed by dichotomy.

If the series involved in a specific problem are rather dense and do not contain too many zero coefficients amongst the admissible index combinations one may not store the indices at all. It is possible to re-establish their values, when necessary, based only on the place of the corresponding term in the coefficient array. This place is determined by the numeration formula, such as (1.3.4) or (1.3.18). The solution of the inverse problem to find the index set by the number of the term in the coefficient array is not needed here since it is sufficient to determine for any index set the next admissible one. Such a tabular method of series representation has been widely used at the Bureau des Longitudes. In spite of its efficiency in particular problems this method is less general compared to methods involving index coding (by means of packing or numerating).

2 The Keplerian Processor

2.1 The Keplerian Processor in Closed Form

The Keplerian processor is intended to be a software package for operating with the functions related to the undisturbed two-body problem. As a rule, one confines oneself only to the case of elliptic motion, which is the most important for practical applications. The Keplerian processor may be developed on the basis of a PS processor or some general CAS. It illustrates the application of these software tools to the derivation of the formulae describing Keplerian motion.

It is well known that the solution of the two-body problem may be presented in three main forms, namely: (1) closed-form expressions with the aid of true or eccentric anomalies, (2) trigonometric series in multiples of the mean anomaly, and (3) power series with respect to time. The third form is valid, of course, only for some finite interval of time. Depending on the choice of any one of these forms we shall speak about the Keplerian processor in closed form, Poisson-series form or Taylor-series form. Recently, quite a new technique for solving the two-body problem has been proposed by E. Brumberg (1992) based on the transformation and expansion theory of elliptic functions. This leads to one more form of the Keplerian processor which may be extremely useful for investigating highly eccentric orbits. This elliptic-function form will also be considered below.

We start with the classic closed-form Keplerian processor. The basic relations for the two-body problem are

$$\mathbf{r} = r(\mathbf{l}\cos\phi + \mathbf{m}\sin\phi) \tag{2.1.1}$$

and

$$\dot{\mathbf{r}} = \frac{na}{\eta}\left[-\mathbf{l}(\sin\phi + e\sin\omega) + \mathbf{m}(\cos\phi + e\cos\omega)\right] . \tag{2.1.2}$$

Here $\mathbf{r}$ and $\dot{\mathbf{r}}$ are the relative coordinates and velocity components in the two-body problem, the radius-vector r and argument of latitude ϕ are the polar coordinates in the plane of the orbit and ω is the argument of the pericentre. The orientation of the orbital plane with respect to the reference plane is characterized by the inclination i and the longitude of the ascending node Ω.

Instead of these two angles one may use the triad of orthonormal vectors **l**, **m** and **k**:

$$\mathbf{l}=\begin{pmatrix}\cos\Omega\\ \sin\Omega\\ 0\end{pmatrix}, \qquad \mathbf{m}=\begin{pmatrix}-\cos i\sin\Omega\\ \cos i\cos\Omega\\ \sin i\end{pmatrix}, \qquad \mathbf{k}=\begin{pmatrix}\sin i\sin\Omega\\ -\sin i\cos\Omega\\ \cos i\end{pmatrix}.$$

Introducing the true anomaly v instead of ϕ,

$$\phi = v + \omega\,,$$

one has

$$r = \frac{a\eta^2}{1+e\cos v}\,, \tag{2.1.3}$$

$$\mathbf{r} = r(\mathbf{P}\cos v + \mathbf{Q}\sin v) \tag{2.1.4}$$

and

$$\dot{\mathbf{r}} = \frac{na}{\eta}\Big[-\mathbf{P}\sin v + \mathbf{Q}(\cos v + e)\Big]\,. \tag{2.1.5}$$

The orthogonal unit vectors **P** and **Q** result from the rotation of the vectors **l** and **m** in the orbital plane by the angle ω:

$$\mathbf{P} = \mathbf{l}\cos\omega + \mathbf{m}\sin\omega\,, \qquad \mathbf{Q} = \frac{\partial\mathbf{P}}{\partial\omega} = -\mathbf{l}\sin\omega + \mathbf{m}\cos\omega\,.$$

Instead of v one often uses the eccentric anomaly g defined by the relations

$$r = a(1 - e\cos g)\,, \tag{2.1.6}$$

$$r\cos v = a(\cos g - e)\,, \tag{2.1.7}$$

$$r\sin v = a\eta\sin g\,. \tag{2.1.8}$$

In all these relations a, e and n stand for the semi-major axis, eccentricity and mean motion, respectively. One has

$$\eta = (1-e^2)^{1/2} \tag{2.1.9}$$

and

$$n^2a^3 = Gm\,,$$

G being the gravitational constant and m the sum of the masses of both bodies. The relationship with time is given by the Kepler equation

$$g - e\sin g = M\,, \tag{2.1.10}$$

the mean anomaly M being a linear function of time,

$$M = M_0 + nt\,. \tag{2.1.11}$$

Instead of (2.1.6)–(2.1.8) one may relate v and g by means of the formulae (Broucke and Cefola, 1973)

$$\tan\frac{v-g}{2} = \frac{e\sin v}{1+\eta+e\cos v} \tag{2.1.12}$$

or

$$\tan\frac{v-g}{2} = \frac{e\sin g}{1+\eta-e\cos g}\,. \tag{2.1.13}$$

The most important operation of the closed-form Keplerian processor is to take quadrature of the type

$$G = \int F\,dM \tag{2.1.14}$$

with

$$F = \sum A(r,\dot r)\,{\cos \atop \sin}\,(jv+kg)\,, \tag{2.1.15}$$

where $A(r,\dot r)$ is a polynomial containing positive and negative powers of its arguments with rational coefficients in e and η. The summation in (2.1.15) is expanded over some finite set of j and k values. Such a problem occurs in determining perturbations in closed form without using series expansions. For example, in the motion of the Earth's artificial satellites the perturbations due to the Earth's oblateness may be found as closed-form expressions in terms of the true anomaly v whereas the main luni-solar perturbations are representable as finite expressions in terms of the eccentric anomaly g and the mean longitude of the disturbing body. In taking into account both these factors one encounters in a second-order theory terms of the general form (2.1.15) containing both v and g.

The algorithm for problem (2.1.14) and (2.1.15), presented below, belongs to Jefferys (1971). This algorithm involves several steps.

1. Using the trigonometric addition formulae one separates the arguments jv and kg in (2.1.15).
2. Using the expressions

$$\cos v = \frac{1}{e}\left(\eta^2\frac{a}{r}-1\right), \qquad \sin v = \frac{\eta}{e}\frac{\dot r}{na} \tag{2.1.16}$$

and

$$\cos g = \frac{1}{e}\left(1-\frac{r}{a}\right), \qquad \sin g = \frac{1}{e}\frac{r}{a}\frac{\dot r}{na} \tag{2.1.17}$$

one easily finds the necessary trigonometric functions of jv and kg in terms of polynomial variables r, $\dot r$, e and η. To do this one may use the recurrence relation:

$$\exp imx = 2\cos x \exp i(m-1)x - \exp i(m-2)x \tag{2.1.18}$$

valid for both $\cos mx$ and $\sin mx$. One may also use the representation of $\cos mx$ and $\sin mx$ with the aid of the hypergeometric polynomials

$$\cos mx = (\cos x)^{\operatorname{mod}(m,2)} F\Big(-\mathrm{E}\left(\tfrac{m}{2}\right), \mathrm{E}\left(\tfrac{m+1}{2}\right), \tfrac{1}{2}, \sin^2 x\Big), \tag{2.1.19}$$

$$\begin{aligned} \sin mx = {} & m \sin x (\cos x)^{\operatorname{mod}(m-1,2)} \times \\ & \times F\Big(-\mathrm{E}\left(\tfrac{m-1}{2}\right), 1+\mathrm{E}\left(\tfrac{m}{2}\right), \tfrac{3}{2}, \sin^2 x\Big). \end{aligned} \tag{2.1.20}$$

Analytical computation of $\cos mx$ and $\sin mx$ by means of (2.1.18) or (2.1.19) and (2.1.20) using some CAS presents no difficulties. As a result, the integrand F of (2.1.14) becomes a polynomial in r and $\dot{r}$ containing only non-negative powers of $\dot{r}$ (otherwise integral (2.1.14) may have singularities). For simplicity, one may put $a = 1$, $n = 1$, restoring these parameters on the basis of dimensional considerations only in the final expression (2.1.14).

3. All powers of $\dot{r}$ beyond the first one are eliminated from F by applying the relation

$$\frac{\dot{r}^2}{n^2 a^2} = -1 + 2\frac{a}{r} - \eta^2 \frac{a^2}{r^2}. \tag{2.1.21}$$

The integrand F now takes the form

$$F = A(r) + B(r)\frac{\dot{r}}{na}, \tag{2.1.22}$$

$A(r)$ and $B(r)$ being polynomials of positive and negative powers in r with power-function coefficients in e and η.

4. The polynomial $A(r)$ is separated into two parts. The first part includes terms with powers less than -1 whereas the remaining terms with powers equal to or greater than -1 form the second part. Hence,

$$A(r) = \frac{a^2}{r^2} P_1\left(\frac{1}{r}\right) + \frac{a}{r} P_2(r), \tag{2.1.23}$$

P_1 and P_2 being polynomials of only non-negative powers in their arguments.

5. The functions r^{-1} in P_1 and r in P_2 are replaced by (2.1.3) and (2.1.6), respectively. The resulting polynomials in powers of cosines are transformed to the cosine polynomials $B_1(v)$ and $B_2(g)$ in multiples of their arguments. Finally, the function $A(r)$ will become

$$A(r) = \frac{a^2}{r^2} B_1(v) + \frac{a}{r} B_2(g). \tag{2.1.24}$$

6. For unperturbed Keplerian motion one has

$$dM = ndt = \frac{r}{a}dg = \frac{1}{\eta}\frac{r^2}{a^2}dv\,. \tag{2.1.25}$$

Therefore, using (2.1.22), (2.1.24) and (2.1.25) one may present integral (2.1.14) in the form

$$G = \frac{1}{a}\int B(r)\,dr + \frac{1}{\eta}\int B_1(v)\,dv + \int B_2(g)\,dg\,. \tag{2.1.26}$$

All three integrals occurring here are performed in an elementary way. But along with power functions of r and sines of multiples of v and g one may meet here the function $\ln r$ (if $B(r)$ contains the term r^{-1}) and the secular terms proportional to v or g (if $B_1(v)$ or $B_2(g)$ contain the constant terms). The secular terms are usually treated separately. But in doing this one most often has to deal with the secular terms in the mean anomaly M. In this respect the secular term proportional to g presents no complications since with the aid of the Kepler equation it may be replaced by the secular term in M and the sine term in g. On the other hand, the secular term proportional to v may be replaced by the expression

$$v = M + f\,, \tag{2.1.27}$$

where the function f, known as the equation of the centre, cannot be expressed in closed form of (2.1.15). Indeed, combining (2.1.12) or (2.1.13) with the Kepler equation (2.1.10), one obtains

$$f = \frac{\eta e \sin v}{1 + e\cos v} + 2\arctan\frac{e\sin v}{1 + \eta + e\cos v} \tag{2.1.28}$$

in terms of the true anomaly v or

$$f = e\sin g + 2\arctan\frac{e\sin g}{1 + \eta - e\cos g} \tag{2.1.29}$$

in terms of the eccentric anomaly g. Neither of these expressions has the form of (2.1.15). One may still avoid the use of infinite series for representing f. As proposed by N. N. Vasiliev (see Brumberg et al., 1989) the function

$$\Phi(r) = \frac{v - g}{\dot{r}} = \frac{2}{\dot{r}}\arctan\frac{e\sin v}{1 + \eta + e\cos v} \tag{2.1.30}$$

as any function of r may be represented in the domain of variation of its argument $a(1-e) \le r \le a(1+e)$ by the uniformly-best-approximation polynomial in r. Indeed, taking into account expressions (2.1.16) it is easy to see that function (2.1.30) depends only on r and takes the values

$$\frac{2}{na}\frac{\eta}{1+\eta+e}\,, \qquad \frac{2}{na}\frac{1+e}{1+\eta+e}$$

in the pericentre and apocentre, respectively. By fixing maximum and minimum powers of the approximation polynomial we can reduce the construction of the approximation to the problem of linear programming. Having constructed such a polynomial for $\Phi(r)$ one finds an expression for $v - g$ in the form of (2.1.15), substituting $\dot{r}$ from (2.1.16) or (2.1.17). Of course, to obtain f one might from the very beginning consider instead of $\Phi(r)$ the function

$$\Psi(r) = \frac{\eta}{e \sin v} f \tag{2.1.31}$$

taking the values

$$\eta \frac{\eta - e + 3}{1 + \eta + e}, \qquad (1 + e) \frac{\eta + e + 3}{1 + \eta + e}$$

at the pericentre and apocentre, respectively. Quite similarly, one may replace the function $\ln r$ by its uniformly-best-approximation polynomial. In such a way integral (2.1.14) will be presented in the form of (2.1.15) adding, possibly, a secular term in M. Another technique to conserve the closed form of G is to express $\ln r$ and f in terms of dilogarithmic functions of a purely imaginary argument (Osácar and Palacián, 1994; Palacián, 1994). So far it is not clear whether this interesting approach, introducing one more type of special function into celestial mechanics, leads to a real simplification.

7. To simplify the resulting expression for integral (2.1.14) it is reasonable to use the relation $\eta^2 + e^2 = 1$ and to choose the most compact form.

Implementation of this algorithm within some universal CAS like MATHEMATICA presents no difficulties.

The Keplerian processor in closed form may be applied to construct, for example, the Earth-satellite compact theory of motion (without using power expansions in eccentricity). The main theoretical results in this domain belong to Deprit and his followers (see the review paper by Deprit and Ferrer, 1989). The very elegant Deprit techniques involve the set of canonical transformations to new variables, which permit the straightforward integration of the resulting equations of motion. But operationally this is nothing other than integration by parts using the closed-form Keplerian processor. Indeed, let us consider the integral

$$G = \int r^{-k-1} F(\phi)\, dt\,, \tag{2.1.32}$$

where k is a positive integer and $F(\phi)$ is a trigonometric polynomial of the argument of latitude. Such integrals occur in the evaluation of the zonal-harmonic perturbations in the Earth-satellite theory of motion. Using the elimination of the parallax technique (Deprit, 1981) one may exclude from the perturbing Hamiltonian all the terms r^{-3}, r^{-4}, ... and consider integrals like

(2.1.32) only for $k = 1$. But this is done in the Jefferys algorithm by reducing general expression (2.1.15) to (2.1.22) with (2.1.24). It is also possible to evaluate (2.1.32) by integration by parts:

$$G = \frac{1}{na^2\eta} \int r^{-k+1} F(\phi)\, d\phi =$$
$$= \frac{1}{na^2\eta} r^{-k+1} F^{-1}(\phi) + (k-1) \frac{e}{na^3\eta^3} \int r^{-k+2} F^{-1}(\phi) \sin v \, d\phi$$

with

$$F^{-1}(\phi) = \int F(\phi)\, d\phi \,. \tag{2.1.33}$$

Repeating this process one has

$$G = \sum_{s=1}^{k} (-1)^{s-1} (1-k)_{s-1} \frac{e^{s-1}}{na^{s+1}\eta^{2s-1}} r^{-k+s} \tilde{F}^{-s}(\phi) \tag{2.1.34}$$

with

$$\tilde{F}^{-1}(\phi) = F^{-1}(\phi), \quad \tilde{F}^{-s}(\phi) = \int \tilde{F}^{-s+1}(\phi) \sin v \, d\phi \,, \quad s > 1 \,. \tag{2.1.35}$$

This means that all integrals (2.1.32) with $k \geq 1$ are performed by parts.

The use of the nodes technique proposed recently (Palacián, 1992) to deal with the tesseral harmonic perturbations in the Earth-satellite theory may also be interpreted as an integration-by-parts process. Indeed, given the integral

$$G = \int r^{-k-1} F(\phi) \Phi(t)\, dt \tag{2.1.36}$$

with trigonometric polynomials $F(\phi)$ and $\Phi(t)$ in multiples of ϕ and some angular variable ωt, respectively, one has

$$G = r^{-k-1} F(\phi) \Phi^{-1}(t) + \int r^{-k-2} F^{*}(\phi) \Phi^{-1}(t)\, dt \tag{2.1.37}$$

with

$$\Phi^{-1}(t) = \int \Phi(t)\, dt \tag{2.1.38}$$

and

$$F^{*}(\phi) = \frac{na}{\eta} \left[(k+1) e \sin v F(\phi) - (1 + e\cos v) \frac{dF(\phi)}{d\phi} \right] . \tag{2.1.39}$$

In the right-hand part of (2.1.37) we meet an integral of the same type as (2.1.36) but the power $(-k-1)$ is now replaced by $(-k-2)$. This augmentation of the negative power of r is characteristic for the use of the nodes technique. Repeating this process one may hope that the integral term in (2.1.37) eventually becomes negligibly small.

More examples of the application of the closed-form Keplerian processor will be encountered below.

2.2 The Keplerian Processor in Poisson-Series Form

The aim of the Keplerian processor in Poisson-series form is to derive by computer the classic expansions of the functions of the elliptic motion. Of course, all these expansions may be found in tabular form (see, for instance, Jarnagin, 1965) and their general terms are given in many textbooks. But for computer calculations one needs specialized software to reproduce typical expansions for the undisturbed motion. Such software is the subject of the Keplerian processor in Poisson-series form.

One of the versions of the Keplerian processor has been described by Broucke (1970, 1974a). Here we describe another version employed for many years in the Institute of Theoretical Astronomy (St. Petersburg). The four main subroutines of this processor called CENTER, KEPLER, HANSEN and COORD give all the necessary two-body-problem expansions starting with the simplest initial formulae.

The subroutine CENTER produces the expansion of the equation of the centre f defined by (2.1.27). The resulting series is a $(1,1)$ Poisson series with one polynomial variable e and one trigonometric argument M. The algorithm is based on the differential equation

$$\frac{dv}{dM} = \frac{(1+e\cos v)^2}{(1-e^2)^{3/2}} \tag{2.2.1}$$

resulting from (2.1.25). The algorithm involves iterations

$$f_k = \int \left\{ (1-e^2)^{-3/2} \left[1 + e\cos(M + f_{k-1})\right]^2 - 1 \right\} dM\,, \tag{2.2.2}$$

$$f_0 = 0\,, \quad k = 1, 2, \ldots, N\,.$$

If we restrict ourselves to N iterations we obtain $f = f_N$ within the terms of e^N inclusive. The realization of (2.2.2) involves such operations of a PS processor as TAYLOR, BINOME and integration with respect to the trigonometric variable M. Every step of this iteration process increases the accuracy of the determination of f by one order with respect to e so that f_k is the exact approximation of f up to the terms e^k inclusive. In evaluating the integrand in (2.2.2) all operations should be performed within this accuracy because the terms beyond e^k are erroneous and involve corresponding errors in f_k (such as the appearance of secular terms with respect to M). One should always keep this in mind when determining Poisson series by iterations. In particular, one must take care in dealing with numerical values of the polynomial variables. If, for instance, the eccentricity e in (2.2.2) is given numerically then to obtain the correct trigonometric series for f it is necessary to attribute to the quantities e and e^2 the analytical orders 1 and 2, respectively, and to perform all operations taking into account these orders of smallness.

The subroutine KEPLER gives a (1,1) Poisson series for the difference $h = g - M$ between the eccentric and mean anomalies. The algorithm involves iterations based on the Kepler equation

$$h_k = e\sin(M + h_{k-1}), \qquad h_0 = 0\,, \quad k = 1, 2, \ldots, N\,. \tag{2.2.3}$$

Here one again uses the operation TAYLOR of a PS processor. After completing N iterations one obtains $h = h_N$ up to the terms e^N inclusive.

The subroutines CENTER and KEPLER serve to initialize the subroutine HANSEN resulting in a (1,1) Poisson series as follows:

$$C \equiv \left(\frac{r}{a}\right)^n \cos mz = C_0^{n,m} + \sum_{k=1}^{\infty} C_k^{n,m} \cos kM \tag{2.2.4}$$

and

$$S \equiv \left(\frac{r}{a}\right)^n \sin mz = \sum_{k=1}^{\infty} S_k^{n,m} \sin kM\,. \tag{2.2.5}$$

Here z stands for the true anomaly v or the eccentric anomaly g. The coefficients $C_k^{n,m}$ and $S_k^{n,m}$ are power series in the eccentricity e starting with degree $|m-k|$. If $z = v$ one uses BINOME to expand $(r/a)^n$ starting with (2.1.3) and then TAYLOR to expand $\cos v$, $\cos mv$ and $\sin mv$ for $v = M + f$, f being given in advance by CENTER. If $z = g$ one uses BINOME to expand $(r/a)^n$ starting with (2.1.6) and then TAYLOR to expand $\cos g$, $\cos mg$ and $\sin mg$ for $g = M + h$, h being given in advance by KEPLER.

If $z = v$ one may obtain the expansions for the Hansen coefficients. Indeed, the Hansen coefficients $X_k^{n,m}(e)$ are the Fourier coefficients of the function

$$\left(\frac{r}{a}\right)^n \exp imv = \sum_{k=-\infty}^{\infty} X_k^{n,m}(e) \exp ikM \tag{2.2.6}$$

and comparison with (2.2.4) and (2.2.5) shows that

$$\begin{aligned} X_0^{n,m} &= C_0^{n,m}\,, \\ X_k^{n,m} &= \tfrac{1}{2}(C_k^{n,m} + S_k^{n,m})\,, \\ X_{-k}^{n,m} &= \tfrac{1}{2}(C_k^{n,m} - S_k^{n,m})\,, \quad k > 0\,. \end{aligned} \tag{2.2.7}$$

The coefficients of the power expansions for the Hansen coefficients are called Newcomb operators,

$$X_k^{n,m}(e) = \sum_{s=0}^{\infty} P_{k-m}^{|k-m|+2s}(n,m)e^{|k-m|+2s}\,. \tag{2.2.8}$$

The subroutine HANSEN gives these Newcomb operators in numerical form.

The version $z = g$ leads to expansions with Bessel coefficients. For example,

$$C_0^{0,m} = -\tfrac{1}{2}e\delta_{m,1}\,,$$

$$C_k^{0,m} = \frac{m}{k}\left[J_{k-m}(ke) - J_{k+m}(ke)\right],$$

$$S_k^{0,m} = \frac{m}{k}\left[J_{k-m}(ke) + J_{k+m}(ke)\right], \quad k > 0$$

and

$$C_0^{-1,0} = 1\,, \qquad C_k^{-1,0} = 2J_k(ke)\,, \quad k > 0\,.$$

Not so long ago some interesting expansions were discovered dealing with the velocity V of elliptic motion (Broucke, 1974b; Jupp, 1975). In virtue of the relation

$$\left(\frac{V}{na}\right)^2 = \frac{1}{\eta^2}\left(1 + 2e\cos v + e^2\right) \tag{2.2.9}$$

all such expansions may be easily derived with the aid of the CENTER, TAYLOR and BINOME subroutines.

The subroutine COORD produces the expansions of the rectangular coordinates $\mathbf{r} = (x, y, z)$ of the two-body problem into (3,3) or (2,3) Poisson series. The three trigonometric arguments are Ω, ω and M (or else Ω, the longitude of the pericentre $\pi = \omega + \Omega$ and the mean longitude $\lambda = M + \pi$). In the (3,3) version the polynomial variables are e, $\sin(i/2)$ and $\cos(i/2)$. The expansion is performed only in powers of e up to some degree N inclusive. There is no expansion with respect to the inclination. In the (2,3) version the polynomial variables are e and $\sin i$. This version is suitable for orbits at a small inclination, and expansion is performed in powers of e and $\sin i$ up to some total order N inclusive. This subroutine is based on (2.1.4), i.e.

$$\left(\frac{r}{a}\right)^n \frac{\mathbf{r}}{a} = \left(\frac{r}{a}\right)^{n+1}\left(\mathbf{P}\cos v + \frac{\partial \mathbf{P}}{\partial \omega}\sin v\right). \tag{2.2.10}$$

Vector $\mathbf{P}$ is of the form

$$\mathbf{P} = \begin{pmatrix} \cos^2\frac{i}{2}\cos(\Omega+\omega) + \sin^2\frac{i}{2}\cos(\Omega-\omega) \\ \cos^2\frac{i}{2}\sin(\Omega+\omega) + \sin^2\frac{i}{2}\sin(\Omega-\omega) \\ \sin\omega\sin i \end{pmatrix}. \tag{2.2.11}$$

In the (3,3) version one should replace $\sin i$ by $2\sin(i/2)\cos(i/2)$ and in the (2,3) version it is necessary to expand $\sin^2(i/2)$ and $\cos^2(i/2)$ in powers of $\sin i$ producing series like the following:

$$\sin^2\frac{i}{2} = \frac{1}{2}\sum_{r=0}^{\infty}(\delta_{r0}-1)\frac{(-\frac{1}{2})_r}{(1)_r}(\sin i)^{2r} = \frac{1}{4}\sum_{r=1}^{\infty}\frac{(\frac{1}{2})_{r-1}}{(2)_{r-1}}(\sin i)^{2r}\,. \tag{2.2.12}$$

The algorithm is based on the application to (2.2.10) of the subroutines CENTER and HANSEN. The final series in the (3,3) version are of the form

$$\frac{r^n}{a^{n+1}}(x+\mathrm{i}y) = \sum_{k=-\infty}^{\infty} X_{k+1}^{n+1,1}(e)\Big\{\cos^2\frac{i}{2}\exp\mathrm{i}\big[\Omega+\omega+(k+1)M\big]+ + \sin^2\frac{i}{2}\exp\mathrm{i}\big[\Omega-\omega-(k+1)M\big]\Big\}\,, \tag{2.2.13}$$

and

$$\frac{r^n}{a^{n+1}}z = \sin i\sum_{k=-\infty}^{\infty} X_{k+1}^{n+1,1}(e)\sin\big[\omega+(k+1)M\big]\,. \tag{2.2.14}$$

If we introduce the longitudes Ω, π and λ reckoned from one and the same point then the sum of all trigonometric indices is equal to 1 in the cosine series for x and the sine series for y and is equal to 0 in the sine series for z.

The subroutines CENTER, KEPLER, HANSEN and COORD permit one to obtain all the classic expansions. For small eccentricities and inclinations it may be useful to have subroutines to convert the trigonometric expansions (2.2.13) and (2.2.14) into (4,0) power expansions. Indeed, with the complex variables

$$K = e\exp\mathrm{i}M\,,\qquad L = \sin i\exp\mathrm{i}(M+\omega) \tag{2.2.15}$$

the general term in (2.2.13) and (2.2.14) is transformed as follows:

$$e^{i_1}(\sin i)^{i_2}\exp\mathrm{i}(k_1\Omega+k_2\omega+k_3M) = K^{j_1}\bar{K}^{j_2}L^{j_3}\bar{L}^{j_4}\exp\mathrm{i}k_1\lambda\,, \tag{2.2.16}$$

the bar denoting the complex-conjugate quantity. The power and trigonometric indices of the general term on the left-hand side of (2.2.16) satisfy the relations (D'Alembert characteristics)

$$i_1 = |k_3-k_2|+2s\,,\qquad i_2 = |k_2-k_1|+2r\,,\qquad s,r\geq 0\,.$$

This enables one to determine the indices j_1, j_2, j_3 and j_4:

$$j_1 = s+\max\{0,k_3-k_2\}\,,\qquad j_2 = s+\max\{0,k_2-k_3\}\,,$$

$$j_3 = r+\max\{0,k_2-k_1\}\,,\qquad j_4 = r+\max\{0,k_1-k_2\}\,.$$

Rewriting (2.2.14) as

$$2\mathrm{i}\frac{r^n}{a^{n+1}}z = \sin i \sum_{k=-\infty}^{\infty} X_{k+1}^{n+1,1}(e)\Big\{\exp\mathrm{i}\big[\omega + (k+1)M\big] - \\ - \exp\mathrm{i}\big[-\omega - (k+1)M\big]\Big\} \qquad (2.2.17)$$

we see that the functions

$$\frac{r^n}{a^{n+1}}(x+\mathrm{i}y)\exp(-\mathrm{i}\lambda)\,, \qquad 2\mathrm{i}\frac{r^n}{a^{n+1}}z$$

may be presented by the purely power (4,0) series with respect to the variables K, $\bar{K}$, L and $\bar{L}$. It is easy to give the general terms of these series explicitly. Indeed, substituting expansion (2.2.8) for the Hansen coefficients one has

$$\frac{r^n}{a^{n+1}}(x+\mathrm{i}y) = \sum_{k=0}^{\infty}\sum_{s=0}^{\infty}\left(1-\tfrac{1}{2}\delta_{k0}\right)e^{k+2s}\times \\ \times\Big\{P_k^{k+2s}\Big[\cos^2\frac{i}{2}\exp\mathrm{i}\big(\Omega+\omega+(k+1)M\big)+ \\ +\sin^2\frac{i}{2}\exp\mathrm{i}\big(\Omega-\omega-(k+1)M\big)\Big]+ \\ +P_{-k}^{k+2s}\Big[\cos^2\frac{i}{2}\exp\mathrm{i}\big(\Omega+\omega-(k-1)M\big)+ \\ +\sin^2\frac{i}{2}\exp\mathrm{i}\big(\Omega-\omega+(k-1)M\big)\Big]\Big\}.$$

Putting $k+2s=m$, changing the order of summation and applying (2.2.12) and (2.2.16) results in

$$\frac{r^n}{a^{n+1}}(x+\mathrm{i}y)\exp(-\mathrm{i}\lambda) = \frac{1}{2}\sum_{m=0}^{\infty}\sum_{s=0}^{\mathrm{E}(\frac{m}{2})}\sum_{r=0}^{\infty}\frac{(-\frac{1}{2})_r}{(1)_r}\left(1-\frac{1}{2}\delta_{2s,m}\right)\times \\ \times\Big[(\delta_{r0}+1)P_{m-2s}^{m}K^{m-s}\bar{K}^{s}L^{r}\bar{L}^{r}+ \\ +\ (\delta_{r0}+1)P_{2s-m}^{m}K^{s}\bar{K}^{m-s}L^{r}\bar{L}^{r}+ \\ +\ (\delta_{r0}-1)P_{2s-m}^{m}K^{m-s}\bar{K}^{s}L^{r-1}\bar{L}^{r+1}+ \\ +\ (\delta_{r0}-1)P_{m-2s}^{m}K^{s}\bar{K}^{m-s}L^{r-1}\bar{L}^{r+1}\Big]. \qquad (2.2.18)$$

In a similar way one obtains

$$2\mathrm{i}\frac{r^n}{a^{n+1}}z = \sin i\sum_{k=0}^{\infty}\sum_{s=0}^{\infty}\left(1-\tfrac{1}{2}\delta_{k0}\right)e^{k+2s}\times \\ \times\Big\{P_k^{k+2s}\Big[\exp\mathrm{i}\big(\omega+(k+1)M\big)-\exp\mathrm{i}\big(-\omega-(k+1)M\big)\Big]+ \\ +P_{-k}^{k+2s}\Big[\exp\mathrm{i}\big(\omega-(k-1)M\big)-\exp\mathrm{i}\big(-\omega+(k-1)M\big)\Big]\Big\}$$

or finally

$$2\mathrm{i}\frac{r^n}{a^{n+1}}z = \sum_{m=0}^{\infty}\sum_{s=0}^{E(\frac{m}{2})}\left(1-\tfrac{1}{2}\delta_{2s,m}\right)\times$$
$$\times\left[(P^m_{m-2s}K^{m-s}\bar{K}^s + P^m_{2s-m}K^s\bar{K}^{m-s})L - \right.$$
$$\left. -(P^m_{m-2s}Kz^s\bar{K}^{m-s} + P^m_{2s-m}K^{m-s}\bar{K}^s)\bar{L}\right]. \quad (2.2.19)$$

Arguments $(n+1, 1)$ of the Newcomb operators in (2.2.18) and (2.2.19) are omitted.

Power expansion of the coordinates of the two-body problem in variables (2.2.15) is obtained here by converting Poisson series (2.2.13) and (2.2.14). An algorithm of the straightforward derivation of such a power expansion using variables similar to (2.2.15) will be described in Chapter 7.

Series (2.2.18) and (2.2.19) may be put in different forms dependent on the choice of the power variables. In using the elements

$$k = e\exp(-\mathrm{i}\pi), \qquad l = \sin i\exp(-\mathrm{i}\Omega) \quad (2.2.20)$$

related to K and L by

$$K = k\exp\mathrm{i}\lambda, \qquad L = l\exp\mathrm{i}\lambda \quad (2.2.21)$$

series (2.2.18) and (2.2.19) take the form of a (4,1) Poisson series with four power variables k, $\bar{k}$, l, $\bar{l}$ and one trigonometric argument λ. If necessary, it is easy to replace the set K, $\bar{K}$, L, $\bar{L}$ or k, $\bar{k}$, l, $\bar{l}$ by the corresponding tetrad of the real variables. Such transformations are easily performed by a PS processor.

2.3 General Terms of the Elliptic-Motion Expansions

Keplerian processors in the Poisson-series form enable one to obtain by computer the most complicated expansions in elliptic motion. But the general terms of such expansions usually remain unknown. In this respect Keplerian processors cannot replace specialists in celestial mechanics trying to derive the general terms of the expansions rather than calculating as many initial terms as possible. Elegant mathematical expansions of the two- and three-body problems may be regarded as the chefs d'oeuvre of analytical celestial mechanics. In this section we give typical examples of the art of celestial mechanics. Most of the expansions given below are, of course, well known and only the derivation technique may differ from that of common textbooks.

We start with the function

$$\gamma(n,x,y,\nu,\alpha,\zeta) = \alpha^n(1-\alpha\zeta^{-1})^x(1-\alpha\zeta)^y(-\zeta)^\nu, \quad (2.3.1)$$

which turns out to be very useful in the expansions related to the two- and three-body problems (a similar function was considered by Brown and Shook, 1933). Here n, x, y and ν are integers and α is a real parameter, which may be assumed always to be less than 1 in virtue of the relation

$$\gamma(n,x,y,\nu,\alpha,\zeta) = \gamma(-n-x-y,\, y,\, x,\, \nu - x + y,\, \alpha^{-1},\, \zeta) . \tag{2.3.2}$$

The argument ζ is usually of the form $\zeta = \exp i\varphi$, φ being some angular variable so that $\bar{\zeta} = \zeta^{-1}$. Evidently,

$$\gamma(n,x,y,\nu,\alpha,\zeta^{-1}) = \gamma(n,y,x,-\nu,\alpha,\zeta) . \tag{2.3.3}$$

Expanding this function into an exponential series one has

$$\gamma(n,x,y,\nu,\alpha,\zeta) = \sum_{\sigma=0}^{\infty}\sum_{\rho=0}^{\infty}(-1)^{\nu}\frac{(-x)_{\rho}(-y)_{\sigma}}{(1)_{\rho}(1)_{\sigma}}\alpha^{n+\sigma+\rho}\zeta^{\nu+\sigma-\rho} =$$

$$= \sum_{\rho=0}^{\infty}\sum_{\sigma=\nu-\rho}^{\infty}(-1)^{\nu}\frac{(-x)_{\rho}(-y)_{\sigma+\rho-\nu}}{(1)_{\rho}(1)_{\sigma+\rho-\nu}}\alpha^{n+2\rho+\sigma-\nu}\zeta^{\sigma} =$$

$$= \sum_{\sigma=-\infty}^{\infty}\sum_{\rho=0}^{\infty}(-1)^{\nu}\frac{(-x)_{\rho+\max\{0,\nu-\sigma\}}(-y)_{\rho+\max\{0,\sigma-\nu\}}}{(1)_{\rho+\max\{0,\nu-\sigma\}}(1)_{\rho+\max\{0,\sigma-\nu\}}}\alpha^{2\rho+n+|\sigma-\nu|}\zeta^{\sigma} .$$

In deriving this expansion we first replaced the index σ by $\sigma+\rho-\nu$, then changed the summation order

$$\sum_{\rho=0}^{\infty}\sum_{\sigma=\nu-\rho}^{\infty} = \sum_{\sigma=-\infty}^{\infty}\sum_{\rho=\max\{0,\nu-\sigma\}}^{\infty}$$

and, lastly, wrote $\rho+\max\{0,\nu-\sigma\}$ instead of ρ. Hence, we finally obtain

$$\gamma(n,x,y,\nu,\alpha,\zeta) = \sum_{\sigma=-\infty}^{\infty}\gamma_{\sigma}(n,x,y,\nu,\alpha)\zeta^{\sigma} \tag{2.3.4}$$

with

$$\gamma_{\sigma}(n,x,y,\nu,\alpha) = (-1)^{\nu}\frac{(-x)_{\max\{0,\nu-\sigma\}}(-y)_{\max\{0,\sigma-\nu\}}}{(1)_{|\sigma-\nu|}}\alpha^{|\sigma-\nu|+n}\times$$
$$\times F(-x+\max\{0,\nu-\sigma\}, -y+\max\{0,\sigma-\nu\}, 1+|\sigma-\nu|, \alpha^2) . \tag{2.3.5}$$

This form enables one to use different transformations of the hypergeometric function. In particular, if α^2 is not so small it may be suitable to apply the Euler transformation

$$F(\alpha,\beta,\gamma,z) = (1-z)^{\gamma-\alpha-\beta}F(\gamma-\alpha,\gamma-\beta,\gamma,z) \tag{2.3.6}$$

to rewrite (2.3.5) as

$$\gamma_\sigma(n,x,y,\nu,\alpha) = (-1)^\nu \frac{(-x)_{\max\{0,\nu-\sigma\}}(-y)_{\max\{0,\sigma-\nu\}}}{(1)_{|\sigma-\nu|}} \frac{\alpha^{|\sigma-\nu|+n}}{(1-\alpha^2)^{-x-y-1}} \times$$
$$\times F(1+x+\max\{0,\sigma-\nu\}, 1+y+\max\{0,\nu-\sigma\}, 1+|\sigma-\nu|, \alpha^2)\,. \tag{2.3.7}$$

Let us remember that if (at least) one of the first two parameters of $F(\alpha,\beta,\gamma,z)$ is a negative integer then the hypergeometric series is, in fact, the hypergeometric polynomial.

For the first time we meet function (2.3.1) in deriving analytic expressions for the Hansen coefficients. From defining relation (2.2.6) it follows that

$$X_s^{n,m} = \frac{1}{2\pi}\int_0^{2\pi} \left(\frac{r}{a}\right)^n \exp i(mv - sM)\, dM\,. \tag{2.3.8}$$

Along with the eccentricity e and parameter η defined by (2.1.9) many elliptic-motion expansions involve the quantity

$$\beta = \frac{e}{1+\eta}\,. \tag{2.3.9}$$

Evidently,

$$\frac{1}{\beta} = \frac{e}{1-\eta}\,, \qquad \beta^2 = \frac{1-\eta}{1+\eta}\,, \qquad 1+\beta^2 = \frac{2}{1+\eta} \tag{2.3.10}$$

and

$$\eta = \frac{1-\beta^2}{1+\beta^2}\,, \qquad e = \frac{2\beta}{1+\beta^2}\,. \tag{2.3.11}$$

Putting now

$$\sigma = \exp iv\,, \qquad \zeta = \exp ig\,, \qquad \tau = \exp iM \tag{2.3.12}$$

and considering that

$$\sigma^2 + \frac{2}{e}\sigma + 1 = \frac{1}{\beta}\sigma(1+\beta\sigma^{-1})(1+\beta\sigma)\,, \tag{2.3.13}$$

$$\zeta^2 - \frac{2}{e}\zeta + 1 = -\frac{1}{\beta}\zeta(1-\beta\zeta^{-1})(1-\beta\zeta) \tag{2.3.14}$$

one may express r/a either in terms of σ or in terms of ζ as follows:

$$\frac{r}{a} = \frac{(1-\beta^2)^2}{1+\beta^2}(1+\beta\sigma^{-1})^{-1}(1+\beta\sigma)^{-1} \tag{2.3.15}$$

or

$$\frac{r}{a} = (1+\beta^2)^{-1}(1-\beta\zeta^{-1})(1-\beta\zeta)\,. \tag{2.3.16}$$

Indeed, (2.3.15) and (2.3.16) are nothing other than (2.1.3) and (2.1.6), respectively. Trigonometric relations (2.1.7) and (2.1.8) involve the relationships between σ and ζ

$$\sigma = \zeta(1-\beta\zeta^{-1})(1-\beta\zeta)^{-1} \tag{2.3.17}$$

and

$$\zeta = \sigma(1+\beta\sigma^{-1})(1+\beta\sigma)^{-1} . \tag{2.3.18}$$

On the other hand, the Kepler equation (2.1.10) may be rewritten in exponential form:

$$\tau = \zeta \exp\left[-\frac{e}{2}(\zeta-\zeta^{-1})\right] . \tag{2.3.19}$$

Changing the independent argument in (2.3.8) by (2.1.25) and making use of (2.3.16), (2.3.17) and (2.3.19) one gets

$$X_s^{n,m} = \frac{1}{2\pi}(1+\beta^2)^{-n-1}\int_0^{2\pi}(1-\beta\zeta^{-1})^{n+1+m}(1-\beta\zeta)^{n+1-m}\zeta^{m-s}\times$$
$$\times \exp\left[\frac{se}{2}(\zeta-\zeta^{-1})\right] dg . \tag{2.3.20}$$

Depending on the order of treating the multipliers in the integrand function one obtains the formulae of Hansen or Hill. In Hansen's approach one starts with the binomial and exponential series to produce

$$(1-\beta\zeta)^{n-m+1}\exp(\nu\beta\zeta) = \sum_{k=0}^{\infty} P_k^{(n)}(m,\nu)\beta^k\zeta^k \tag{2.3.21}$$

with

$$\nu = \frac{s}{1+\beta^2} . \tag{2.3.22}$$

The coefficients of (2.3.21) are polynomials in ν:

$$P_k^{(n)}(m,\nu) = \frac{\nu^{k-q}}{(1)_{k-q}}\sum_{r=0}^{q}\frac{(-n+m-1)_{q-r}}{(1)_{q-r}}\frac{\nu^r}{(1+k-q)_r} \tag{2.3.23}$$

with

$$q = \begin{cases} \min\{k, n-m+1\} , & n-m+1 \geq 0 , \\ k , & n-m+1 < 0 . \end{cases}$$

These polynomials differ only in the factor $(-1)^k$ from the Laguerre polynomials $L_k^{\alpha}(\nu)$:

$$P_k^{(n)}(m,\nu) = (-1)^k L_k^{n-m+1-k}(\nu) , \tag{2.3.24}$$

or taking into account the relationship

$$L_k^\alpha(\nu) = \frac{(\alpha+1)_k}{(1)_k}\Phi(-k,\alpha+1,\nu) \tag{2.3.25}$$

between the Laguerre polynomials and the confluent hypergeometric functions $\Phi(\alpha,\gamma,\nu)$ they may also be expressed in the form

$$P_k^{(n)}(m,\nu) = \frac{(-n+m-1)_k}{(1)_k}\Phi(-k,n-m+2-k,\nu)\,. \tag{2.3.26}$$

Two other multipliers in (2.3.20) result in

$$(1-\beta\zeta^{-1})^{n+m+1}\exp(-\nu\beta\zeta^{-1}) = \sum_{j=0}^{\infty} P_j^{(n)}(-m,-\nu)\beta^j\zeta^{-j}\,. \tag{2.3.27}$$

Substituting (2.3.23) and (2.3.27) into (2.3.20) one has

$$X_s^{n,m} = (1+\beta^2)^{-n-1}\sum_{k=0}^{\infty}\sum_{j=0}^{\infty} P_k^{(n)}(m,\nu)P_j^{(n)}(-m,-\nu)\beta^{k+j}\times$$
$$\times\frac{1}{2\pi}\int_0^{2\pi}\zeta^{k-j+m-s}\,dg\,.$$

Replacing k by $k+j-m+s$ here, changing the order of summation

$$\sum_{j=0}^{\infty}\;\sum_{k=-j+m-s}^{\infty} = \sum_{k=-\infty}^{\infty}\;\sum_{j=\max\{0,-k+m-s\}}^{\infty}$$

and replacing j by $j+\max\{0,-k+m-s\}$ one gets

$$X_s^{n,m} = (1+\beta^2)^{-n-1}\sum_{k=-\infty}^{\infty}\sum_{j=0}^{\infty} P_{j+max\{0,k-m+s\}}^{(n)}(m,\nu)\times$$
$$\times P_{j+\max\{0,-k+m-s\}}^{(n)}(-m,-\nu)\beta^{2j+|k-m+s|}\frac{1}{2\pi}\int_0^{2\pi}\zeta^k\,dg\,.$$

The integral term is evidently equal to $2\pi\delta_{k0}$ so that the first sum actually reduces to a single term, and we finally get

$$X_s^{n,m} = \beta^{|m-s|}(1+\beta^2)^{-n-1}\sum_{j=0}^{\infty}P_{j+\max\{0,-m+s\}}^{(n)}(m,\nu)\times$$
$$\times P_{j+\max\{0,m-s\}}^{(n)}(-m,-\nu)\beta^{2j}\,. \tag{2.3.28}$$

This is the Hansen-type formula, expressing the coefficients $X_s^{n,m}$ in terms of β. In Hill's approach one combines together the first three multipliers in (2.3.20) to get

$$X_s^{n,m} = (1+\beta^2)^{-n-1}\frac{(-1)^{m-s}}{2\pi}\int_0^{2\pi}\gamma(0,n+1+m,n+1-m,m-s,\beta,\zeta)\times$$
$$\times \exp\left[\frac{se}{2}(\zeta-\zeta^{-1})\right]\,dg$$

with the function (2.3.1). Using expansion (2.3.4) here and the expansion

$$\exp\left[\frac{se}{2}(\zeta-\zeta^{-1})\right] = \sum_{j=-\infty}^{\infty} J_j(se)\zeta^j \tag{2.3.29}$$

involving the Bessel coefficients, one obtains

$$X_s^{n,m} = (1+\beta^2)^{-n-1}\sum_{k=-\infty}^{\infty}\sum_{j=-\infty}^{\infty}\gamma_k(0,n+1+m,n+1-m,m-s,\beta)\times$$
$$\times J_j(se)\frac{(-1)^{m-s}}{2\pi}\int_0^{2\pi}\zeta^{k+j}\,dg\,.$$

The last integral survives only for $j=-k$. Taking into account that

$$J_{-k}(se) = (-1)^k J_k(se)$$

we finally have the Hill-type formula

$$X_s^{n,m} = (1+\beta^2)^{-n-1}\times$$
$$\times\sum_{k=-\infty}^{\infty}(-1)^{k+m-s}\gamma_k(0,n+1+m,n+1-m,m-s,\beta)J_k(se) \tag{2.3.30}$$

with coefficients (2.3.5). Formula (2.3.28) shows directly that $X_s^{n,m}$ is of the order $|m-s|$ with respect to e (or β). This follows, of course, from (2.3.30) also. Indeed, the general term in (2.3.30) has the order

$$|m-s-k|+|k|\geq|m-s|\,.$$

But in using (2.3.30) we have to sum $|m-s|+1$ terms (for $k=0,1,\ldots,m-s$ in the case $m-s\geq 0$ and for $k=m-s,m-s+1,\ldots,0$ in the case $m-s\leq 0$) having the same order of magnitude $|m-s|$ in e. In this respect (2.3.28) may be more advantageous for numerical computation. To derive (2.2.8) in analytical form it is sufficient to replace β by e in (2.3.28) using a CAS such as MATHEMATICA, for instance. The same aim may be achieved, of course, by using recurrence relations (see, for instance, Hughes, 1981) or by applying some Keplerian processor but we are interested here in the general terms of elliptic-motion expansions. As is well known, the Hansen coefficients for $s = 0$ may be expressed in closed form. Indeed, from (2.3.28) we have

$$X_0^{n,m} = \beta^{|m|}(1+\beta^2)^{-n-1}\sum_{s=0}^{\infty} P_s^{(n)}(|m|,0)P_{s+|m|}^{(n)}(-|m|,0)\beta^{2s}$$

or

$$X_0^{n,m} = \frac{(-n-|m|-1)_{|m|}}{(1)_{|m|}}\beta^{|m|}(1+\beta^2)^{-n-1}\times$$
$$\times F(-n+|m|-1,-n-1,1+|m|,\beta^2). \qquad (2.3.31)$$

This formula is convenient when $|m|-1 > n \geq -1$. It also shows that $X_0^{n,m} = 0$ for $-1 > n \geq -|m|-1$. Using the quadratic transformation formula

$$F(2\alpha, 2\alpha+1-\gamma,\gamma,z) = (1+z)^{-2\alpha}F\left(\alpha,\alpha+\tfrac{1}{2},\gamma,\frac{4z}{(1+z)^2}\right) \qquad (2.3.32)$$

one obtains from (2.3.31)

$$X_0^{n,m} = \frac{(-n-|m|-1)_{|m|}}{(1)_{|m|}}\left(\frac{e}{2}\right)^{|m|}\times$$
$$\times F\left(\frac{|m|-n-1}{2},\frac{|m|-n}{2},1+|m|,e^2\right), \qquad (2.3.33)$$

which results in the closed-form expression for $n \geq |m|-1$ and is more compact than (2.3.31). Applying here Euler transformation (2.3.6) one gets

$$X_0^{n,m} = \frac{(-n-|m|-1)_{|m|}}{(1)_{|m|}}\left(\frac{e}{2}\right)^{|m|}(1-e^2)^{n+\frac{3}{2}}\times$$
$$\times F\left(\frac{|m|+n+2}{2},\frac{|m|+n+3}{2},1+|m|,e^2\right), \qquad (2.3.34)$$

which is convenient for obtaining the closed-form expression for $n < -|m|-1$.

Expansions of elliptic motion in multiples of v or g also often occur in celestial-mechanics perturbation theory. The general terms of these expansions are much simpler than the Hansen coefficients of the (e, M) expansions (Brumberg E. and Fukushima, 1994). Indeed, consider the expansions

$$\left(\frac{r}{a}\right)^n \exp imv = \sum_{s=-\infty}^{\infty} Y_s^{n,m}\exp isv \qquad (2.3.35)$$

and

$$\left(\frac{r}{a}\right)^n \exp imv = \sum_{s=-\infty}^{\infty} Z_s^{n,m}\exp isg\,. \qquad (2.3.36)$$

From (2.3.15) in terms of σ it follows that

$$\left(\frac{r}{a}\right)^n \exp imv = \frac{(1-\beta^2)^{2n}}{(1+\beta^2)^n}\gamma(0,-n,-n,m,\beta,-\sigma) \tag{2.3.37}$$

and hence

$$Y_s^{n,m} = (-1)^{|m-s|}\frac{(n)_{|m-s|}}{(1)_{|m-s|}}\beta^{|m-s|}\frac{(1-\beta^2)^{2n}}{(1+\beta^2)^n}\times$$
$$\times F(n, n+|m-s|, 1+|m-s|, \beta^2)\,. \tag{2.3.38}$$

On the other hand, in terms of ζ from (2.3.16) we have

$$\left(\frac{r}{a}\right)^n \exp imv = (1+\beta^2)^{-n}(-1)^m\gamma(0,n+m,n-m,m,\beta,\zeta)\,, \tag{2.3.39}$$

involving

$$Z_s^{n,m} = \frac{(-n-m)_{\max\{0,m-s\}}(-n+m)_{\max\{0,s-m\}}}{(1)_{|m-s|}}\beta^{|m-s|}(1+\beta^2)^{-n}\times$$
$$\times F(-n-m+\max\{0,m-s\}, -n+m+\max\{0,s-m\},$$
$$1+|m-s|, \beta^2)\,. \tag{2.3.40}$$

Expressions (2.3.38) and (2.3.40) may always be reduced to finite (closed-form) polynomials by repeated application of the Gauss relations for the adjacent hypergeometric functions or just by using the Euler transformation (2.3.6). Indeed, in this case instead of (2.3.38) and (2.3.40) one has, respectively,

$$Y_s^{n,m} = (-1)^{|m-s|}\frac{(n)_{|m-s|}}{(1)_{|m-s|}}\beta^{|m-s|}\frac{1-\beta^2}{(1+\beta^2)^n}\times$$
$$\times F(1-n,\ 1+|m-s|-n,\ 1+|m-s|,\ \beta^2) \tag{2.3.41}$$

and

$$Z_s^{n,m} = \frac{(-n-m)_{\max\{0,m-s\}}(-n+m)_{\max\{0,s-m\}}}{(1)_{|m-s|}}\beta^{|m-s|}\frac{(1-\beta^2)^{2n+1}}{(1+\beta^2)^n}\times$$
$$\times F(1+n+m+\max\{0,s-m\}, 1+n-m+\max\{0,m-s\},$$
$$1+|m-s|, \beta^2)\,. \tag{2.3.42}$$

Formulae (2.3.38) and (2.3.40), first obtained in a slightly different form by Brown and Shook (1933), enable one to derive in a straightforward manner the well-known classical expansions. Indeed, noticing that

$$F(1, 1+|s|, 1+|s|, \beta^2) = \frac{1}{1-\beta^2}\,,$$

$$F(2, 2+|s|, 1+|s|, \beta^2) = \frac{1+\beta^2}{(1-\beta^2)^3}\,\frac{1+|s|\eta}{1+|s|}$$

we find that

$$Y_s^{1,0} = (-1)^s \eta \beta^{|s|}\,, \qquad Y_s^{2,0} = (-1)^s \eta \beta^{|s|}(1 + |s|\eta) \tag{2.3.43}$$

and

$$Z_s^{-1,0} = \frac{1}{\eta}\beta^{|s|}\,, \qquad Z_s^{-2,0} = \frac{1}{\eta^3}\beta^{|s|}(1 + |s|\eta)\,. \tag{2.3.44}$$

The difference $v - g$ between the true anomaly v and the eccentric anomaly g may now be expressed by a very simple series in multiples of v or g. From (2.1.25) it follows that

$$\frac{dg}{dv} = \frac{1}{\eta}\frac{r}{a} = \frac{1}{\eta}\left(Y_0^{1,0} + 2\sum_{k=1}^{\infty} Y_k^{1,0}\cos kv\right),$$

and after integration

$$v - g = 2\sum_{k=1}^{\infty}\frac{(-1)^{k-1}}{k}\beta^k \sin kv\,. \tag{2.3.45}$$

On the other hand,

$$\frac{dv}{dg} = \eta\frac{a}{r} = \eta\left(Z_0^{-1,0} + 2\sum_{k=1}^{\infty} Z_k^{-1,0}\cos kg\right),$$

and after integration

$$v - g = 2\sum_{k=1}^{\infty}\frac{1}{k}\beta^k \sin kg\,. \tag{2.3.46}$$

The equation of the centre $v - M$ may also be expressed quite easily in terms of v or g. Again from (2.1.25) one gets

$$\frac{dM}{dv} = \frac{1}{\eta}\frac{r^2}{a^2} = \frac{1}{\eta}\left(Y_0^{2,0} + 2\sum_{k=1}^{\infty} Y_k^{2,0}\cos kv\right)$$

or

$$v - M = 2\sum_{k=1}^{\infty}\frac{(-1)^{k-1}}{k}\beta^k(1 + k\eta)\sin kv\,. \tag{2.3.47}$$

In terms of g in virtue of the Kepler equation (2.1.10) and expansion (2.3.46) one has

$$v - M = 2\sum_{k=1}^{\infty}\frac{1}{k}\beta^k\left(1 + \frac{\delta_{k,1}}{1+\beta^2}\right)\sin kg\,. \tag{2.3.48}$$

Finally, from (2.3.45) and (2.3.47) we have

$$g - M = 2\eta \sum_{k=1}^{\infty} (-1)^{k-1} \beta^k \sin kv \tag{2.3.49}$$

while in terms of g we simply have the Kepler equation (2.1.10). In a similar way one can get

$$\left(\frac{r}{a}\right)^{-2} = Z_0^{-2,0} + 2\sum_{k=1}^{\infty} Z_k^{-2,0} \cos kg =$$

$$= \frac{1}{\eta^3}\left(1 + 2\sum_{k=1}^{\infty} \beta^k (1 + k\eta) \cos kg\right) \tag{2.3.50}$$

and so on. In terms of M all these expansions become more complicated since their coefficients are now represented by an infinite series (Hansen or Bessel coefficients). For example,

$$\frac{dv}{dM} = \eta \frac{a^2}{r^2} = \eta \sum_{k=0}^{\infty} (2 - \delta_{k0}) X_k^{-2,0}(e) \cos kM\,,$$

and considering from (2.3.34) that $X_0^{-2,0} = \eta^{-1}$ we have

$$v - M = 2\eta \sum_{k=1}^{\infty} \frac{1}{k} X_k^{-2,0}(e) \sin kM\,. \tag{2.3.51}$$

Similarly,

$$\frac{dg}{dM} = \frac{a}{r} = \sum_{k=0}^{\infty} (2 - \delta_{k0}) X_k^{-1,0}(e) \cos kM\,,$$

and considering from (2.3.33) that $X_0^{-1,0} = 1$ we have

$$g - M = 2\sum_{k=1}^{\infty} \frac{1}{k} X_k^{-1,0}(e) \sin kM\,. \tag{2.3.52}$$

The expressions for the general terms of the expansions for the simple functions of the eccentric anomaly g are usually given with the aid of Bessel coefficients. But in such cases Hansen coefficients reduce automatically to Bessel coefficients when we apply formulae (2.3.30) and (2.3.5). For example, for $n = -1$, $m = 0$ the sum in (2.3.30) will consist of a single term with $k = -s$ so that

$$X_s^{-1,0}(e) = (-1)^s J_{-s}(se) = J_s(se)\,,$$

and expansion (2.3.52) can be written in the familiar form

$$g - M = 2\sum_{k=1}^{\infty} \frac{1}{k} J_k(ke) \sin kM\,. \tag{2.3.53}$$

Nowadays it is possible to combine the tools of Keplerian processors with the classical expansions dealt with by universal CAS like MATHEMATICA or MAPLE.

2.4 The Keplerian Processor in Taylor-Series Form

Expansions of the coordinates of the two-body problem in series in powers of time are of less use than the other two forms of the solution. However, these expansions are actually applied in determining orbits from observations and in various problems of astrodynamics dealing with the representation of motion for fairly short intervals of time. They also occur in theoretical investigations based on Taylor series and their modifications such as series of polynomials, etc. We consider here two versions of the Taylor expansions in the two-body problem. The first version involves the representation

$$\mathbf{r}(\tau) = \mathbf{r}_0 F(\tau) + \dot{\mathbf{r}}_0 G(\tau) \tag{2.4.1}$$

with $\tau = \sqrt{Gm}(t - t_0)$ and $\mathbf{r}_0$, $\dot{\mathbf{r}}_0$ being the coordinates and velocity components at the initial moment of time t_0. Here and below in this section the dot denotes differentiation with respect to τ. For all three types of Keplerian motion, with the exception only of the degenerate case of rectilinear motion, the functions $F(\tau)$ and $G(\tau)$ are holomorphic in τ:

$$F(\tau) = \sum_{n=0}^{\infty} f_n \frac{\tau^n}{(1)_n}, \qquad G(\tau) = \sum_{n=0}^{\infty} g_n \frac{\tau^n}{(1)_n} \tag{2.4.2}$$

with

$$f_0 = 1, \qquad g_0 = 0, \qquad f_1 = 0, \qquad g_1 = 1. \tag{2.4.3}$$

In a universal treatment of all three types of Keplerian motion one often makes use of the so-called local invariants

$$\mu = \frac{1}{r^3}, \qquad \sigma = \frac{\mathbf{r}\dot{\mathbf{r}}}{r^2}, \qquad \epsilon = \frac{1}{r^2}\left(\dot{\mathbf{r}}^2 - \frac{1}{r}\right).$$

The local invariants satisfy the equations

$$\dot{\mu} = -3\mu\sigma, \qquad \dot{\sigma} = \epsilon - 2\sigma^2, \qquad \dot{\epsilon} = -\sigma(\mu + 2\epsilon). \tag{2.4.4}$$

The coefficients f_n and g_n represent polynomials with integer coefficients with respect to the local invariants (2.4.3) taken at the initial moment of time. Most simply they are determined from the recurrence relations

$$f_n = \dot{f}_{n-1} - \mu g_{n-1} \tag{2.4.5}$$

and

$$g_n = f_{n-1} + \dot{g}_{n-1} \tag{2.4.6}$$

with initial values (2.4.3). In taking derivatives of f_n and g_n with respect to μ, σ and ϵ one should use relations (2.4.4). The first few coefficients are as follows:

$$\begin{aligned}
f_2 &= -\mu\,, & g_2 &= 0\,,\\
f_3 &= 3\sigma\mu\,, & g_3 &= -\mu\,,\\
f_4 &= -15\sigma^2\mu + 3\epsilon\mu + \mu^2\,, & g_4 &= 6\sigma\mu\,,\\
f_5 &= 105\sigma^3\mu - \sigma(45\epsilon\mu + 15\mu^2)\,, & g_5 &= -45\sigma^2\mu + 9\epsilon\mu + \mu^2\,,\\
&\ldots
\end{aligned}$$

The algorithm based on recurrence relations (2.4.5) and (2.4.6) was realized with the aid of FORMAC (Sconzo et al., 1965). This was one of the first CAS applications in celestial mechanics. The theory of the local invariants is investigated in the first volume of the treatise by Stumpff (1959). Stumpff (1943–1947) also suggested an algorithm to derive the general terms of the series (2.4.2) without using recurrence relations (2.4.5) and (2.4.6), but the relevant formulae are too cumbersome. The technique of the F and G series was generalized for the N-body problem by Papadakos (1983).

Consider now another algorithm to develop the Taylor expansions in the two-body problem (Brumberg, 1963). Rewriting solution (2.1.4) for the elliptic case in the form

$$\mathbf{r} = \mathbf{P}X + \mathbf{Q}Y\,, \tag{2.4.7}$$

$$X = a(\cos g - e)\,, \qquad Y = a\eta \sin g \tag{2.4.8}$$

we expand the orbital coordinates X and Y in powers of the mean anomaly M:

$$X = a\sum_{k=0}^{\infty} a_k M^{2k}\,, \qquad Y = a\sum_{k=0}^{\infty} b_k M^{2k+1}\,. \tag{2.4.9}$$

This representation is not so general as (2.4.2) but the coefficients a_k and b_k depend only on the eccentricity and hence have much simpler form. The functions $X^* = X/a$ and $Y^* = Y/a$ satisfy the differential equations

$$\begin{aligned}
(1 - e^2 - eX^*)\frac{dX^*}{dM} &= -\frac{1}{\eta}Y^*\,,\\
(1 - e^2 - eX^*)\frac{dY^*}{dM} &= \eta(X^* + e)\,.
\end{aligned} \tag{2.4.10}$$

Substitution of (2.4.9) results in the recurrence relations

$$\begin{aligned}
(2k+1)(1-e)b_k &= e\sum_{j=1}^{k}(2j-1)b_{j-1}a_{k-j+1} + \eta a_k\,,\\
(k+1)(1-e)a_{k+1} &= e\sum_{j=1}^{k} j a_j a_{k-j+1} - \frac{1}{2\eta}b_k
\end{aligned} \tag{2.4.11}$$

with initial values

$$a_0 = 1 - e\,, \qquad b_0 = \frac{1+e}{\eta}\,, \qquad a_1 = -\frac{1}{2(1-e)^2}\,. \tag{2.4.12}$$

Relations (2.4.11) imply the general form of the coefficients a_k and b_k

$$a_k = \frac{(-1)^k \tilde{a}_k}{(1)_{2k}(1-e)^{3k-1}}\,, \qquad b_k = \frac{(-1)^k \eta \tilde{b}_k}{(1)_{2k+1}(1-e)^{3k+1}}\,, \tag{2.4.13}$$

$\tilde{a}_k$ and $\tilde{b}_k$ $(k = 1, 2, \ldots)$ being polynomials of degree $k-1$ in e with positive integer coefficients

$$\tilde{a}_k = \sum_{j=0}^{k-1} a_j^{(k)} e^j\,, \qquad \tilde{b}_k = \sum_{j=0}^{k-1} b_j^{(k)} e^j\,. \tag{2.4.14}$$

To calculate these polynomials one may use the recurrence relations resulting from (2.4.11)

$$\begin{aligned} \tilde{a}_k &= -e \sum_{j=1}^{k-1} \frac{(-2k+1)_{2j-1}}{(1)_{2j-1}} \tilde{a}_j \tilde{a}_{k-j} + \tilde{b}_{k-1}\,, \\ \tilde{b}_k &= e \sum_{j=1}^{k-1} \frac{(-2k)_{2j}}{(1)_{2j}} \tilde{b}_j \tilde{a}_{k-j} + \tilde{a}_k\,. \end{aligned} \tag{2.4.15}$$

The first few polynomials calculated by means of these formulae are as follows:

$$\begin{aligned} \tilde{a}_1 &= 1\,, \\ \tilde{a}_2 &= 1 + 3e\,, \\ \tilde{a}_3 &= 1 + 24e + 45e^2\,, \\ \tilde{a}_4 &= 1 + 117e + 1107e^2 + 1575e^3\,, \\ \tilde{a}_5 &= 1 + 498e + 15066e^2 + 85410e^3 + 99225e^4\,, \\ &\ldots \end{aligned}$$

and

$$\begin{aligned} \tilde{b}_1 &= 1\,, \\ \tilde{b}_2 &= 1 + 9e\,, \\ \tilde{b}_3 &= 1 + 54e + 225e^2\,, \\ \tilde{b}_4 &= 1 + 243e + 4131e^2 + 11025e^3\,, \\ \tilde{b}_5 &= 1 + 1008e + 50166e^2 + 457200e^3 + 893025e^4\,, \\ &\ldots\,. \end{aligned}$$

A more extended table of these polynomials is given by Kuzmin (1980). The coefficients of polynomials (2.4.14) may be computed also with the aid of the closed-form expressions

$$a_j^{(k+1)} = \sum_{i=0}^{j} (i+1) \frac{(-3k-2)_{j-i}}{2^i (1)_{j-i}} d_{ik} \tag{2.4.16}$$

and

$$b_j^{(k)} = \sum_{i=0}^{j} \frac{(-3k-1)_{j-i}}{2^i (1)_{j-i}} d_{ik} \tag{2.4.17}$$

with

$$d_{ik} = \sum_{s=0}^{E(\frac{i}{2})} \frac{(-1)^s (i+1-2s)^{i+2k+1}}{(1)_s (1)_{i+1-s}} . \tag{2.4.18}$$

2.5 The Keplerian Processor with the Aid of Elliptic Functions

As mentioned above, one more version of the Keplerian processor may be developed with the aid of elliptic-function theory (Brumberg E., 1992). Elliptic functions themselves act only as an intermediate step to introduce more convenient variables than the eccentricity e and mean anomaly M. Indeed, regarding e as the modulus k of Jacobi elliptic functions and using the well-known trigonometric expansions for elliptic functions with closed-form (rational) coefficients with respect to the Jacobi nome $q = q(k)$ one may replace the classic (e, M) trigonometric expansions by the (q, w) expansions in multiples of the new (elliptic) anomaly w, which are related linearly to the argument u of elliptic functions. Even for large eccentricities the nome q remains rather small (for instance, $q = 0.2622\ldots$ for $k^2 = 0.99$). Not only are the coefficients of the (q, w) expansions much more compact compared to the (e, M) expansions but so are these trigonometric expansions themselves and they may be easily used in perturbation theory.

The closed-form representation of the two-body-problem solution in terms of elliptic functions results from the transformation

$$k = e \tag{2.5.1}$$

and

$$\operatorname{am} u = g + \frac{\pi}{2} \tag{2.5.2}$$

for the modulus k and argument u of the Jacobi elliptic functions (Brumberg E., 1992). Hence,

$$\sin g = -\operatorname{cn} u\,, \qquad \cos g = \operatorname{sn} u\,, \tag{2.5.3}$$

and for the orbital coordinates X, Y defined by (2.4.8) and for the radius-vector r one has

$$X = a(\operatorname{sn} u - k)\,, \qquad Y = -ak' \operatorname{cn} u \tag{2.5.4}$$

and

$$r = a(1 - k \operatorname{sn} u)\,, \tag{2.5.5}$$

where the complementary modulus $k' = (1-k^2)^{1/2}$ now replaces the quantity η introduced above by (2.1.9). From (2.1.7) and (2.1.8) it follows that

$$\sin v = -\frac{k' \operatorname{cn} u}{1 - k \operatorname{sn} u}\,, \qquad \cos v = \frac{\operatorname{sn} u - k}{1 - k \operatorname{sn} u}\,. \tag{2.5.6}$$

These relations may be rewritten as

$$\exp \mathrm{i}v = \frac{a}{r}(\operatorname{sn} u - k - \mathrm{i}k' \operatorname{cn} u) \tag{2.5.7}$$

or

$$\exp \mathrm{i}v = \frac{1}{\operatorname{dn}^2 u}(k' - \mathrm{i}k \operatorname{cn} u)(k' \operatorname{sn} u - \mathrm{i} \operatorname{cn} u)\,.$$

In terms of elliptic functions the Kepler equation takes the form

$$\operatorname{am} u + k \operatorname{cn} u = M + \frac{\pi}{2}\,. \tag{2.5.8}$$

Relationships between M, g, v and u may be expressed in differential form as follows:

$$dM = (1 - k \operatorname{sn} u) \operatorname{dn} u\, du\,, \tag{2.5.9}$$

$$dg = \operatorname{dn} u\, du \tag{2.5.10}$$

and

$$dv = k' \frac{1 + k \operatorname{sn} u}{\operatorname{dn} u}\, du\,. \tag{2.5.11}$$

One benefit of introducing elliptic functions is that they make it possible to obtain the closed-form solution of many modelling problems of celestial mechanics (the theory of intermediate orbits, typical resonance schemes, rotation models, etc.). But the main advantage in dealing with elliptic functions to construct analytical theories of motion is the possibility of using their expansions in fast converging trigonometric series with rational coefficients with respect to q, k, k', K and E. Here q is the Jacobi nome determined by

$$q = \exp\left(-\frac{\pi K'}{K}\right), \qquad K' = K(k')\,, \tag{2.5.12}$$

and $K = K(k)$ and $E = E(k)$ are complete elliptic integrals of the first and second kinds, respectively. The standard trigonometric expansions listed below may be found in many textbooks and handbooks such as those by Whittaker and Watson (1935), Magnus and Oberhettinger (1949), Erdélyi (1954), Abramowitz and Stegun (1965), Byrd and Friedman (1971), Gradshteyn and Ryzhik (1980). On applying these to the elliptic two-body problem we see that these standard expansions involve the following ones (Brumberg E., 1992):

$$\sin g = \frac{2\pi}{kK}\sum_{m=1}^{\infty}(-1)^{m+1}\frac{q^{m-\frac{1}{2}}}{1+q^{2m-1}}\sin(2m-1)w\,, \tag{2.5.13}$$

$$\cos g = \frac{2\pi}{kK}\sum_{m=1}^{\infty}(-1)^{m+1}\frac{q^{m-\frac{1}{2}}}{1-q^{2m-1}}\cos(2m-1)w\,, \tag{2.5.14}$$

$$g = w + 2\sum_{m=1}^{\infty}\frac{(-1)^m}{m}\frac{q^m}{1+q^{2m}}\sin 2mw \tag{2.5.15}$$

and

$$\sin v = \frac{\pi^2}{kK^2}\sum_{m=1}^{\infty} m\frac{q^{\frac{m}{2}}}{1+q^m}\sin mw\,, \tag{2.5.16}$$

$$\cos v = \frac{E-K}{kK} + \frac{\pi^2}{kK^2}\sum_{m=1}^{\infty} m\frac{q^{\frac{m}{2}}}{1-q^m}\cos mw\,, \tag{2.5.17}$$

$$v = w + 4\sum_{m=1}^{\infty}\frac{1}{m}\frac{q^{\frac{m}{2}}}{1+q^m}\sin mw\,, \tag{2.5.18}$$

where w stands for

$$w = \frac{\pi u}{2K} - \frac{\pi}{2}\,. \tag{2.5.19}$$

We shall call w the elliptic anomaly although this quantity differs from the elliptic anomaly introduced by Janin and Bond (see Bond and Janin, 1981; Nacozy, 1977; Ferrandiz et al., 1987). Expressions (2.5.13)–(2.5.15) follow directly from substituting the standard series for $\operatorname{sn} u$, $\operatorname{cn} u$ and $\operatorname{am} u$ into (2.5.3) and (2.5.2), respectively. Expressions (2.5.6) may be rewritten in the form

$$\sin v = -\frac{k'}{k}\frac{d}{du}\frac{1+k\operatorname{sn} u}{\operatorname{dn} u}\,, \tag{2.5.20}$$

$$\cos v = -k\operatorname{sn}^2 u - \frac{d}{du}\frac{(1+k\operatorname{sn} u)\operatorname{cn} u}{\operatorname{dn} u}\,. \tag{2.5.21}$$

Substituting into (2.5.20), (2.5.21) and (2.5.11) the standard series for $1/\operatorname{dn} u$, $\operatorname{sn} u/\operatorname{dn} u$, $\operatorname{cn} u/\operatorname{dn} u$, $\operatorname{sn} u \operatorname{cn} u/\operatorname{dn} u$ and $\operatorname{sn}^2 u$ one obtains expansions (2.5.16)–(2.5.18). By analogy with expansions (2.2.6), (2.3.35) and

(2.3.36) in multiples of the mean, true and eccentric anomalies, respectively, one can consider expansions in multiples of the elliptic anomaly

$$\left(\frac{r}{a}\right)^n \exp imv = \sum_{s=-\infty}^{\infty} B_s^{n,m}(q) \exp isw . \tag{2.5.22}$$

The coefficients $B_s^{n,m}(q)$, known as elliptic Hansen coefficients (Brumberg E. and Fukushima, 1994), may be presented in closed form as rational functions of k, k', q, K and E (although such a form may turn out to be too cumbersome in practice and a q-power-series representation might be preferable). They may be calculated by recurrence relations (Brumberg E. and Fukushima, 1994) or by analytical formulae. Such analytical formulae are obtained most simply by proceeding from (2.3.35) or (2.3.36). Proceeding with (2.3.36) one finds from (2.5.13) and (2.5.14) with definition (2.3.12) that

$$\zeta = \frac{2\pi}{kK} \sum_{m=-\infty}^{\infty} (-1)^{\frac{|2m-1|-1}{2}} \frac{q^{\left(1-\frac{1}{2}\,\mathrm{sgn}(2m-1)\right)|2m-1|}}{1-q^{2|2m-1|}} \exp i(2m-1)w . \tag{2.5.23}$$

By raising ζ or $\bar{\zeta} = \zeta^{-1}$ to a positive integer power with the aid of some CAS one can obtain the expansion of ζ^s for arbitrary (positive or negative) s:

$$\zeta^s = \sum_{k=-\infty}^{\infty} (-1)^k A_k^{(s)} \exp i(s+2k)w , \quad A_{-k}^{(s)} = A_k^{(-s)} . \tag{2.5.24}$$

Substituting this expansion into (2.3.36) and comparing the result with (2.5.22) one has

$$B_s^{n,m} = \sum_{k=-\infty}^{\infty} (-1)^k Z_{s-2k}^{n,m} A_k^{(s-2k)} . \tag{2.5.25}$$

This formula was derived by Brumberg E. (1992), where the coefficients $A_k^{(s)}$ were determined by means of the expansion (Brumberg, 1992; Klioner, 1992)

$$\exp i(s \,\mathrm{am}\, u) = \sum_{k=-\infty}^{\infty} A_k^{(s)} \exp i(s+2k)\frac{\pi u}{2K} \tag{2.5.26}$$

and the standard series

$$\mathrm{am}\, u = \frac{\pi u}{2K} + \sum_{m=1}^{\infty} \frac{2}{m} \frac{q^m}{1+q^{2m}} \sin m\frac{\pi u}{K} . \tag{2.5.27}$$

In virtue of the relation $\zeta = -\mathrm{i} \exp(\mathrm{i}\,\mathrm{am}\, u)$ expansions (2.5.24) and (2.5.26) are equivalent. On the other hand, proceeding from (2.3.35) one finds from (2.5.16) and (2.5.17) with definition (2.3.12) that

$$\sigma = \frac{E-K}{kK} + \frac{\pi^2}{kK^2} \sum_{m=-\infty}^{\infty} |m| \frac{q^{(1-\frac{1}{2}\operatorname{sgn} m)|m|}}{1-q^{2|m|}} \exp imw\,. \tag{2.5.28}$$

Again, for any integer s one can obtain

$$\sigma^s = \sum_{k=-\infty}^{\infty} C_k^{(s)} \exp ikw\,, \quad C_{-k}^{(s)} = C_k^{(-s)} \tag{2.5.29}$$

and then

$$B_s^{n,m} = \sum_{k=-\infty}^{\infty} Y_k^{n,m} C_s^{(k)}\,. \tag{2.5.30}$$

In virtue of the evident relation $B_s^{n,m} = B_{-s}^{n,-m}$, valid for all four types of elliptic-motion expansions (2.2.6), (2.3.35), (2.3.36) and (2.5.22), one can always assume that $m \geq 0$. Comparing (2.5.25) and (2.5.30) it is seen that for $n \leq 0$ it is preferable to use (2.5.30) because it reduces to a finite sum with $m+n \leq k \leq m-n$. This results from representation (2.3.37). For $n \geq m$ formula (2.5.25) is preferable, reducing as it does to a finite sum with $s-n \leq 2k \leq s+n$. This follows from representation (2.3.39). For $0 < n < m$ both representations (2.5.25) and (2.5.30) are not so convenient, since they involve infinite sums, but this case is not of practical importance (for example, it does not occur in expanding the disturbing function). For this case it might be more advantageous to proceed from the representation

$$\left(\frac{r}{a}\right)^n \exp imv = (1+\beta^2)^{-n}(1-\beta\zeta^{-1})^{2n}\zeta^n\sigma^{m-n}\,, \tag{2.5.31}$$

which involves only raising to positive powers.

Hence, the coefficients $B_s^{n,m}$ may always be presented in closed form with respect to k, k', q, K and E. Expanding the rational functions of q in powers of q one may get the q-power-series representation of $B_s^{n,m}(q)$. The (q,w) expansions (2.5.22) have been compared by Brumberg E. and Fukushima (1994) with the classical (e,M), (e,v) and (e,g) expansions (2.2.6), (2.3.35) and (2.3.36), respectively, in three respects:

1. applicability to highly eccentric orbits;
2. compactness of the coefficients;
3. compactness of the trigonometric series involved.

The conclusion is that in all respects the (q,w) expansions are at least as good as the (e,v) expansions for $n \leq 0$ or the (e,g) expansions for $n \geq m \geq 0$ and have doubtless advantages compared with the (e,M) expansions. From the operational point of view the derivation of the (q,w) expansions presents no difficulties. Proceeding with the very simple expansions (2.5.15) and (2.5.18) one may develop a Keplerian (q,w) processor just along the same lines as the Keplerian (e,M) processor described in Section 2.2 .

The (e, M) expansions are widespread because they enable one to express the coordinates of the two-body problem as explicit functions of time. But this is not of primary importance for constructing analytical theories of motion. As we shall see below, the integration of the equations of perturbations may be performed using the Hansen device to interrelate different angular arguments. If one wants to use instead of M some other anomaly in canonical equations of motion this may be easily achieved by the technique of Bond and Janin (1981).

The relationship of w with time may be easily deduced by solving the Kepler equation (2.5.8). Substituting expansions (2.5.13) and (2.5.27) onto the left-hand side one has

$$w + \sum_{m=1}^{\infty} d_m \sin mw = M\,. \tag{2.5.32}$$

The coefficients d_m have order $O(q^{m/2})$ and are determined by

$$d_m = (-1)^{E(\frac{m+1}{2})} \frac{2q^{\frac{m}{2}}}{1+q^m} \Delta_m\,, \qquad \Delta_m = \begin{cases} \dfrac{2}{m}\,, & m \text{ is even,} \\ \dfrac{\pi}{K}\,, & m \text{ is odd.} \end{cases} \tag{2.5.33}$$

Equation (2.5.32) belongs to the Lagrange implicit type and may be easily solved with the aid of some CAS by applying the Deprit technique (Deprit, 1979) or the straightforward INVERSE technique described in Section 1.2 . The solution of (2.5.32) given by Brumberg E. (1992) and Klioner (1992) has the form

$$w = M + \sum_{m=1}^{\infty} c_m \sin mM \tag{2.5.34}$$

with the initial terms

$$\begin{aligned}
c_1 &= -d_1 + \tfrac{1}{8}d_1^3 - \tfrac{1}{2}d_1 d_2 - \tfrac{1}{192}d_1^5 + \tfrac{1}{24}d_1^3 d_2 - \tfrac{1}{8}d_1^2 d_3 \\
&\quad + \tfrac{1}{4}d_1 d_2^2 - \tfrac{1}{2}d_2 d_3 + \ldots\,, \\
c_2 &= -d_2 + \tfrac{1}{2}d_1^2 - \tfrac{1}{6}d_1^4 + d_1^2 d_2 - d_1 d_3 + \ldots\,, \\
c_3 &= -d_3 - \tfrac{3}{8}d_1^3 + \tfrac{3}{2}d_1 d_2 + \tfrac{27}{128}d_1^5 - \tfrac{27}{16}d_1^3 d_2 + \tfrac{9}{4}d_1^2 d_3 + \\
&\quad + \tfrac{9}{8}d_1 d_2^2 - \tfrac{3}{2}d_1 d_4 + \ldots\,, \\
c_4 &= -d_4 + \tfrac{1}{3}d_1^4 - 2d_1^2 d_2 + 2d_1 d_3 + d_2^2 + \ldots\,, \\
c_5 &= -d_5 - \tfrac{125}{384}d_1^5 + \tfrac{125}{48}d_1^3 d_2 - \tfrac{25}{8}d_1^2 d_3 - \tfrac{25}{8}d_1 d_2^2 + \tfrac{5}{2}d_1 d_4 + \\
&\quad + \tfrac{5}{2}d_2 d_3 + \ldots\,, \\
&\ldots\,.
\end{aligned} \tag{2.5.35}$$

The expansion of any given function of w may be performed with the aid of equation (2.5.32) without obtaining solution (2.5.34) itself.

Let us note that (2.5.32) enables one to express the equation of the centre $f = v - M$ as the (q, w) series. Indeed, combining together (2.5.18) and (2.5.32) one obtains (Brumberg E., 1992)

$$v - M = \sum_{m=1}^{\infty} \nabla_m \frac{q^{\frac{m}{2}}}{1+q^m} \sin mw \qquad (2.5.36)$$

with the coefficients

$$\nabla_m = \frac{4}{m}\Big|2 - \operatorname{mod}(m+2,4)\Big| + (-1)^{\frac{m-1}{2}} \operatorname{mod}(m,2)\frac{2\pi}{K} \qquad (2.5.37)$$

or

$$\nabla_m = \begin{cases} \dfrac{4}{m} + \dfrac{2\pi}{K}, & m = 4s-3 \\ \dfrac{8}{m}, & m = 4s-2 \\ \dfrac{4}{m} - \dfrac{2\pi}{K}, & m = 4s-1 \\ 0, & m = 4s \end{cases} \qquad (2.5.38)$$

for $s = 1, 2, \ldots$.

Before concluding this section let us consider the case of very large eccentricities, say $k^2 = 0.99$. As mentioned above, the corresponding value of q is $q = 0.2622\ldots$ and this value is not small enough to ensure compact analytical expansions. There is no problem in applying the Landen transformation numerically, which involves the new Jacobi nome $q_1 = q^2$. But in analytical manipulation the advantages of the Landen transformation are not so certain. The Landen transformation from modulus k to k_1 is

$$k_1 = \frac{1-k'}{1+k'}, \qquad k_1' = \frac{2\sqrt{k'}}{1+k'} \qquad (2.5.39)$$

or

$$k = \frac{2\sqrt{k_1}}{1+k_1}, \qquad k' = \frac{1-k_1}{1+k_1}. \qquad (2.5.40)$$

The Jacobi functions of modulus k and argument u

$$\begin{aligned} &\operatorname{sn} u = \sin\varphi, \\ &\operatorname{cn} u = \cos\varphi, \\ &\delta = \operatorname{dn} u = (1 - k^2\sin^2\varphi)^{\frac{1}{2}}, \\ &\varphi = \operatorname{am} u \end{aligned} \qquad (2.5.41)$$

are transformed therewith to functions of modulus k_1 and argument u_1 by means of the relations

$$\operatorname{sn} u_1 = \sin\varphi_1 = \frac{1+k'}{\delta}\sin\varphi\cos\varphi, \qquad (2.5.42)$$

$$\operatorname{cn} u_1 = \cos\varphi_1 = \frac{1}{\delta}\left[1 - (1+k')\sin^2\varphi\right], \tag{2.5.43}$$

$$\operatorname{dn} u_1 = \delta_1 = (1 - k_1 \sin^2\varphi_1)^{\frac{1}{2}} = \frac{1}{\delta}\left[1 - (1-k')\sin^2\varphi\right], \tag{2.5.44}$$

$$\varphi_1 = \operatorname{am} u_1, \qquad u_1 = (1+k')u. \tag{2.5.45}$$

One may add to this the transformation for the elliptic integral of the second kind $E(\varphi, k)$,

$$E(\varphi_1, k_1) = \frac{2}{1+k'}\left[E(\varphi, k) + k'u\right] - \frac{1-k'}{\delta}\sin\varphi\cos\varphi. \tag{2.5.46}$$

The inverse transformation may be expressed as follows:

$$\sin 2\varphi = (\delta_1 + k_1\cos\varphi_1)\sin\varphi_1, \tag{2.5.47}$$

$$\cos 2\varphi = (\delta_1 + k_1\cos\varphi_1)\cos\varphi_1 - k_1, \tag{2.5.48}$$

$$\delta = \frac{1}{1+k_1}\left(\delta_1 + k_1\cos\varphi_1\right) = \frac{1-k_1}{\delta_1 - k_1\cos\varphi_1}, \tag{2.5.49}$$

$$u = \tfrac{1}{2}(1+k_1)u_1$$

and

$$E(\varphi, k) = -\frac{1}{2}(1-k_1)u_1 + \frac{1}{1+k_1}E(\varphi_1, k_1) + \frac{k_1}{1+k_1}\sin\varphi_1. \tag{2.5.50}$$

Instead of (2.5.48) one may use either of two relations:

$$\sin^2\varphi = \tfrac{1}{2}(1 - \delta_1\cos\varphi_1 + k_1\sin^2\varphi_1) \tag{2.5.51}$$

or

$$\cos^2\varphi = \tfrac{1}{2}(1 + \delta_1\cos\varphi_1 - k_1\sin^2\varphi_1). \tag{2.5.52}$$

Hence, the inverse transformation involves irrationalities when expressing $\operatorname{sn} u$ and $\operatorname{cn} u$ in terms of functions of u_1 and k_1. This is the price of introducing the new modulus k_1. From (2.5.45) and (2.5.46) it follows that

$$\begin{aligned} K_1 &= K(k_1) = \frac{1+k'}{2}K, \\ K_1' &= K(k_1') = (1+k')K', \\ K &= (1+k_1)K_1 \end{aligned} \tag{2.5.53}$$

and

$$\begin{aligned} E_1 &= E(k_1) = \frac{1}{1+k'}(E + k'K), \\ E &= -(1-k_1)K_1 + \frac{2}{1+k_1}E_1. \end{aligned} \tag{2.5.54}$$

Functions (2.5.41) are developable in multiples of $\pi u/2K$ or w determined by (2.5.19) with rational coefficients in q defined by (2.5.12). Functions (2.5.42)–(2.5.45) are expanded into the same series in multiples of $\pi u_1/2K_1$ or

$$w_1 = \frac{\pi u_1}{2K_1} - \frac{\pi}{2} = 2w + \frac{\pi}{2} \tag{2.5.55}$$

with rational coefficients with respect to

$$q_1 = \exp\left(-\frac{\pi K_1'}{K_1}\right) = q^2 . \tag{2.5.56}$$

When applied to the coordinates of the two-body problem these (q_1, w_1) series for functions (2.5.42)–(2.5.45) result in Landen expansions in multiples of $w_1/2$ with coefficients of the form $A(q_1)+\sqrt{q_1}B(q_1)$, where $A(q_1)$ and $B(q_1)$ are rational functions of q_1. Therefore, the Landen transformation retains the elliptic anomaly w as the trigonometric argument and changes the form of the coefficients to deal with two q_1-series instead of one q-series. To find the relevant expansions it is necessary to relate the amplitude functions φ and φ_1. From (2.5.42) and (2.5.43) we get

$$\tan(\varphi_1 - \varphi) = k' \tan\varphi . \tag{2.5.57}$$

An equation of this type often occurs in the classical two-body problem. For example, the relationship between the eccentric and true anomalies satisfies the same equation:

$$\tan\frac{g}{2} = \left(\frac{1-e}{1+e}\right)^{\frac{1}{2}} \tan\frac{v}{2}, \tag{2.5.58}$$

and comparing this with (2.3.46) for $\varphi = v/2$, $\varphi_1 = (g+v)/2$ one can write the solution of (2.5.57) in the form

$$\varphi_1 = 2\varphi + \sum_{m=1}^{\infty} \frac{(-1)^m}{m} k_1^m \sin 2m\varphi . \tag{2.5.59}$$

With respect to $\varphi = \operatorname{am} u$ this relation may be regarded as an equation of type (2.5.32), to be solved for φ or its functions by means of the Deprit technique or INVERSE operation (Section 1.2). In such a way one can get the Landen expansions for $\operatorname{am} u$, $\operatorname{sn} u$ and $\operatorname{cn} u$ but in the inversion of (2.5.59) one has to deal with k_1 as the small parameter, which actually may not be so small.

It is possible to put (2.5.57) in an explicit form with respect to φ. From (2.5.51) and (2.5.52) we have

$$\tan\varphi = \frac{(1+k_1)\sin\varphi_1}{\delta_1 + \cos\varphi_1}, \tag{2.5.60}$$

but this relation is not particularly convenient for analytical manipulation. Equations (2.5.47) and (2.5.48) involve

$$\sin(2\varphi - \varphi_1) = k_1 \sin\varphi_1\,, \tag{2.5.61}$$

$$\cos(2\varphi - \varphi_1) = \delta_1\,. \tag{2.5.62}$$

From (2.5.61) one gets

$$2\varphi = \varphi_1 + \arcsin(k_1 \sin\varphi_1)\,. \tag{2.5.63}$$

Expanding $\arcsin(k_1 \sin\varphi_1)$ in multiples of φ_1 one has

$$2\varphi = \varphi_1 + \sum_{m=1}^{\infty} a_m(k_1)\sin(2m-1)\varphi_1\,, \tag{2.5.64}$$

where coefficients $a_m(k_1)$ of the order $(\sqrt{q_1})^{2m-1}$ are expressed in closed form with the aid of the complete elliptic integrals K_1 and E_1 (Klioner, 1992). Having the (q_1, w_1) expansions for $\operatorname{am} u_1$, $\operatorname{sn} u_1$ and $\operatorname{cn} u_1$ one may obtain such an expansion for $\sin(2m-1)\varphi_1$ using (2.1.18) or (2.1.20), resulting in the Landen series for $\operatorname{am} u$. But the simplest way is to replace $\arcsin(k_1 \sin\varphi_1)$ by its expansion into a hypergeometric series. Then relation (2.5.64) takes the form

$$2 \operatorname{am} u = \operatorname{am} u_1 + k_1 \operatorname{sn} u_1 F\left(\tfrac{1}{2}, \tfrac{1}{2}, \tfrac{3}{2}, k_1^2 \operatorname{sn}^2 u_1\right)\,, \tag{2.5.65}$$

permitting one to obtain the Landen series for $\operatorname{am} u$ in a more straightforward manner. Then the Landen series for $\operatorname{sn} u$ and $\operatorname{cn} u$ may be obtained by the SIN and COS operations of Section 1.2.

Finally, we may conclude that for small eccentricities a Keplerian processor based on (e, M) expansions (2.2.13) and (2.2.14) is quite adequate. For moderate and large eccentricities it is appropriate to construct (q, w) expansions, replacing in (2.2.13) and (2.2.14) the Hansen coefficients $X_{k+1}^{n+1,1}(e)$ and the mean anomaly M by the elliptic Hansen coefficients $B_{k+1}^{n+1,1}(q)$ and the elliptic anomaly w, respectively. For very large eccentricities ($e^2 > 0.95$) the Landen expansions may be used.

To conclude this section we list below the standard trigonometric expansions, which are useful to construct an elliptic-function Keplerian processor:

$$\operatorname{sn} u = \frac{2\pi}{kK}\sum_{m=1}^{\infty}\frac{q^{m-\frac{1}{2}}}{1-q^{2m-1}}\sin(2m-1)\frac{\pi u}{2K}\,, \tag{2.5.66}$$

$$\operatorname{cn} u = \frac{2\pi}{kK}\sum_{m=1}^{\infty}\frac{q^{m-\frac{1}{2}}}{1+q^{2m-1}}\cos(2m-1)\frac{\pi u}{2K}\,, \tag{2.5.67}$$

$$\operatorname{dn} u = \frac{\pi}{2K} + \frac{2\pi}{K}\sum_{m=1}^{\infty}\frac{q^{m}}{1+q^{2m}}\cos m\frac{\pi u}{K}\,, \tag{2.5.68}$$

$$\operatorname{sn}^2 u = \frac{K-E}{k^2 K} - \frac{2\pi^2}{k^2 K^2} \sum_{m=1}^{\infty} m \frac{q^m}{1-q^{2m}} \cos m\frac{\pi u}{K}, \tag{2.5.69}$$

$$\operatorname{sn} u \operatorname{cn} u = \frac{2\pi^2}{k^2 K^2} \sum_{m=1}^{\infty} m \frac{q^m}{1+q^{2m}} \sin m\frac{\pi u}{K}, \tag{2.5.70}$$

$$\operatorname{sn} u \operatorname{dn} u = \frac{\pi^2}{k K^2} \sum_{m=1}^{\infty} (2m-1) \frac{q^{m-\frac{1}{2}}}{1+q^{2m-1}} \sin(2m-1)\frac{\pi u}{2K}, \tag{2.5.71}$$

$$\operatorname{cn} u \operatorname{dn} u = \frac{\pi^2}{k K^2} \sum_{m=1}^{\infty} (2m-1) \frac{q^{m-\frac{1}{2}}}{1-q^{2m-1}} \cos(2m-1)\frac{\pi u}{2K}, \tag{2.5.72}$$

$$\operatorname{dn}^2 u = \frac{E}{K} + \frac{2\pi^2}{K^2} \sum_{m=1}^{\infty} m \frac{q^m}{1-q^{2m}} \cos m\frac{\pi u}{K}, \tag{2.5.73}$$

$$\frac{1}{\operatorname{dn} u} = \frac{\pi}{2k'K} + \frac{2\pi}{k'K} \sum_{m=1}^{\infty} (-1)^m \frac{q^m}{1+q^{2m}} \cos m\frac{\pi u}{K}, \tag{2.5.74}$$

$$\frac{\operatorname{sn} u}{\operatorname{dn} u} = \frac{2\pi}{kk'K} \sum_{m=1}^{\infty} (-1)^{m+1} \frac{q^{m-\frac{1}{2}}}{1+q^{2m-1}} \sin(2m-1)\frac{\pi u}{2K}, \tag{2.5.75}$$

$$\frac{\operatorname{cn} u}{\operatorname{dn} u} = \frac{2\pi}{kK} \sum_{m=1}^{\infty} (-1)^{m+1} \frac{q^{m-\frac{1}{2}}}{1-q^{2m-1}} \cos(2m-1)\frac{\pi u}{2K}, \tag{2.5.76}$$

$$\frac{\operatorname{sn} u \operatorname{cn} u}{\operatorname{dn} u} = \frac{4\pi}{k^2 K} \sum_{m=1}^{\infty} \frac{q^{2m-1}}{1-q^{4m-2}} \sin(2m-1)\frac{\pi u}{K}, \tag{2.5.77}$$

$$\frac{1}{\operatorname{dn}^2 u} = \frac{E}{k'^2 K} + \frac{2\pi^2}{k'^2 K^2} \sum_{m=1}^{\infty} (-1)^m m \frac{q^m}{1-q^{2m}} \cos m\frac{\pi u}{K}. \tag{2.5.78}$$

Recall that the standard expansion for $\operatorname{am} u$ was given in (2.5.27).

2.6 Functions Involving the Coordinates of Two Bodies

All the algorithms given above are related to functions that depend only on the coordinates of one celestial body. Strictly speaking, the expansions related to functions involving the coordinates of two bodies moving on the Keplerian orbits are beyond the framework of the Keplerian processor and should be considered in perturbation theory. Nevertheless, some of these expansions are related so closely to the Keplerian-processor algorithms that it is reasonable to treat the corresponding subroutines together with these algorithms. The following two subroutines are of primary importance.

The subroutine COSINE produces the expansion of the cosine of the angle between the directions from the central body to the moving bodies. The algorithm is based on the elementary formula

$$\cos H = \frac{xx' + yy' + zz'}{rr'}, \tag{2.6.1}$$

where the prime relates to the coordinates of the disturbing body. The quantities x/r, y/r and z/r and the corresponding primed quantities are obtained by the subroutine COORD in the form of (2.2.13) and (2.2.14). Their substitution into (2.6.1) results in a Poisson series with six trigonometric arguments Ω, Ω', ω, ω', M and M'. Depending on the specific version of COORD one may have four, five or six power variables as follows:

$$(a) \quad e, \quad e', \quad \sin i, \quad \sin i'\,;$$

$$(b) \quad e, \quad e', \quad \sin i, \quad \sin\frac{i'}{2}, \quad \cos\frac{i'}{2}\,;$$

$$(c) \quad e, \quad e', \quad \sin\frac{i}{2}, \quad \cos\frac{i}{2}, \quad \sin i'\,;$$

$$(d) \quad e, \quad e', \quad \sin\frac{i}{2}, \quad \cos\frac{i}{2}, \quad \sin\frac{i'}{2}, \quad \cos\frac{i'}{2}\,.$$

The trigonometric functions of half arguments are taken into account rigorously, whereas e, e', $\sin i$, $\sin i'$ are treated as small expansion parameters. The eccentricities e, e' and anomalies M, M' may be replaced by nomes q, q' and elliptic anomalies w, w', as indicated in the previous section.

In expanding the perturbing function, the negative powers of the mutual distance between bodies and other functions occurring in the right-hand members of the equations of motion one needs the series for Legendre polynomials $P_n(\cos H)$, Gegenbauer polynomials $C_n^{k/2}(\cos H)$ and trigonometric functions such as $\cos nH$ and $\sin nH$. If the expansion for $\cos H$ is available then the series for all these functions are obtained by the subroutine GBAUER, which is based on the recurrence relation for the Gegenbauer polynomials

$$C_n^{k/2}(x) = \frac{2n+k-2}{n} x C_{n-1}^{k/2}(x) - \frac{n+k-2}{n} C_{n-2}^{k/2}(x)\,. \tag{2.6.2}$$

For $n = 2$ the initial values are

$$C_0^{k/2}(x) = 1\,, \qquad C_1^{k/2}(x) = kx\,.$$

Putting $k = 1$ one obtains the Legendre polynomials $P_n(x) = C_n^{1/2}(x)$. Comparison with (2.1.18) shows that recurrence (2.6.2) with $k = 2$ may still be used for obtaining $\cos nx$ or $\sin nx$ provided that x in the first coefficient of the right-hand member is replaced by $\cos x$. In this case $C_m^1(x)$ $(m = n-2, n-1, n)$ stands for $\cos mx$ or $\sin mx$. Hence the initial values are 1 and $\cos x$ for determining $\cos nx$ and 0 and $\sin x$ for determining $\sin nx$. The initial expansion for x in the case of the Gegenbauer polynomials or for

$\cos x$ in the case of the sine and cosine functions determines the type of the resulting expansion.

The simplest example of the application of these two subroutines is given by expanding the satellite disturbing function due to the third-body attraction. The expansion of the disturbing function in powers of the ratio of the semi-major axes has the form

$$R = \frac{Gm'a^2}{a'^3} \sum_{n=2}^{\infty} \left(\frac{a}{a'}\right)^{n-2} R_n \tag{2.6.3}$$

with

$$R_n = \left(\frac{r}{a}\right)^n \left(\frac{a'}{r'}\right)^{n+1} P_n(\cos H) . \tag{2.6.4}$$

The expansiòn of R_n can be easily obtained under one of the four forms indicated in the subroutine COSINE. Note that the fourth form (d) in the (q, w) version enables one to obtain the expansion of the perturbing function valid for large eccentricities and for any value of the inclination.

Other techniques for expanding the perturbing function and different applications of the Keplerian processor will be discussed below.

3 Quasi-polynomial Systems

3.1 The N-Planet Problem in Polynomial Form

Numerical or analytical investigation of systems of differential equations is usually greatly facilitated if the right-hand sides of the equations are polynomials with respect to the unknown functions (polynomial systems). Such systems occur quite naturally in celestial mechanics in dealing with solutions in the vicinity of some known particular solution, in investigating the long-term evolution of the planetary orbits for small eccentricities and inclinations (the secular system) and so on. Moreover, the differential equations of celestial mechanics may be reduced in typical cases to polynomial systems by introducing some extra unknown variables. To illustrate this let us consider the problem of the heliocentric motion of N planets.

In heliocentric rectangular coordinates the motion of N planets is described by the system

$$\ddot{\mathbf{r}}_i = -G(m_0 + m_i)\frac{\mathbf{r}_i}{r_i^3} + \sum_{j=1}^{N}{}^{(i)} Gm_j \left(\frac{\mathbf{r}_j - \mathbf{r}_i}{\Delta_{ij}^3} - \frac{\mathbf{r}_j}{r_j^3} \right), \quad i = 1, 2, \ldots, N \tag{3.1.1}$$

with the generally accepted designations for heliocentric position vectors $\mathbf{r}_i$, mutual distances Δ_{ij}, mass of the Sun m_0 and planetary masses m_i. For numerical integration of this system one often uses the Steffensen technique. This method involves the representation of the coordinates of the planets as finite segments of the Taylor series in powers of time, the coefficients being determined by the recurrence relations resulting from (3.1.1). But before that, this system is reduced to a form containing only the terms of (at most) the second degree with respect to the unknown functions and their derivatives. This is achieved by considering the quantities r_i, Δ_{ij}, $s_i = r_i^{-3}$ and $s_{ij} = \Delta_{ij}^{-3}$ as the extra unknown variables. As a result, system (3.1.1) takes the form

$$\dot{\mathbf{r}}_i = \mathbf{r}'_i, \tag{3.1.2}$$

$$\dot{\mathbf{r}}'_i = -G(m_0 + m_i)s_i\mathbf{r}_i + \sum_{j=1}^{N}{}^{(i)} Gm_j \left[s_{ij}(\mathbf{r}_j - \mathbf{r}_i) - s_j\mathbf{r}_j \right], \tag{3.1.3}$$

$$r_i\dot{r}_i = \mathbf{r}_i\mathbf{r}'_i\,, \tag{3.1.4}$$

$$r_i\dot{s}_i = -3s_i\dot{r}_i\,, \tag{3.1.5}$$

$$\Delta_{ij}\dot{\Delta}_{ij} = (\mathbf{r}_i - \mathbf{r}_j)(\mathbf{r}'_i - \mathbf{r}'_j)\,, \tag{3.1.6}$$

$$\Delta_{ij}\dot{s}_{ij} = -3s_{ij}\dot{\Delta}_{ij}\,. \tag{3.1.7}$$

Hence, the original system (3.1.1), which is equivalent to a system of $6N$ equations of the first order, is replaced by a new system consisting of $6N$ equations (3.1.2) and (3.1.3), $2N$ equations (3.1.4) and (3.1.5) and $N(N-1)$ equations (3.1.6) and (3.1.7). The total order of the system is equal to $N(N+7)$. In such a form the equations of planetary motion were integrated using the Steffensen technique by, for example, Broucke (1971b).

System (3.1.1) may also be reduced to a polynomial system. Introducing the extra variables

$$u_i = r_i^{-1}\,, \qquad u_{ij} = \Delta_{ij}^{-1} \tag{3.1.8}$$

one may write (3.1.1) in the form

$$\dot{\mathbf{r}}_i = \mathbf{r}'_i\,, \tag{3.1.9}$$

$$\dot{\mathbf{r}}'_i = -G(m_0 + m_i)u_i^3\mathbf{r}_i + \sum_{j=1}^{N}{}^{(i)} Gm_j\left[u_{ij}^3(\mathbf{r}_j - \mathbf{r}_i) - u_j^3\mathbf{r}_j\right]\,, \tag{3.1.10}$$

$$\dot{u}_i = -u_i^3\mathbf{r}_i\mathbf{r}'_i\,, \tag{3.1.11}$$

$$\dot{u}_{ij} = -u_{ij}^3\,(\mathbf{r}_j - \mathbf{r}_i)(\mathbf{r}'_j - \mathbf{r}'_i)\,, \tag{3.1.12}$$

which consists of $6N$ equations (3.1.9) and (3.1.10), N equations (3.1.11) and $N(N-1)/2$ equations (3.1.12) with the total order of the system equal to $N(N+13)/2$. The maximum degree of the polynomials on the right-hand side of this system is 5. Equations (3.1.9)–(3.1.12) are less convenient than (3.1.2)–(3.1.7) when one uses the Steffensen technique but they are well adapted for the general algorithm, given below, for solving polynomial systems with the aid of a Taylor series.

3.2 Kustaanheimo–Stiefel (KS) Variables

The introduction of the KS variables into celestial mechanics resulted in the elaboration of a set of numerical and analytical techniques that constitute a new domain known as linear and regular celestial mechanics (Stiefel and Scheifele, 1971). The KS variables are remarkable in three aspects since they (1) ensure the spatial regularization of the equations of motion, (2) enable one to treat uniformly all three types of Keplerian motion, and (3) transform the equations of the two-body problem into a linear form. In relation to the content of this chapter the last feature is of particular importance. If we take the variations of the KS variables with respect to their Keplerian values as the unknown variables, the right-hand sides of the equations of motion will be holomorphic functions of these variables and the resulting polynomial systems can be treated by perturbation-theory methods. Of course, it should be noted that the reducibility of the two-body problem to the linear oscillator form and its application in perturbation theory have been used even in classical celestial mechanics (recall, for example, the equations by Clairaut and Laplace and their application by Adams to construct the theory of motion of the Moon), but the introduction of the KS variables enabled the realization of this idea in a more complete form.

The standard equations of the perturbed two-body problem have the form

$$\ddot{\mathbf{x}} + \frac{Gm}{r^3}\mathbf{x} = \frac{\partial R}{\partial \mathbf{x}} + \mathbf{F}\,, \tag{3.2.1}$$

where for the sake of convenience the potential and non-potential disturbing forces on the right-hand side are given separately. Here and in the next section the position vector $\mathbf{r} = (x,\, y,\, z)$ is designated by $\mathbf{x} = (x_1,\, x_2,\, x_3)$. The KS variables are the components of the four-dimensional vector $\mathbf{u}$ of the parametric space related to the three-dimensional vector $\mathbf{x}$ of the physical space by the KS transformation

$$\mathbf{x} = L(\mathbf{u})\mathbf{u} \tag{3.2.2}$$

with

$$L(\mathbf{u}) = \begin{pmatrix} u_1 & -u_2 & -u_3 & u_4 \\ u_2 & u_1 & -u_4 & -u_3 \\ u_3 & u_4 & u_1 & u_2 \\ u_4 & -u_3 & u_2 & -u_1 \end{pmatrix}. \tag{3.2.3}$$

In virtue of (3.2.2)

$$r = |\mathbf{x}| = |\mathbf{u}|^2\,. \tag{3.2.4}$$

The KS transformation is accompanied by replacement of the physical time t by the fictitious time s:

$$\frac{dt}{ds} = r\,. \tag{3.2.5}$$

Therefore,

$$\dot{\mathbf{x}} = \frac{2}{r} L(\mathbf{u}) \frac{d\mathbf{u}}{ds} \tag{3.2.6}$$

and, in particular,

$$|\dot{\mathbf{x}}|^2 = \frac{4}{r} \left| \frac{d\mathbf{u}}{ds} \right|^2 . \tag{3.2.7}$$

In formulae like (3.2.2) or (3.2.6) containing vectors of different dimensions it is always assumed that the three-dimensional vectors are converted to four dimensions by adding a zero fourth component. Therefore, formulae (3.2.6) involve the non-holonomic relationship for the KS variables

$$u_4 \frac{du_1}{ds} - u_3 \frac{du_2}{ds} + u_2 \frac{du_3}{ds} - u_1 \frac{du_4}{ds} = 0\,. \tag{3.2.8}$$

If $L^T(\mathbf{u})$ denotes the transposed matrix of (3.2.3) then

$$L^T(\mathbf{u}) \frac{\partial R}{\partial \mathbf{x}} = \frac{1}{2} \frac{\partial R}{\partial \mathbf{u}}\,. \tag{3.2.9}$$

The equations of motion (3.2.1) expressed in the KS variables and referred to the fictitious time s take the form

$$\frac{d^2\mathbf{u}}{ds^2} + \frac{1}{2} h_K \mathbf{u} = \frac{1}{2} |\mathbf{u}|^2 \left(\frac{1}{2} \frac{\partial R}{\partial \mathbf{u}} + L^T(\mathbf{u}) \mathbf{F} \right) . \tag{3.2.10}$$

Here h_K represents the energy of the Keplerian motion (taken with the opposite sign):

$$h_K = \frac{Gm}{r} - \frac{1}{2} |\dot{\mathbf{x}}|^2 . \tag{3.2.11}$$

This quantity is constant for the Keplerian motion and changes in the perturbed problem in accordance with the equation

$$\frac{dh_K}{ds} = - \left(\frac{\partial R}{\partial \mathbf{u}} + 2L^T(\mathbf{u}) \mathbf{F} \right) \frac{d\mathbf{u}}{ds}\,. \tag{3.2.12}$$

In virtue of (3.2.7) relation (3.2.11) may be put into the form

$$h_K |\mathbf{u}|^2 + 2 \left| \frac{d\mathbf{u}}{ds} \right|^2 = Gm\,. \tag{3.2.13}$$

Reduced to the first-order equations the system of 10 equations (3.2.10), (3.2.12) and (3.2.5) with two relationships (3.2.8) and (3.2.13) is equivalent to the six first-order equations of the original system (3.2.1). Instead of equation

(3.2.12) one may use (3.2.13), reducing the total order of the system to 9. Such a system has been used, in particular, by Rössler (Stiefel et al., 1966) for the planetary three-body problem.

In some cases it is appropriate to replace h_K by

$$h = h_K + R\,. \tag{3.2.14}$$

Equations (3.2.10) and (3.2.12) are thus replaced respectively by the following ones:

$$\frac{d^2\mathbf{u}}{ds^2} + \frac{1}{2}h\mathbf{u} = \frac{1}{4}\frac{\partial}{\partial \mathbf{u}}\left(|\mathbf{u}|^2 R\right) + \frac{1}{2}|\mathbf{u}|^2 L^T(\mathbf{u})\mathbf{F} \tag{3.2.15}$$

and

$$\frac{dh}{ds} = |\mathbf{u}|^2 \frac{\partial R}{\partial t} - 2\frac{d\mathbf{u}}{ds}\left(L^T(\mathbf{u})\mathbf{F}\right)\,. \tag{3.2.16}$$

In the case of conservative potential perturbations when $\mathbf{F} = 0$ and $\dfrac{\partial R}{\partial t} = 0$ the quantity h represents the energy constant and the harmonic oscillator frequency on the left-hand side of (3.2.15) is the constant quantity. In the general case in virtue of (3.2.14) equations (3.2.15) differ from (3.2.10) by the presence of the disturbing function R itself (not only in the form of the derivatives). In accordance with the original equations (3.2.1) the disturbing function R is determined up to an additive constant (and even arbitrary additive function of t alone) and this feature may be used to choose a non-Keplerian intermediate orbit.

Equations (3.2.10) involve the equation for the radius-vector

$$\frac{d^2 r}{ds^2} + 2h_K r = Gm + r\left(\frac{1}{2}\frac{\partial R}{\partial \mathbf{u}} + L^T(\mathbf{u})\mathbf{F}\right)\mathbf{u}\,. \tag{3.2.17}$$

This equation is not related to the KS variables and its right-hand side can be represented in terms of the old variables:

$$\frac{d^2 r}{ds^2} + 2h_K r = Gm + r\left(\frac{\partial R}{\partial \mathbf{x}} + \mathbf{F}\right)\mathbf{x}\,. \tag{3.2.18}$$

Inversion of the KS transformation is performed with the aid of any one of the following sets of relations:

$$\begin{aligned} u_1^2 + u_4^2 &= \tfrac{1}{2}(r + x_1)\,,\\ u_2 &= \frac{x_2 u_1 + x_3 u_4}{r + x_1}\,,\\ u_3 &= \frac{x_3 u_1 - x_2 u_4}{r + x_1} \end{aligned} \tag{3.2.19}$$

or

$$u_2^2 + u_3^2 = \tfrac{1}{2}(r - x_1)\,,$$

$$u_1 = \frac{x_2 u_2 + x_3 u_3}{r - x_1}\,, \qquad (3.2.20)$$

$$u_4 = \frac{x_3 u_2 - x_2 u_3}{r - x_1}\,.$$

The first set is suitable for $x_1 \geq 0$ whereas the second one is suitable for the case $x_1 \leq 0$. Evidently, one of the components of $\mathbf{u}$ is chosen arbitrarily. Inversion of (3.2.6) results in

$$\frac{d\mathbf{u}}{ds} = \tfrac{1}{2} L^T(\mathbf{u})\dot{\mathbf{x}}\,. \qquad (3.2.21)$$

Relations (3.2.19)–(3.2.21) serve to determine the initial values of $\mathbf{u}$ and $d\mathbf{u}/ds$ from the given initial values of $\mathbf{x}$ and $\dot{\mathbf{x}}$.

The fictitious time s is well adapted for uniformly treating all three types of Keplerian motion. If the undisturbed motion belongs to the elliptic type and if the perturbations are not so strong as to change the type of motion, then it might be better to take as an independent argument the quantity g introduced by the equation

$$\frac{dt}{dg} = \frac{r}{\sqrt{2(h_K + \gamma R)}}\,. \qquad (3.2.22)$$

Here γ denotes a numerical factor taking values 0 and 1. For $\gamma = 0$ the quantity g coincides with the eccentric anomaly. For $\gamma = 1$ one obtains the so-called generalized eccentric anomaly widely used by Stiefel and Scheifele (1971). Equations (3.2.10) and (3.2.15) referred to g take the form

$$\frac{d^2\mathbf{u}}{dg^2} + \tfrac{1}{4}\mathbf{u} = \frac{1}{2(h_K + \gamma R)} \Bigg\{ \tfrac{1}{2}\gamma R \mathbf{u} + r\Big(\tfrac{1}{4}\frac{\partial R}{\partial \mathbf{u}} + \tfrac{1}{2} L^T(\mathbf{u})\mathbf{F}\Big) +$$

$$+ \left[\Big((1-\gamma)\frac{\partial R}{\partial \mathbf{u}} + 2L^T(\mathbf{u})\mathbf{F}\Big)\frac{d\mathbf{u}}{dg} - \gamma \frac{r}{\sqrt{2(h_K + \gamma R)}}\frac{\partial R}{\partial t}\right]\frac{d\mathbf{u}}{dg}\Bigg\}\,. \qquad (3.2.23)$$

In virtue of (3.2.13) one has

$$h_K + \gamma R = \frac{Gm + \gamma r R}{r + 4\left|\dfrac{d\mathbf{u}}{dg}\right|^2} \qquad (3.2.24)$$

with the differential equation

$$\frac{d(h_K+\gamma R)}{dg} = -\left[(1-\gamma)\frac{\partial R}{d\mathbf{u}} + 2L^T(\mathbf{u})\mathbf{F}\right]\frac{d\mathbf{u}}{dg} + \gamma\frac{r}{\sqrt{2(h_K+\gamma R)}}\frac{\partial R}{\partial t}. \tag{3.2.25}$$

Considering that the Keplerian energy h_K is related to the semi-major axis a by $h_K = Gm/2a$ one may replace the equation (3.2.12) for h_K by

$$\frac{da}{dg} = \frac{2a^2}{Gm}\left(\frac{\partial R}{\partial \mathbf{u}} + 2L^T(\mathbf{u})\mathbf{F}\right)\frac{d\mathbf{u}}{dg}. \tag{3.2.26}$$

Equations (3.2.23) in the undisturbed case correspond to the equations of a harmonic oscillator with constant frequency. As in the case of equations (3.2.15), for $\gamma \neq 0$ equations (3.2.23) contain the disturbing function R explicitly, extending the possibilities for choosing an intermediate solution.

Equations (3.2.22) and (3.2.23) involve different sets of elements given by Stiefel and Scheifele (1971). But the consideration of these equations in the "coordinate" form may also be of great interest. They may be solved both by traditional techniques with secular terms and by Gylden-type techniques valid for very long time intervals. Some other aspects of the KS transformation, in particular, its representation in quaternion form and the sets of new linearizing transformations, are considered by Deprit et al. (1994).

3.3 Hansen Coordinates and Euler Parameters

In classical celestial mechanics the transformation formulae from one rectangular coordinate system to another are written with the aid of the Euler angles. As a result, the right-hand sides of the differential equations will contain the trigonometric functions of these angles, which are not very convenient from the point of view of computation. The right-hand sides of the differential equations of celestial mechanics take a more symmetrical form, approaching polynomial systems, and become more convenient for computation if the three Euler angles are replaced by four Euler parameters that are the components of a four-dimensional unit vector. Such a modification has been initiated by Musen (see, for example, Musen, 1963). In this section we consider the application of the Euler parameters to derive the equations of the perturbed motion in Hansen coordinates.

Let x_1, x_2, x_3 and x'_1, x'_2, x'_3 be two rectangular coordinate systems. Their mutual orientation is given by three Euler angles ψ, φ, θ (Fig. 2). Transformation between two systems is described by the equations

$$\mathbf{x} = \Lambda\mathbf{x}' \tag{3.3.1}$$

and

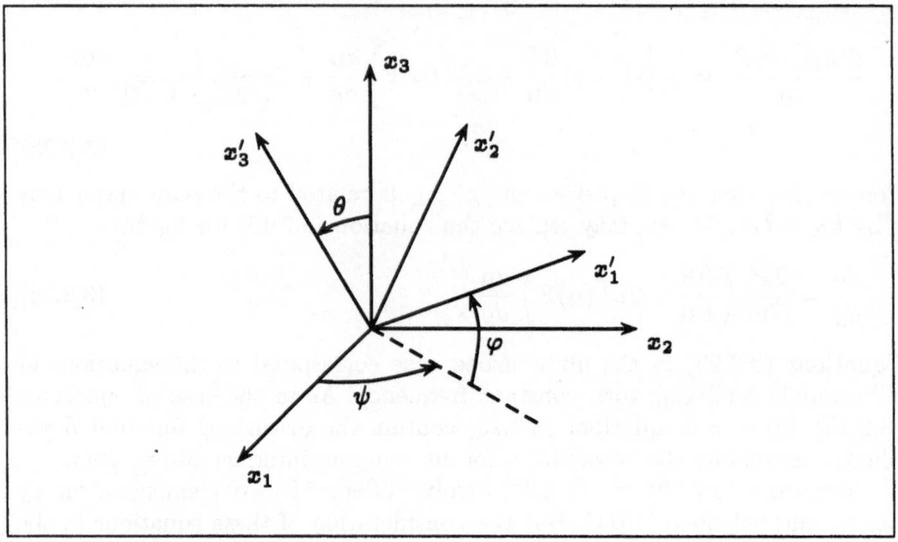

Fig. 2. Euler angles

$$\mathbf{x}' = \Lambda^T \mathbf{x} \tag{3.3.2}$$

with

$$\Lambda = D_3(-\psi)\, D_1(-\theta)\, D_3(-\varphi) \tag{3.3.3}$$

and

$$\Lambda^T = D_3(\varphi)\, D_1(\theta)\, D_3(\psi) \tag{3.3.4}$$

for the transposed matrix Λ^T. Here we use standard designations for the respective unit rotations about the axes x_1, x_2 and x_3:

$$D_1(\alpha) = \begin{pmatrix} 1 & 0 & 0 \\ 0 & \cos\alpha & \sin\alpha \\ 0 & -\sin\alpha & \cos\alpha \end{pmatrix}, \tag{3.3.5}$$

$$D_2(\beta) = \begin{pmatrix} \cos\beta & 0 & -\sin\beta \\ 0 & 1 & 0 \\ \sin\beta & 0 & \cos\beta \end{pmatrix}, \tag{3.3.6}$$

$$D_3(\gamma) = \begin{pmatrix} \cos\gamma & \sin\gamma & 0 \\ -\sin\gamma & \cos\gamma & 0 \\ 0 & 0 & 1 \end{pmatrix}. \tag{3.3.7}$$

The Euler parameters u_α $(\alpha = 1, 2, 3, 4)$ are determined by the relations

$$u_1 = \sin\frac{\theta}{2}\cos\frac{\psi - \varphi}{2},$$

$$u_2 = \sin\frac{\theta}{2}\,\sin\frac{\psi-\varphi}{2}\,,$$
$$u_3 = \cos\frac{\theta}{2}\,\sin\frac{\psi+\varphi}{2}\,,$$
$$u_4 = \cos\frac{\theta}{2}\,\cos\frac{\psi+\varphi}{2}\,. \tag{3.3.8}$$

Evidently, $|\mathbf{u}| = 1$, $\mathbf{u}$ being the four-dimensional vector with components u_α. The elements of matrix (3.3.3) are the simple quadratic functions of the Euler parameters. Indeed, introducing three vectors

$$\boldsymbol{\alpha} = \begin{pmatrix} u_1^2 - u_2^2 - u_3^2 + u_4^2 \\ 2(u_1u_2 + u_3u_4) \\ 2(u_1u_3 - u_2u_4) \end{pmatrix}, \tag{3.3.9}$$

$$\boldsymbol{\beta} = \begin{pmatrix} 2(u_1u_2 - u_3u_4) \\ -u_1^2 + u_2^2 - u_3^2 + u_4^2 \\ 2(u_2u_3 + u_1u_4) \end{pmatrix} \tag{3.3.10}$$

and

$$\boldsymbol{\gamma} = \begin{pmatrix} 2(u_1u_3 + u_2u_4) \\ 2(u_2u_3 - u_1u_4) \\ -u_1^2 - u_2^2 + u_3^2 + u_4^2 \end{pmatrix}, \tag{3.3.11}$$

one may rewrite the elements of matrix (3.3.3) as

$$\Lambda_{i1} = \alpha_i\,, \qquad \Lambda_{i2} = \beta_i\,, \qquad \Lambda_{i3} = \gamma_i\,, \quad i = 1,2,3. \tag{3.3.12}$$

This means that the components of the vectors $\boldsymbol{\alpha}$, $\boldsymbol{\beta}$ and $\boldsymbol{\gamma}$ form the first, second and third columns of the matrix $\Lambda(\mathbf{u})$, respectively.

Consider now again the equations (3.2.1) for the perturbed two-body problem. Let x_1', x_2', x_3' be the system of Hansen coordinates such that the disturbed motion always satisfies the equations

$$Z = \dot{Z} = 0\,. \tag{3.3.13}$$

Direct and inverse transformations (3.3.1) and (3.3.2) may be put into the form

$$\mathbf{x} = X\boldsymbol{\alpha} + Y\boldsymbol{\beta} + Z\boldsymbol{\gamma} \tag{3.3.14}$$

and

$$X = \boldsymbol{\alpha}\mathbf{x}\,, \qquad Y = \boldsymbol{\beta}\mathbf{x}\,, \qquad Z = \boldsymbol{\gamma}\mathbf{x}\,. \tag{3.3.15}$$

Under transformation (3.3.14) with (3.3.13) system (3.2.1) reduces to the system of the fourth order

$$\ddot{X} + \frac{Gm}{r^3}X = \frac{\partial R}{\partial X} + \alpha\mathbf{F}\,,$$
$$\ddot{Y} + \frac{Gm}{r^3}Y = \frac{\partial R}{\partial Y} + \beta\mathbf{F}\,, \tag{3.3.16}$$

which represents the motion in the Hansen plane, and the system of the third order for the Euler angles describing the motion of this plane. Using the widely accepted denotations

$$\psi = \Omega\,, \qquad \varphi = -\sigma\,, \qquad \theta = i \tag{3.3.17}$$

one has

$$\frac{di}{dt} = c^{-1}\left[-\frac{1}{\sin i}\left(\frac{\partial R}{\partial \Omega} + \cos i\,\frac{\partial R}{\partial \sigma}\right) + (X\cos\sigma + Y\sin\sigma)\,\gamma\mathbf{F}\right], \tag{3.3.18}$$

$$\sin i\frac{d\Omega}{dt} = c^{-1}\left[\frac{\partial R}{\partial i} + (Y\cos\sigma - X\sin\sigma)\,\gamma\mathbf{F}\right] \tag{3.3.19}$$

and

$$\frac{d\sigma}{dt} = \cos i\,\frac{d\Omega}{dt} \tag{3.3.20}$$

with

$$c = X\dot{Y} - Y\dot{X}\,. \tag{3.3.21}$$

The disturbing function R, which depends on the coordinates x_1, x_2, x_3 and time t, should be considered in equations (3.3.16) and (3.3.18)–(3.3.20) as a function of X, Y, i, Ω, σ and t.

In terms of the Euler parameters u_α defined by (3.3.8) with (3.3.17) system (3.3.18)–(3.3.20) may be replaced by the system of four differential equations of the first order

$$\dot{u}_1 = \frac{1}{4}c^{-1}\Big[-(u_3^2+u_4^2)\frac{\partial R}{\partial u_2} + (u_2u_3 - u_1u_4)\frac{\partial R}{\partial u_3} + $$
$$+ (u_1u_3 + u_2u_4)\frac{\partial R}{\partial u_4} + 2(u_4X - u_3Y)\gamma\mathbf{F}\Big]\,, \tag{3.3.22}$$

$$\dot{u}_2 = \frac{1}{4}c^{-1}\Big[(u_3^2+u_4^2)\frac{\partial R}{\partial u_1} - (u_1u_3 + u_2u_4)\frac{\partial R}{\partial u_3} + $$
$$+ (u_2u_3 - u_1u_4)\frac{\partial R}{\partial u_4} + 2(u_3X + u_4Y)\gamma\mathbf{F}\Big]\,, \tag{3.3.23}$$

$$\dot{u}_3 = \frac{1}{4}c^{-1}\Big[(u_1u_4 - u_2u_3)\frac{\partial R}{\partial u_1} + (u_1u_3 + u_2u_4)\frac{\partial R}{\partial u_2} - $$
$$- (u_1^2+u_2^2)\frac{\partial R}{\partial u_4} - 2(u_2X - u_1Y)\gamma\mathbf{F}\Big]\,, \tag{3.3.24}$$

$$\dot{u}_4 = \frac{1}{4}c^{-1}\Big[-(u_1u_3+u_2u_4)\frac{\partial R}{\partial u_1} + (u_1u_4 - u_2u_3)\frac{\partial R}{\partial u_2} + \\ + (u_1^2+u_2^2)\frac{\partial R}{\partial u_3} - 2(u_1X+u_2Y)\gamma\mathbf{F}\Big] . \tag{3.3.25}$$

One may add to this system the equation for the quantity $q = c^{-1}$:

$$\dot{q} = q^2\Big[Y\Big(\frac{\partial R}{\partial X} + \alpha\mathbf{F}\Big) - X\Big(\frac{\partial R}{\partial Y} + \beta\mathbf{F}\Big)\Big] \tag{3.3.26}$$

with the possible substitution of

$$Y\frac{\partial R}{\partial X} - X\frac{\partial R}{\partial Y} = \frac{1}{2}\left(-u_2\frac{\partial R}{\partial u_1} + u_1\frac{\partial R}{\partial u_2} - u_4\frac{\partial R}{\partial u_3} + u_3\frac{\partial R}{\partial u_4}\right) . \tag{3.3.27}$$

In equations (3.3.16) and (3.3.22)–(3.3.25) R should be regarded as a function of X, Y, u_1, u_2, u_3, u_4 and t.

Equations analogous to (3.3.22)–(3.3.25) have been derived in a different manner by Deprit (1976).

With respect to a specific problem, the system (3.3.16) describing the motion in the Hansen plane may be subjected to further transformations. In particular, the left-hand sides of these equations are easily reduced to a linear form by introducing the parabolic Levi–Civita coordinates and a new independent argument of the eccentric anomaly type (the planar case of the KS transformation considered above). The final reduction of (3.3.16) and (3.3.22)–(3.3.26) to a polynomial form depends on the specific structure of R and $\mathbf{F}$.

It is to be noted that the Euler parameters are also suitable for problems dealing with the rotation of a body around its centre of mass. The well-known kinematic Euler equations

$$\begin{aligned} \omega_1 &= \dot{\psi}\sin\varphi\sin\theta + \dot{\theta}\cos\varphi , \\ \omega_2 &= \dot{\psi}\cos\varphi\sin\theta - \dot{\theta}\sin\varphi , \\ \omega_3 &= \dot{\psi}\cos\theta + \dot{\varphi} \end{aligned} \tag{3.3.28}$$

relate the components of the angular rotation velocity $\boldsymbol{\omega}$ to the Euler angles ψ, φ, θ determining the orientation of the principal axes of inertia of the body. Expressed in terms of the Euler parameters (3.3.8) these equations take the form

$$\dot{\mathbf{u}} = \tfrac{1}{2}\Omega(\boldsymbol{\omega})\mathbf{u} \tag{3.3.29}$$

with

$$\Omega(\boldsymbol{\omega}) = \begin{pmatrix} 0 & \omega_3 & -\omega_2 & \omega_1 \\ -\omega_3 & 0 & \omega_1 & \omega_2 \\ \omega_2 & -\omega_1 & 0 & \omega_3 \\ -\omega_1 & -\omega_2 & -\omega_3 & 0 \end{pmatrix} . \tag{3.3.30}$$

The angular momentum **M** of the forces acting on the body may also be written in a fairly symmetrical form. Indeed, in terms of the Euler angles the components of **M** are determined by

$$
\begin{aligned}
M_1 &= \left(\frac{\partial U}{\partial \psi} - \cos\theta \frac{\partial U}{\partial \varphi}\right)\frac{\sin\varphi}{\sin\theta} + \cos\varphi \frac{\partial U}{\partial \theta}\,, \\
M_2 &= \left(\frac{\partial U}{\partial \psi} - \cos\theta \frac{\partial U}{\partial \varphi}\right)\frac{\cos\varphi}{\sin\theta} - \sin\varphi \frac{\partial U}{\partial \theta}\,, \\
M_3 &= \frac{\partial U}{\partial \varphi}\,,
\end{aligned}
\tag{3.3.31}
$$

U being the force function. In terms of the Euler parameters the same relations take the form

$$
\begin{aligned}
M_1 &= \frac{1}{2}\left(u_4\frac{\partial U}{\partial u_1} + u_3\frac{\partial U}{\partial u_2} - u_2\frac{\partial U}{\partial u_3} - u_1\frac{\partial U}{\partial u_4}\right), \\
M_2 &= \frac{1}{2}\left(-u_3\frac{\partial U}{\partial u_1} + u_4\frac{\partial U}{\partial u_2} + u_1\frac{\partial U}{\partial u_3} - u_2\frac{\partial U}{\partial u_4}\right), \\
M_3 &= \frac{1}{2}\left(u_2\frac{\partial U}{\partial u_1} - u_1\frac{\partial U}{\partial u_2} + u_4\frac{\partial U}{\partial u_3} - u_3\frac{\partial U}{\partial u_4}\right),
\end{aligned}
\tag{3.3.32}
$$

which may be advantageous for computer calculations.

4 Algorithms to Solve Polynomial Systems

4.1 Taylor Expansions

In most of the cases that actually occur solution of the polynomial systems is sought either in the form of Taylor expansions in powers of an independent argument or in pure trigonometric form with respect to some linear functions of this argument. We start with the first form. Let $x_1(t), \ldots, x_N(t)$ be the functions satisfying the differential polynomial system

$$\frac{dx_i}{dt} = F_i(x), \quad i = 1, 2, \ldots, N, \tag{4.1.1}$$

$F_i(x)$ being power series with constant coefficients

$$F_i(x) = \sum_{j=0}^{J} \sum F^{(i)}_{j_1 \ldots j_N} x_1^{j_1} \ldots x_N^{j_N}. \tag{4.1.2}$$

The internal sum is taken over all non-negative values $j_1, \ldots, j_N$, satisfying the condition

$$j_1 + \ldots + j_N = j. \tag{4.1.3}$$

Given the initial conditions $x_0^{(i)} = x_i(0)$ the problem is to find the approximate solution of (4.1.1) in the form of the finite segments of the Taylor series

$$x_i = \sum_{k=0}^{M} x_k^{(i)} t^k. \tag{4.1.4}$$

The parameters of this problem are the number N of the variables x_i, the maximum degree J of the polynomials $F_i(x)$, the total number MAX of the terms in all polynomials $F_i(x)$ and the maximum degree M of the resulting polynomials of the solution. One way to solve this problem is to try to apply the Steffensen technique to determine the coefficients $x_k^{(i)}$ by means of recurrence relations. But its realization strongly depends on the specific form of the right-hand sides of equations (4.1.1). The algorithm given below does not depend on the specific form of the right-hand sides, it being quite universal in this respect. In applying this algorithm to some specific system (4.1.1) it may be more computer time consuming than the Steffensen technique and cannot be recommended, for example, for numerical integration of equations (4.1.1) by means of Taylor series. Nevertheless, the simplicity and universality of this algorithm make it convenient for solving equations (4.1.1) with various right-hand sides.

The algorithm involves the straightforward substitution of solution (4.1.4), already known within the terms of degree $k-1$ inclusive with respect to t ($k = 1, 2, \dots, M$) into right-hand sides (4.1.2). This substitution demands only that some simple software manipulate polynomials. As a result, one finds the coefficients $F_{k-1}^{(i)}$ of the $(k-1)$-degree terms of the Taylor expansions of the right-hand sides $F_i(x)$ in powers of t. Then the terms of degree k in solution (4.1.4) are determined from

$$kx_k^{(i)} = F_{k-1}^{(i)} \tag{4.1.5}$$

and the process is repeated until the value $k = M$ is attained.

To ensure the universal and compact computer representation of right-hand sides (4.1.2) one can numerate the sets of indices $j_1, \dots, j_N$ and store not the indices themselves in packed form but the relevant numbers of sets. The solution of the direct and inverse problems in the numeration technique was considered in Section 1.3. In using this technique the right-hand sides of (4.1.1) are completely given by three arrays as follows:

AMOUNT(I), $I = 1, \dots, N$, numbers of terms on the right-hand sides (4.1.2);

COEFF(L), $L = 1, \dots, MAX$, coefficients $F_{j_1 \dots j_N}^{(i)}$ for $i = 1, \dots, N$;

IEX(L), $L = 1, \dots, MAX$, numbers of sets of indices $j_1, \dots, j_N$ according to numeration function (1.3.4). The elements COEFF(L) and IEX(L) with equal values of L relate to one and the same term determining, respectively, its coefficient and the set of indices $j_1, \dots, j_N$.

The initial values and the results obtained may be described by the array $X(k, i)$, $k = 0, 1, \dots, M$, $i = 1, 2, \dots, N$, so that the element $X(k, i)$ corresponds to the coefficient $x_k^{(i)}$. The values $X(0, i)$, $i = 1, 2, \dots, N$, are the given initial values of series (4.1.4) for $t = 0$. The remaining elements of X, i.e. the coefficients of series (4.1.4), are determined consecutively. The process of calculation may be arranged in different ways. In the simplest version the elements of IEX(L), $L = 1, \dots, MAX$, are looked through one by one, the stored numbers are converted into the corresponding index sets and the segments of (4.1.4) up to degree $k-1$ inclusive are substituted into the expression $x_1^{j_1} \dots x_N^{j_N}$. The coefficient obtained in t^{k-1} is multiplied by the relevant coefficients of the array COEFF(L), $L = 1, \dots, MAX$, and the results are summed for all the terms of one and the same polynomial $F_i(x)$ (the boundary values of the elements of COEFF and IEX for different polynomials are determined with the aid of the elements of AMOUNT). Formula (4.1.5) results in coefficients in t^k and the process is repeated again. To substitute

(4.1.4) into (4.1.2) one can of course use faster algorithms, for example, the Horner technique for polynomials in several variables, arranging beforehand the terms of the polynomials $F_i(x)$ in a suitable order.

To obtain the Taylor expansions is not always the final aim in solving (4.1.1). Using these expansions one can find more compact expansions in Chebyshev polynomials (Lanczos, 1956) or with the aid of different summation methods for constructing series of polynomials with a larger domain of convergence, including the series of polynomials in the three-body problem, which converge for any real moment of time (Brumberg, 1963).

4.2 Normalization and Trigonometric Expansions

Consider now the problem of solving secular polynomial systems in trigonometric form. Such systems are often written in Hamiltonian form by constructing the normalizing Birkhoff transformation with the techniques of Hori or Deprit. The corresponding computation algorithms may be found, for instance, in the papers by Deprit et al. (1969) and Markeev and Sokolsky (1976, 1978). Here we give the algorithm to solve the secular system based directly on Birkhoff normalization without involving the Hamiltonian form of the system.

The secular system will be taken here in the form occurring in the general planetary theory:

$$\dot{\alpha} = \mathrm{i}\mathcal{N}\left[A\alpha + \Phi(\alpha, \bar{\alpha})\right] . \tag{4.2.1}$$

This form is typical for many of the problems of celestial mechanics. Here α and $\bar{\alpha}$ are complex conjugate vectorial variables, the components $\alpha_1, \ldots, \alpha_N$ and $\bar{\alpha}_1, \ldots, \bar{\alpha}_N$ being Laplace-type elements as in (2.2.20). $\mathcal{N}$ stands for the diagonal matrix of the mean motions $n_1, \ldots, n_N$. A is a real $N \times N$ square matrix. $\Phi(\alpha, \bar{\alpha})$ are N-dimensional vectorial functions expanded in series of α and $\bar{\alpha}$ and containing, generally speaking, only forms of odd degree in these variables starting with the third-degree terms. The coefficients of these series are real. The matrix $\mathcal{N}$ is indicated explicitly so as to be able to deal in A and Φ with dimensionless quantities. Of course, one should add to (4.2.1) the conjugate equation for $\bar{\alpha}$ so that the total order of the secular system is $2N$. In general planetary theory N is an even number (twice the number of the planets) and the variables $\alpha_1, \ldots, \alpha_N$ are split into two groups (eccentric and oblique variables) interrelated only by the non-linear terms in (4.2.1). Therefore, the matrix A is of block-diagonal form, which is also typical for the equations of celestial mechanics. But this feature is not used in the algorithm given below.

Let c_j and $S_{\cdot j}$ $(j = 1, 2, \ldots, N)$ be the eigenvalues and eigenvectors of the matrix $\mathcal{N}A$. In the case typical for celestial mechanics they are real. The numbers c_j are assumed to be incommensurable (the non-resonant case) but one of them may vanish (the zero frequency for the oblique variables). In virtue of the defining relation

$$\mathcal{N}AS_{.j} = c_j S_{.j} \tag{4.2.2}$$

one has

$$S^{-1}\mathcal{N}AS = c\,. \tag{4.2.3}$$

Here c is the diagonal matrix of the eigennumbers $c_1, \ldots, c_N$ and $S = \|S_{ij}\|$, S_{ij} being the component i of the vector $S_{.j}$.

The first step of the algorithm is to perform the linear transformation to the new variables x:

$$\alpha = Sx\,. \tag{4.2.4}$$

It is appropriate to use the block-diagonal structure of S to manipulate polynomials with a smaller number of variables. As a result of transformation (4.2.4) equation (4.2.1) takes the form

$$\dot{x} = \mathrm{i}\left[cx + \mathcal{N}R_1(x, \bar{x})\right] \tag{4.2.5}$$

with

$$R_1 = \mathcal{N}^{-1}S^{-1}\mathcal{N}\Phi\,. \tag{4.2.6}$$

Constructing the series $R_1(x, \bar{x})$ with the given series $\Phi(\alpha, \bar{\alpha})$ is the most cumbersome operation of the algorithm. Complementing (4.2.5) by the conjugate equation one may rewrite the resulting secular system in the form of a single equation

$$\dot{X} = \mathrm{i}[\mathcal{P}X + \mathcal{N}R(X)]\,. \tag{4.2.7}$$

Here

$$X = \begin{pmatrix} x \\ \bar{x} \end{pmatrix}, \qquad \mathcal{P} = \begin{pmatrix} c & 0 \\ 0 & -c \end{pmatrix}, \qquad R = \begin{pmatrix} R_1 \\ R_2 \end{pmatrix} \tag{4.2.8}$$

with $R_2 = -\bar{R}_1$. The diagonal matrix $\mathcal{N}$ of order N should be considered in (4.2.7) as the block multiplier for the vectors and matrices composed of blocks of order N.

Birkhoff normalization entails constructing the formal power series

$$X = Y + \Gamma(Y)\,, \tag{4.2.9}$$

reducing (4.2.7) to the system

$$\dot{Y} = \mathrm{i}[\mathcal{P}Y + \mathcal{N}F(Y)] \tag{4.2.10}$$

with the power series F admitting the straightforward integration of this system. Just as in (4.2.8), here we have

$$Y = \begin{pmatrix} u \\ \bar{u} \end{pmatrix}, \quad \Gamma = \begin{pmatrix} \Gamma_1 \\ \Gamma_2 \end{pmatrix}, \quad F = \begin{pmatrix} F_1 \\ F_2 \end{pmatrix}, \tag{4.2.11}$$

where u is the N-vector of the new variables, Γ_1 and F_1 are N-vector functions holomorphic in u and $\bar{u}$. Hence, $\Gamma_2 = \bar{\Gamma}_1$ and $F_2 = -\bar{F}_1$. The right-hand sides of (4.2.7) and (4.2.10) are related by

$$\mathcal{N}F = (E + \Gamma_Y)^{-1}(\mathcal{N}R + \mathcal{P}\Gamma - \Gamma_Y \mathcal{P}Y) \tag{4.2.12}$$

with the unit matrix E and the Jacobi matrix Γ_Y composed of the derivatives of the components of Γ with respect to the variables Y. Equation (4.2.12) serves to determine Γ and F. In virtue of the relation

$$(E + \Gamma_Y)^{-1} = E - (E + \Gamma_Y)^{-1}\Gamma_Y \tag{4.2.13}$$

this equation may be put into the form

$$\mathcal{N}F = \mathcal{P}\Gamma - \Gamma_Y \mathcal{P}Y + \mathcal{N}U, \tag{4.2.14}$$

where the vector U, composed, like (4.2.11), of two N-vectors U_1 and U_2 (with $U_2 = -\bar{U}_1$), is determined by the relation

$$\mathcal{N}U = (E + \Gamma_Y)^{-1}[\mathcal{N}R + \Gamma_Y(\Gamma_Y \mathcal{P}Y - \mathcal{P}\Gamma)]. \tag{4.2.15}$$

U may be split into two parts,

$$U = U^* + U^+, \tag{4.2.16}$$

with U^+ containing all the terms of U, enabling us to integrate without secular terms the partial-derivative equation

$$\Gamma_Y \mathcal{P}Y - \mathcal{P}\Gamma = \mathcal{N}U^+. \tag{4.2.17}$$

In virtue of this separation of U (4.2.15) takes the form

$$U = R - \mathcal{N}^{-1}\Gamma_Y \mathcal{N}U^* \tag{4.2.18}$$

and the right-hand side F of (4.2.10) becomes

$$F = U^*. \tag{4.2.19}$$

Equations (4.2.16)–(4.2.18) enable one to realize Birkhoff normalization by iteration. Indeed, let U and Γ be determined up to terms of degree m, inclusive, with respect to Y $(m = 2, 3, \ldots)$. Then the $(m+1)$-degree terms of U can be found from (4.2.18). These terms are split in accordance with (4.2.16) into two groups and then the $(m+1)$-degree terms of Γ are found by integrating (4.2.17). The secular system of general planetary theory contains only forms of odd degree, and each step of the iteration process increases the accuracy of the determination of U and Γ by two orders. The iterations start with $U = R$.

Let us now give the actual computational formulae. The most complicated operation of the iteration process is to substitute the series

$$x = u + \Gamma_1(u, \bar{u}) \tag{4.2.20}$$

into the power series $R_1(x, \bar{x})$. The resulting power series in u and $\bar{u}$ will be accurate up to one or two degrees more compared with the initial series (4.2.20). Then one finds the correction terms due to the lower-degree terms of U and Γ:

$$U_1 = R_1 - \mathcal{N}^{-1}\left(\frac{\partial \Gamma_1}{\partial u}\mathcal{N}U_1^* - \frac{\partial \Gamma_1}{\partial \bar{u}}\mathcal{N}\bar{U}_1^*\right). \tag{4.2.21}$$

Thus all the components U_{1i} $(i = 1, 2, \ldots, N)$ of the vector U_1 will be represented by the power series

$$U_{1i} = \sum U_{pq}^{(1i)} \prod_{j=1}^{N} u_j^{p_j} \bar{u}_j^{q_j}, \tag{4.2.22}$$

where p and q are multi-indices with components p_j and q_j $(j = 1, \ldots, N)$, respectively. All the terms of U_{1i} with indices satisfying the relation

$$p_j = q_j + \delta_{ij} \tag{4.2.23}$$

will be attributed to U_{1i}^* so that

$$U_{1i}^* = u_i \sum{}^* U_{pq}^{(1i)} \prod_{j=1}^{N} (u_j \bar{u}_j)^{q_j}. \tag{4.2.24}$$

The asterisk at the summation sign indicates that this summation is taken only over critical values (4.2.23). The equation

$$\frac{\partial \Gamma_1}{\partial u} cu - \frac{\partial \Gamma_1}{\partial \bar{u}} c\bar{u} - c\Gamma_1 = \mathcal{N}U_1^+ \tag{4.2.25}$$

with given coefficients $U_{pq}^{(1i)}$ on the right-hand side enables one to find the coefficients $\Gamma_{pq}^{(1i)}$ of the power expansion of Γ_{1i}

$$\Gamma_{pq}^{(1i)} = \frac{n_i U_{pq}^{(1i)}}{\sum\limits_{j=1}^{N} (p_j - q_j - \delta_{ij}) c_j}. \tag{4.2.26}$$

All the divisors in (4.2.26) are non-zero. Indeed, the eigenvalues c_j are incommensurable by assumption and the right-hand sides (4.2.24) do not contain the critical terms satisfying (4.2.23). Moreover, the terms of the secular system of the general planetary theory have indices such that

$$\sum_{j=1}^{N}(p_j - q_j) = 1\,. \tag{4.2.27}$$

Therefore, in spite of one zero eigenvalue the divisors of (4.2.26) may vanish only under condition (4.2.23), i.e. only for the critical terms. In reality, if for instance $c_N = 0$ then the divisors in (4.2.26) might vanish only for

$$p_j = q_j + \delta_{ij}\,, \quad j = 1, 2, \ldots, N-1\,.$$

In virtue of (4.2.27) these relations also imply that

$$p_N - q_N = 1 - \sum_{j=1}^{N-1}(p_j - q_j)$$

or

$$p_N = q_N + \delta_{iN}\,,$$

again resulting in critical values (4.2.23).

Hence, the iterative algorithm enables one to construct formal series like (4.2.22) for U_{1i} and Γ_{1i} with arbitrary accuracy. To solve this problem one should have a processor to manipulate multiple power series including, in particular, the subroutines for linear substitution (4.2.4) and TAYLOR substitution (4.2.20). If the number N is not so small it is appropriate again to replace index packing by the numeration technique.

The normalization process is completed without any difficulties. In virtue of (4.2.10), (4.2.19) and (4.2.24) the final equations for the variables u_i are of the form

$$\dot{u}_i = \mathrm{i}u_i\left[c_i + n_i \sum {}^{*}U_{pq}^{(1i)} \prod_{j=1}^{N}(u_j\bar{u}_j)^{q_j}\right]. \tag{4.2.28}$$

In combination with the conjugate equations we get

$$u_i\bar{u}_i = \mathrm{const}\,. \tag{4.2.29}$$

Hence, equations (4.2.28) are reduced to the linear equations

$$\dot{u}_i = \mathrm{i}(c_i + \delta c_i)u_i \tag{4.2.30}$$

with real constants δc_i. The solution of equations (4.2.30) is

$$u_i = \xi_i \exp \mathrm{i}\,\omega_i\,, \tag{4.2.31}$$

where ω_i are linear functions of t:

$$\omega_i = (c_i + \delta c_i)t + \tau_i\,. \tag{4.2.32}$$

ξ_i and τ_i are real constants of integration. Therefore, the corrections to the eigenvalues will be

$$\delta c_i = n_i \sum{}^{*} U_{pq}^{(1i)} \prod_{j=1}^{N} \xi_j^{2q_j} . \tag{4.2.33}$$

The asterisk reminds us again that the summation is performed over critical values (4.2.23).

5 The Satellite Disturbing Function

5.1 Expansions in Spherical Functions

The main perturbations in satellite problems are caused by the influence of the non-sphericity of the primary body and by the attraction of the third body. In both cases the expansion of the corresponding disturbing function in a spherical-function series may be presented in the form

$$R = \sum_{k=k_0}^{\infty} \sum_{j=0}^{k} R_{kj} \,. \tag{5.1.1}$$

For the non-sphericity perturbations the expression of R_{kj} is as follows:

$$R_{kj} = Gm_0 \frac{r_0^k}{r^{k+1}} \left(C_{kj} \cos j\lambda' + S_{kj} \sin j\lambda' \right) P_{kj}(\sin \beta') \,, \tag{5.1.2}$$

where m_0 is the mass of the primary body, r_0 is its mean equatorial radius, C_{kj} and S_{kj} are numerical coefficients characterizing its gravitational structure, and r, λ' and β' are the spherical coordinates of the satellite P in the body-fixed coordinate system (see Fig. 3).

The right ascension λ' is reckoned in the equatorial plane of the figure from some moving axis x' and β' is the satellite declination with respect to this plane. In the general case $k_0 = 1$. But if the origin of the body-fixed coordinate system coincides with the centre of mass of the primary body then the summation over k in (5.1.1) starts with $k_0 = 2$.

For the case of the perturbations due to the third body P' the standard expansion in Legendre polynomials for the reciprocal of the mutual distance between P and P' with subsequent use of the addition formula for the Legendre polynomials results in the expression

$$R_{kj} = Gm'(2-\delta_{j0}) \frac{(1)_{k-j}}{(1)_{k+j}} \frac{r^k}{r'^{k+1}} P_{kj}(\sin \beta) P_{kj}(\sin \beta') \cos j(\lambda - \lambda') . \tag{5.1.3}$$

Here m' is the mass of the disturbing body P'. r, λ, β and r', λ', β' are the spherical coordinates of P and P', respectively, in some inertial coordinate system (Fig. 4). The summation over k in (5.1.1) starts in this case with $k_0 = 2$.

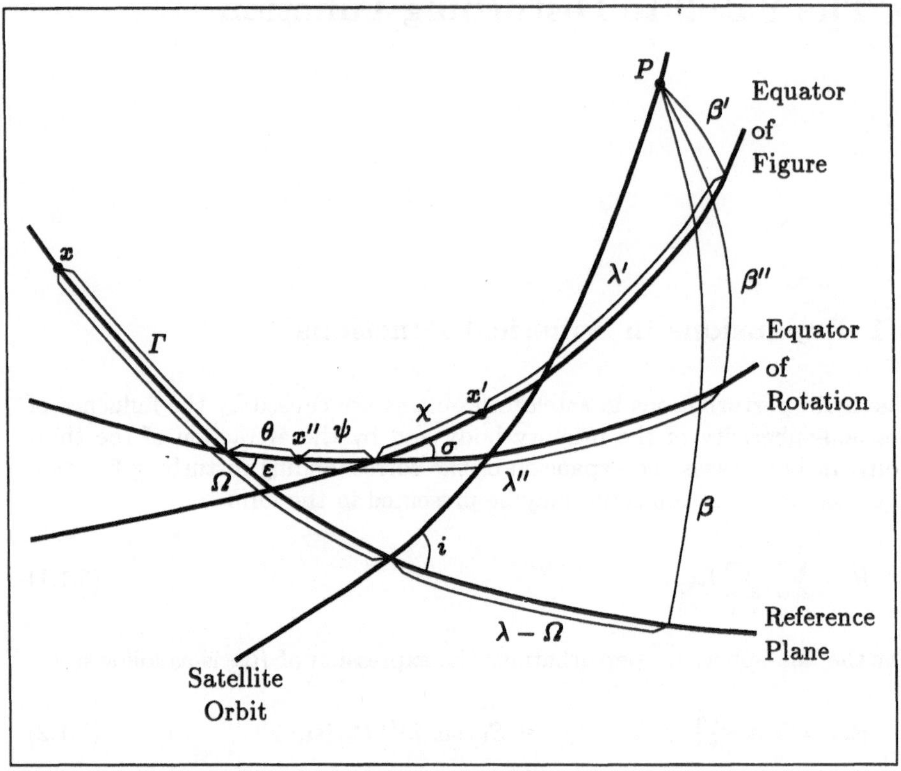

Fig. 3. Coordinate systems and the satellite orbit

In both cases the summation over k in (5.1.1) is actually performed up to some fixed value K. For (5.1.2) this value is determined by the decrease of the absolute values of the coefficients C_{kj} and S_{kj} with increasing k. For (5.1.3) this value depends on the smallness of the ratio r/r'. In both cases K is chosen taking into account the type of satellite orbit and the accuracy of the calculation demanded.

Expressions (5.1.2) and (5.1.3) contain the associated Legendre functions $P_{kj}(t)$. In celestial-mechanics problems where t is the real argument belonging to the interval $(-1, 1)$ they are defined with the aid of the Legendre polynomials $P_k(t)$ by means of the relation (Ferrers' definition)

$$P_{kj}(t) = (1 - t^2)^{\frac{j}{2}} \frac{d^j P_k(t)}{dt^j}, \quad 0 \le j \le k. \tag{5.1.4}$$

Introducing the rectangular coordinates

$$\begin{aligned} x &= r \cos\lambda \cos\beta , \\ y &= r \sin\lambda \cos\beta , \\ z &= r \sin\beta \end{aligned} \tag{5.1.5}$$

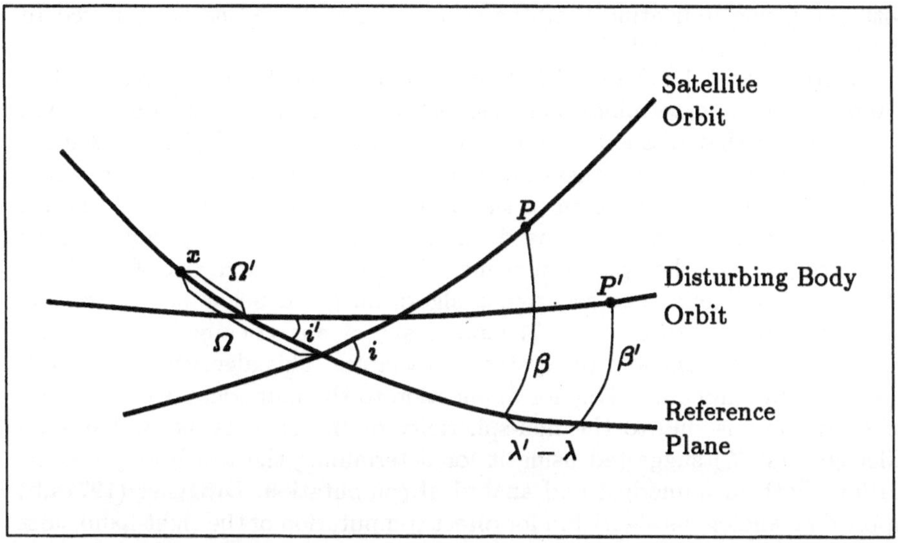

Fig. 4. Orbits of the satellite and disturbing body

one may determine for $0 \leq j \leq k$ solid spherical harmonics as follows:

$$X_{kj} = r^k \, P_{kj}(\sin\beta) \, \exp \mathrm{i} j\lambda \,, \tag{5.1.6}$$

$$Y_{kj} = r^{-k-1} \, P_{kj}(\sin\beta) \, \exp \mathrm{i} j\lambda \,. \tag{5.1.7}$$

Evidently,

$$X_{kj} = r^{2k+1} Y_{kj} \,. \tag{5.1.8}$$

The functions X_{kj} are homogeneous polynomials of degree k with respect to x, $\mathrm{i}y$ and z. Separating the real and imaginary parts in (5.1.6) and (5.1.7) we have

$$X_{kj} = W_{kj} + \mathrm{i} Z_{kj} \tag{5.1.9}$$

and

$$Y_{kj} = U_{kj} + \mathrm{i} V_{kj} \,. \tag{5.1.10}$$

Then (5.1.2) and (5.1.3) take the form

$$R_{kj} = G m_0 r_0^k \left(C_{kj} U'_{kj} + S_{kj} V'_{kj} \right) \tag{5.1.11}$$

and

$$R_{kj} = G m' (2 - \delta_{j0}) \frac{(1)_{k-j}}{(1)_{k+j}} \left(W_{kj} U'_{kj} + Z_{kj} V'_{kj} \right) \tag{5.1.12}$$

with the prime indicating that the corresponding functions are referred to the variables x', y' and z'.

Expressions (5.1.11) and (5.1.12) demonstrate that the problem of analytical or numerical determination of the disturbing function and its derivatives is reduced to that of solid spherical harmonics (5.1.6) and (5.1.7). General analytical expressions for these harmonics in terms of the rectangular coordinates and different systems of the orbital elements are now well known. Some of these expressions will be given in Section 5.3 . But any change of the reference plane or the system of elements necessitates extra transformations of these expressions. The recurrence algorithm given below enables one to find functions (5.1.6) and (5.1.7) numerically or analytically for any choice of the reference plane and the system of variables. This algorithm was developed by Cunningham (1970) for application to the numerical determination of perturbations due to the non-sphericity of the primary body. Later on Giacaglia (1975) suggested using it for determining the third-body perturbations both in numerical and analytical computation. Drożyner (1977a,b) applied an analogous algorithm for direct computation of the right-hand sides of the equations of satellite motion.

5.2 The Recurrence Determination of Solid Spherical Harmonics

For negative values of the second index definitions (5.1.4), (5.1.6) and (5.1.7) are complemented as follows:

$$P_{k,-j}(t) = (-1)^j \frac{(1)_{k-j}}{(1)_{k+j}} P_{kj}(t) , \tag{5.2.1}$$

$$X_{k,-j} = (-1)^j \frac{(1)_{k-j}}{(1)_{k+j}} \bar{X}_{kj} , \tag{5.2.2}$$

$$Y_{k,-j} = (-1)^j \frac{(1)_{k-j}}{(1)_{k+j}} \bar{Y}_{kj} , \tag{5.2.3}$$

the bar denoting the complex conjugate quantity. In addition, in all the recurrences below one should put for $j > k$

$$P_{kj}(t) = 0 , \qquad X_{kj} = 0 , \qquad Y_{kj} = 0 \quad (j > k) . \tag{5.2.4}$$

Let the values of the indices k and j be written in the form of a triangle table (Fig. 5). Functions (5.1.6) and (5.1.7) are determined separately by means of recurrences of the same type in the diagonal and vertical directions as indicated by arrows. The diagonal recurrences enable one to determine successively the sectorial functions X_{kk}, Y_{kk} $(k = 1, 2, \ldots)$ on the basis of only one preceding value. The initial data for this process are

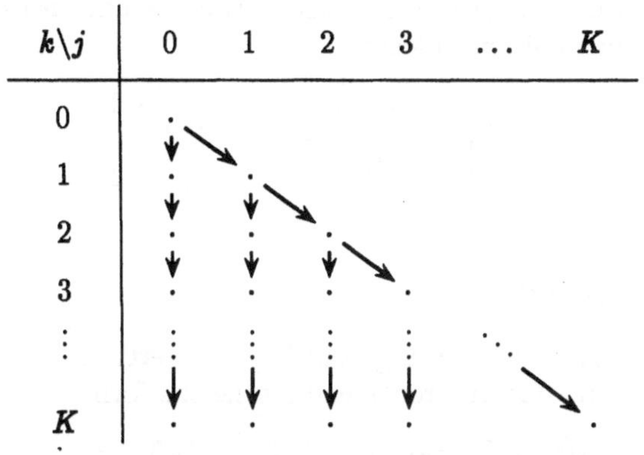

Fig. 5. Recurrences for solid spherical harmonics

$$X_{00} = 1\,, \qquad Y_{00} = \frac{1}{r}\,. \tag{5.2.5}$$

The vertical recurrences enable one to determine for each fixed j on the basis of two preceding values (in the particular case $k = j + 1$ on the basis of only one value) all the subsequent tesseral functions ($k = j + 1, j + 2, \ldots$).

The diagonal formulae result from the relation

$$P_{k+1,j}(t) = (2k+1)(1-t^2)^{\frac{1}{2}} P_{k,j-1}(t) + P_{k-1,j}(t)\,. \tag{5.2.6}$$

Using (5.1.6) one has

$$X_{k+1,k+1} = (2k+1)(x+\mathrm{i}y)X_{kk}\,. \tag{5.2.7}$$

In virtue of (5.1.8)

$$Y_{k+1,k+1} = (2k+1)\frac{1}{r^2}(x+\mathrm{i}y)Y_{kk}\,. \tag{5.2.8}$$

The vertical formulae may be most easily derived from the recurrence relation for the Legendre associated functions

$$(k-j+1)P_{k+1,j}(t) = (2k+1)tP_{kj}(t) - (k+j)P_{k-1,j}(t)\,. \tag{5.2.9}$$

In terms of (5.1.6) one gets

$$(k-j+1)X_{k+1,j} = (2k+1)zX_{kj} - (k+j)r^2X_{k-1,j}\,. \tag{5.2.10}$$

Hence, in virtue of (5.1.8) we have

$$(k-j+1)Y_{k+1,j} = (2k+1)\frac{z}{r^2}Y_{kj} - \frac{k+j}{r^2}Y_{k-1,j}\,. \tag{5.2.11}$$

For the real functions W_{kj} and Z_{kj} the vertical recurrences are of form (5.2.10) and the diagonal ones become

$$\begin{pmatrix} W_{k+1,k+1} \\ Z_{k+1,k+1} \end{pmatrix} = (2k+1)\left[x\begin{pmatrix} W_{kk} \\ Z_{kk} \end{pmatrix} + y\begin{pmatrix} -Z_{kk} \\ W_{kk} \end{pmatrix}\right] \tag{5.2.12}$$

with the initial values

$$W_{00} = 1\,, \qquad Z_{00} = 0\,. \tag{5.2.13}$$

Similarly, for the real functions U_{kj} and V_{kj} the vertical recurrences retain form (5.2.11) and the diagonal recurrences take the form

$$\begin{pmatrix} U_{k+1,k+1} \\ V_{k+1,k+1} \end{pmatrix} = \frac{2k+1}{r^2}\left[x\begin{pmatrix} U_{kk} \\ V_{kk} \end{pmatrix} + y\begin{pmatrix} -V_{kk} \\ U_{kk} \end{pmatrix}\right] \tag{5.2.14}$$

with the initial values

$$U_{00} = \frac{1}{r}\,, \qquad V_{00} = 0\,. \tag{5.2.15}$$

In numerical calculations the coefficients of formulae (5.2.7), (5.2.8), (5.2.10)–(5.2.12) and (5.2.14) are numbers. In analytical computations they are presented by some expansion obtained, for example, by a Keplerian processor. In any case these formulae enable us to determine efficiently functions (5.1.6) and (5.1.7) in the appropriate form.

As far as the derivatives of these functions with respect to the coordinates x, y, z are concerned they may be computed either by means of the analogous recurrences (as is done in the papers by Drożyner cited above) or by means of the original functions themselves. In the latter case one needs to have only some extra values of the functions (5.1.6) and (5.1.7) for $k = K+1$ and $j = 0, 1, \ldots, K+1$. Indeed, the well-known representation

$$Y_{kj} = \frac{(-1)^k}{(1)_{k-j}}\left(\frac{\partial}{\partial x} + \mathrm{i}\frac{\partial}{\partial y}\right)^j \left(\frac{\partial}{\partial z}\right)^{k-j}\frac{1}{r} \tag{5.2.16}$$

gives

$$Y_{k+1,j} = -\frac{1}{k-j+1}\frac{\partial}{\partial z}Y_{kj} \tag{5.2.17}$$

and

$$Y_{k+1,j+1} = -\left(\frac{\partial}{\partial x} + \mathrm{i}\frac{\partial}{\partial y}\right)Y_{kj}\,. \tag{5.2.18}$$

In virtue of the Laplace equation one has

$$\frac{\partial^2}{\partial z^2}\frac{1}{r} = -\left(\frac{\partial}{\partial x} + \mathrm{i}\frac{\partial}{\partial y}\right)\left(\frac{\partial}{\partial x} - \mathrm{i}\frac{\partial}{\partial y}\right)\frac{1}{r}\,.$$

Therefore,

$$Y_{k+1,j-1} = \frac{1}{(k-j+2)(k-j+1)}\left(\frac{\partial}{\partial x} - \mathrm{i}\frac{\partial}{\partial y}\right)Y_{kj}\,. \tag{5.2.19}$$

Solving (5.2.17)–(5.2.19) explicitly in terms of the derivatives, one obtains

$$\frac{\partial}{\partial x}Y_{kj} = -\frac{1}{2}Y_{k+1,j+1} + \frac{1}{2}(k-j+2)(k-j+1)Y_{k+1,j-1}\,, \tag{5.2.20}$$

$$\frac{\partial}{\partial y}Y_{kj} = \frac{\mathrm{i}}{2}Y_{k+1,j+1} + \frac{\mathrm{i}}{2}(k-j+2)(k-j+1)Y_{k+1,j-1} \tag{5.2.21}$$

and

$$\frac{\partial}{\partial z}Y_{kj} = -(k-j+1)Y_{k+1,j}\,. \tag{5.2.22}$$

Using (5.1.8) it is easy to derive from (5.2.20)–(5.2.22) the corresponding relations for the derivatives of X_{kj}:

$$\begin{aligned}\frac{\partial}{\partial x}X_{kj} = &-\frac{1}{2r^2}X_{k+1,j+1} + (2k+1)\frac{x}{r^2}X_{kj} + \\ &+\frac{1}{2r^2}(k-j+2)(k-j+1)X_{k+1,j-1}\,,\end{aligned} \tag{5.2.23}$$

$$\begin{aligned}\frac{\partial}{\partial y}X_{kj} = &\frac{\mathrm{i}}{2r^2}X_{k+1,j+1} + (2k+1)\frac{y}{r^2}X_{kj} + \\ &+\frac{\mathrm{i}}{2r^2}(k-j+2)(k-j+1)X_{k+1,j-1}\end{aligned} \tag{5.2.24}$$

and

$$\frac{\partial}{\partial z}X_{kj} = -\frac{k-j+1}{r^2}X_{k+1,j} + (2k+1)\frac{z}{r^2}X_{kj}\,. \tag{5.2.25}$$

For $j = 0$ in accordance with (5.2.2) and (5.2.3) one has on the right-hand sides of (5.2.20), (5.2.21), (5.2.23) and (5.2.24)

$$(k+2)(k+1)\begin{pmatrix}X_{k+1,-1}\\ Y_{k+1,-1}\end{pmatrix} = -\begin{pmatrix}\bar{X}_{k+1,1}\\ \bar{Y}_{k+1,1}\end{pmatrix}.$$

Expressed in terms of real variables relations (5.2.20) and (5.2.21) take the form

$$\frac{\partial}{\partial x}\begin{pmatrix}U_{kj}\\ V_{kj}\end{pmatrix} = -\frac{1}{2}\begin{pmatrix}U_{k+1,j+1}\\ V_{k+1,j+1}\end{pmatrix} + \frac{1}{2}(k-j+2)(k-j+1)\begin{pmatrix}U_{k+1,j-1}\\ V_{k+1,j-1}\end{pmatrix} \tag{5.2.26}$$

and

$$\frac{\partial}{\partial y}\begin{pmatrix} U_{kj} \\ V_{kj} \end{pmatrix} = \frac{1}{2}\begin{pmatrix} -V_{k+1,j+1} \\ U_{k+1,j+1} \end{pmatrix} + \frac{1}{2}(k-j+2)(k-j+1)\begin{pmatrix} -V_{k+1,j-1} \\ U_{k+1,j-1} \end{pmatrix} \tag{5.2.27}$$

with the substitution for $j = 0$

$$(k+2)(k+1)\begin{pmatrix} U_{k+1,-1} \\ V_{k+1,-1} \end{pmatrix} = \begin{pmatrix} -U_{k+1,1} \\ V_{k+1,1} \end{pmatrix}.$$

Relations (5.2.23) and (5.2.24) may be rewritten in a similar manner:

$$\frac{\partial}{\partial x}\begin{pmatrix} W_{kj} \\ Z_{kj} \end{pmatrix} = -\frac{1}{2r^2}\begin{pmatrix} W_{k+1,j+1} \\ Z_{k+1,j+1} \end{pmatrix} + (2k+1)\frac{x}{r^2}\begin{pmatrix} W_{kj} \\ Z_{kj} \end{pmatrix} + \\ + \frac{1}{2r^2}(k-j+2)(k-j+1)\begin{pmatrix} W_{k+1,j-1} \\ Z_{k+1,j-1} \end{pmatrix} \tag{5.2.28}$$

and

$$\frac{\partial}{\partial y}\begin{pmatrix} W_{kj} \\ Z_{kj} \end{pmatrix} = \frac{1}{2r^2}\begin{pmatrix} -Z_{k+1,j+1} \\ W_{k+1,j+1} \end{pmatrix} + (2k+1)\frac{y}{r^2}\begin{pmatrix} W_{kj} \\ Z_{kj} \end{pmatrix} + \\ + \frac{1}{2r^2}(k-j+2)(k-j+1)\begin{pmatrix} -Z_{k+1,j-1} \\ W_{k+1,j-1} \end{pmatrix} \tag{5.2.29}$$

with the substitution for $j = 0$

$$(k+2)(k+1)\begin{pmatrix} W_{k+1,-1} \\ Z_{k+1,-1} \end{pmatrix} = \begin{pmatrix} -W_{k+1,1} \\ Z_{k+1,1} \end{pmatrix}.$$

The derivatives with respect to z retain form (5.2.22) for functions U_{kj} and V_{kj} and form (5.2.25) for functions W_{kj} and Z_{kj}.

5.3 The Disturbing Function and Its Derivatives

In this section we shall deal with the motion of a satellite in an inertial planetocentric coordinate system x, y, z with an arbitrary plane of reference xy. The equations of the satellite motion will be

$$\ddot{x} = -Gm_0\frac{x}{r^3} + \frac{\partial R_0}{\partial x} + \frac{\partial R'}{\partial x} \tag{5.3.1}$$

with two similar equations for y and z. Here R_0 is the disturbing function (5.1.1), (5.1.2) due to the non-sphericity of the central body and R' is the disturbing function (5.1.1), (5.1.3) caused by the attraction of the third body. The coupling effect between these two main sources of perturbation is neglected here.

Consider first the function R_0. According to (5.1.11) the recurrence algorithm given above should be performed in the body-fixed coordinate system

x', y', z' with the plane $x'y'$ being the equator of the figure. The transformation between the systems x, y, z and x', y', z' involves a third system x'', y'', z'', where $x''y''$ is the equator of rotation (Fig. 3). The relevant canonical variables are known as Serret–Andoyer variables (Deprit and Elipe, 1993). Using relations (3.3.1)–(3.3.4) consecutively for the transformations from (x, y, z) to (x'', y'', z'') and then from (x'', y'', z'') to (x', y', z') one finally has

$$\begin{pmatrix} x \\ y \\ z \end{pmatrix} = \Lambda \begin{pmatrix} x' \\ y' \\ z' \end{pmatrix} \tag{5.3.2}$$

with

$$\begin{aligned} \Lambda &= D_3(-\Gamma)\, D_1(-\varepsilon)\, D_3(-\theta)\, D_3(-\psi)\, D_1(-\sigma)\, D_3(-\chi) = \\ &= D_3(-\Gamma)\, D_1(-\varepsilon)\, D_3(-\theta-\psi)\, D_1(-\sigma)\, D_3(-\chi)\,. \end{aligned} \tag{5.3.3}$$

The unit rotations are defined by (3.3.5)–(3.3.7) and the corresponding Euler angles are shown in Fig. 3. Hence,

$$\begin{pmatrix} \dfrac{\partial R_0}{\partial x} \\ \dfrac{\partial R_0}{\partial y} \\ \dfrac{\partial R_0}{\partial z} \end{pmatrix} = \Lambda \begin{pmatrix} \dfrac{\partial R_0}{\partial x'} \\ \dfrac{\partial R_0}{\partial y'} \\ \dfrac{\partial R_0}{\partial z'} \end{pmatrix} . \tag{5.3.4}$$

The inverse transformation of (5.3.2) is given by

$$\begin{pmatrix} x' \\ y' \\ z' \end{pmatrix} = \Lambda^T \begin{pmatrix} x \\ y \\ z \end{pmatrix} \tag{5.3.5}$$

with the transposed matrix

$$\Lambda^T = D_3(\chi) D_1(\sigma) D_3(\theta+\psi) D_1(\varepsilon) D_3(\Gamma)\,. \tag{5.3.6}$$

The determination of the right-hand sides of equations (5.3.1) in numerically integrating these equations involves four steps:

1. the computation of x', y' and z' by means of (5.3.5) from the given values of x, y and z;
2. the construction of recurrence algorithm (5.2.11) and (5.2.14) in terms of x', y' and z';
3. the determination of the derivatives of R_0 with respect to x', y' and z' by means of (5.2.22), (5.2.26) and (5.2.27);
4. the application of (5.3.4) to obtain the final expressions for the right-hand sides.

In constructing an analytical theory of motion the coordinates $x,\ y,\ z$ are presented in closed form with respect to $r,\ v,\ i,\ \omega$ and Ω or in the form of the infinite series in $a,\ e,\ i,\ \omega,\ \Omega,\ M$ (with possible replacement of e and M by the nome q and elliptic anomaly w). These expressions may be derived by a Keplerian processor. Then the four steps just mentioned are performed using a PS processor. The explicit form of the satellite disturbing function R_0 is well known in terms of the Keplerian elements. We give below expressions for the general terms of the relevant expansions, although actually all these expansions may now be obtained by the recurrence algorithm.

The general terms of these expansions are easily derived from the transformation law of the spherical surface harmonics under rotation of the coordinate system. In passing from the $(x,\ y,\ z)$ to $(x'',\ y'',\ z'')$ system these harmonics are transformed as follows:

$$P_{kj}(\sin\beta)\exp \mathrm{i}j\lambda = \sum_{l=-k}^{k} \mathrm{i}^{|j-l|} A_{kjl}(\varepsilon) P_{kl}(\sin\beta'')\exp \mathrm{i}\big[l(\lambda''+\theta)+j\Gamma\big] \tag{5.3.7}$$

and

$$P_{kj}(\sin\beta'')\exp \mathrm{i}j\lambda'' = \sum_{l=-k}^{k} \mathrm{i}^{-|j-l|} A_{kjl}(\varepsilon) P_{kl}(\sin\beta)\exp \mathrm{i}[l(\lambda-\Gamma)-j\theta]\,. \tag{5.3.8}$$

These transformation formulae given in the mathematical literature on special functions in various forms are of great use in celestial mechanics (see the examples given by Giacaglia (1980)). We put them in the form used by Brumberg et al. (1971) to underline the inclination coefficients A_{kjl} generalizing the inclination coefficients of Kaula F_{kjl}. The inclination coefficients represent a set of special functions of much importance for celestial mechanics. These functions are simpler than the Hansen coefficients since they are expressed in terms of elementary functions and hypergeometric polynomials. Therefore, these functions satisfy rather simple recurrence relations and transformations based on hypergeometric-function theory. The explicit expressions for A_{kjl} and F_{kjl} are as follows:

$$\begin{aligned} A_{kjl}(\varepsilon) = {} & \frac{(1)_{k-l}}{(1)_{k-j}} \frac{(-k-j)_{\max\{0,j-l\}}(-k+j)_{\max\{l-j\}}}{(1)_{|j-l|}} \times \\ & \times \left(\sin\frac{\varepsilon}{2}\right)^{|j-l|} \left(\cos\frac{\varepsilon}{2}\right)^{|j+l|} F\Big(\tfrac{1}{2}|j-l| + \tfrac{1}{2}|j+l| - k, \\ & 1+k+\tfrac{1}{2}|j-l|+\tfrac{1}{2}|j+l|,\ 1+|j-l|,\ \sin^2\frac{\varepsilon}{2}\Big)\,, \end{aligned} \tag{5.3.9}$$

$$F_{kjl}(\varepsilon) = (-1)^{E(\frac{k-j}{2})+\max\{0,k-j-2l\}} \frac{2^{k-2l}\left(\frac{1}{2}\right)_{k-l}}{(1)_l} A_{k,j,k-2l}(\varepsilon)\,. \tag{5.3.10}$$

The functions F_{kjl} occur naturally in applying (5.3.7) to the transformation from the $(x,\ y,\ z)$ system to the orbital system involving the usual Keplerian elements. In this case one should put in (5.3.7) $\varepsilon = i$ (the inclination of the orbit), $\Gamma = \Omega$ (the longitude of the node), $\theta = \omega$ (the argument of the pericentre), $\lambda'' = v$ (the true anomaly) and $\beta'' = 0$. Taking into account that $P_{kl}(0)$ vanishes for odd values of $k - l$ and

$$P_{k,k-2p}(0) = (-1)^p 2^{k-2p} \frac{\left(\frac{1}{2}\right)_{k-p}}{(1)_p} \tag{5.3.11}$$

one has

$$P_{kj}(\sin\beta) \exp ij\lambda = \mathrm{i}^{j-k+2\mathrm{E}\left(\frac{k-j}{2}\right)} \sum_{p=0}^{k} F_{kjp}(i) \exp \mathrm{i}\left[(k-2p)(v+\omega) + j\Omega\right]. \tag{5.3.12}$$

This relation was first given by Kaula (1961) with very complicated expressions for the inclination coefficients. In dealing with functions (5.3.9) one may use the relation

$$A_{k,-j,-l}(\varepsilon) = \frac{(1)_{k+l}\,(1)_{k-j}}{(1)_{k-l}\,(1)_{k+j}} A_{kjl}(\varepsilon) \tag{5.3.13}$$

or in terms of the Kaula coefficients

$$F_{k,-j,k-l}(\varepsilon) = (-1)^k \frac{(1)_{k-j}}{(1)_{k+j}} F_{kjl}(\varepsilon).$$

Now, applying (5.3.8) twice for the transformation from the $(x',\ y',\ z')$ system to the $(x'',\ y'',\ z'')$ system and then to the $(x,\ y,\ z)$ system one has

$$\begin{aligned} P_{kj}(\sin\beta') \exp \mathrm{i}\lambda' = \sum_{l=-k}^{k} \sum_{s=-k}^{k} \mathrm{i}^{-|j-l|-|l-s|} A_{kjl}(\sigma) A_{kls}(\varepsilon) \times \\ \times P_{ks}(\sin\beta) \exp \mathrm{i}\left[s(\lambda - \Gamma) - l(\theta + \psi) - j\chi\right]. \end{aligned} \tag{5.3.14}$$

Of course, if one of the inclination angles, say σ, vanishes then $A_{kjl}(0) = \delta_{jl}$, and there remains only one summation in (5.3.14). Substituting expansion (5.3.12) into the right-hand side of (5.3.14) and using the evident relations

$$|l - s| = s - l + 2\max\{0, l - s\},$$

$$|j - l| = l - j + 2\max\{0, j - l\}$$

we obtain

$$\begin{aligned} P_{kj}(\sin\beta') \exp \mathrm{i}\lambda' = \\ = \mathrm{i}^{j-k} \sum_{l=-k}^{k} \sum_{s=-k}^{k} \sum_{p=0}^{k} (-1)^{E\left(\frac{k-s}{2}\right)+\max\{0,j-l\}+\max\{0,l-s\}} A_{kjl}(\sigma) A_{kls}(\varepsilon) \times \\ \times F_{ksp}(i) \exp \mathrm{i}\left[(k-2p)(v+\omega) + s(\Omega - \Gamma) - l(\theta + \psi) - j\chi\right]. \end{aligned} \tag{5.3.15}$$

Finally, considering that for $j \le k$

$$\mathrm{i}^{j-k} = (-1)^{\mathrm{E}(\frac{k-j}{2})} \exp\left(-\mathrm{i}\nabla_{kj}\frac{\pi}{2}\right), \qquad \nabla_{kj} = \mathrm{mod}(k-j,\ 2) \tag{5.3.16}$$

we get for (5.1.10) in the $(x',\ y',\ z')$ system

$$Y'_{kj} = r^{-k-1}(-1)^{\mathrm{E}(\frac{k-j}{2})} \sum_{l=-k}^{k} \sum_{s=-k}^{k} \sum_{p=0}^{k} (-1)^{\mathrm{E}(\frac{k-s}{2})} \times$$
$$\times (-1)^{\max\{0,l-s\}+\max\{0,j-l\}} A_{kjl}(\sigma) A_{kls}(\varepsilon) F_{ksp}(i) \times$$
$$\times \exp\mathrm{i}\left[(k-2p)(v+\omega) + s(\Omega - \Gamma) - l(\theta+\psi) - j\chi - \nabla_{kj}\frac{\pi}{2}\right]. \tag{5.3.17}$$

Changing here the symbol exp i to cos and sin one obtains the expressions for U'_{kj} and V'_{kj}, respectively, occurring in (5.1.11). The closed-form function (5.3.17) in terms of r and v may be easily expanded in (e, M) or (q, w) series by applying (2.2.6) or (2.5.22), respectively.

It may be useful to change the order of summation in (5.1.1) with (5.1.10) and (5.1.11) to have in the first line a four-argument trigonometric expansion. Indeed, by replacing the summation indices k and p by new indices m and t

$$k = |m| + 2t, \qquad p = t - \min\{0, m\}, \qquad k - 2p = m \tag{5.3.18}$$

and by changing the order of summation

$$\sum_{k=k_0}^{K} \sum_{p=0}^{k} = \sum_{m=-K}^{K} \sum_{t=t^*}^{\mathrm{E}(\frac{K-|m|}{2})},$$

where

$$t^* = \max\left\{0, \mathrm{E}\left(\frac{\max\{k_0, j\} - |m| + 1}{2}\right)\right\}, \tag{5.3.19}$$

one has

$$R_0 = \sum_{m=-K}^{K} \sum_{j=0}^{K} \sum_{l=-K}^{K} \sum_{s=-K}^{K} (A_{mjsl}\cos\Psi_{mjsl} + B_{mjsl}\sin\Psi_{mjsl}) \tag{5.3.20}$$

with

$$\Psi_{mjsl} = m(v+\omega) - j\chi + s(\Omega - \Gamma) - l(\theta + \psi)$$

and

$$\left.\begin{matrix} A_{mjsl} \\ B_{mjsl} \end{matrix}\right\} = Gm_0 \sum_{t=t^*}^{\mathrm{E}(\frac{K-|m|}{2})} (-1)^{\mathrm{E}(\frac{k-j}{2})+\mathrm{E}(\frac{k-s}{2})+\max\{0,l-s\}+\max\{0,j-l\}} \times$$
$$\times \frac{r_0^k}{r^{k+1}} A_{kls}(\varepsilon) A_{kjl}(\sigma) F_{ksp}(i) \begin{pmatrix} (1-\nabla_{kj})C_{kj} - \nabla_{kj}S_{kj} \\ (1-\nabla_{kj})S_{kj} + \nabla_{kj}C_{kj} \end{pmatrix} \tag{5.3.21}$$

with $k = k(t)$ and $p = p(t)$ determined by (5.3.18). There is no problem in replacing in (5.3.20) and (5.3.21) the functions of r and v by (e, M) or (q, w) series. In doing this we shall have one extra argument and the corresponding extra summation in (5.3.20) and an extra factor (a Hansen or elliptic Hansen coefficient) in (5.3.21). The order of smallness of the general term (5.3.21) with respect to the inclination angles is, evidently, equal to

$$O\left(\left(\sin\frac{i}{2}\right)^{|m-s|}\left(\cos\frac{i}{2}\right)^{|m+s|}\left(\sin\frac{\varepsilon}{2}\right)^{|l-s|}\left(\sin\frac{\sigma}{2}\right)^{|j-l|}\right).$$

In using non-Keplerian elements, for instance, Laplace-type elements, the explicit form of (5.3.17) becomes much more complicated. At the same time this does not affect the derivation of such expansions with the aid of the recurrence algorithm. To do this it is sufficient to find with the aid of a Keplerian processor the relevant expansions for the rectangular coordinates. In such a way it is possible, for example, to expand R_0 in terms of the canonical variables of Serret–Andoyer (Babaev and Krasinsky, 1978).

Consider now the function R'. In representation (5.1.12) the functions W_{kj} and Z_{kj}, dependent on the coordinates x, y, z of a satellite, are separated from the functions U'_{kj} and V'_{kj}, which are dependent on the coordinates x', y', z' of the third body. In numerically integrating equations (5.3.1) recurrence algorithm (5.2.10) and (5.2.12) with further application of (5.2.25), (5.2.28) and (5.2.29) results in needed derivatives of W_{kj} and Z_{kj} with respect to x, y and z. Recurrence algorithm (5.2.11) and (5.2.14) applied to the coordinates x', y' and z' of the third body gives the functions U'_{kj} and V'_{kj}. In analytical determination of the functions W_{kj} and Z_{kj} the coordinates x, y and z are expressed with the aid of a Keplerian processor as the closed form expression in r, v, i, ω and Ω or as the infinite series in a, e, i, ω, Ω and M (or else a, q, i, ω, Ω and w). If the disturbing body is assumed to move in an elliptic orbit then its coordinates may be expressed by means of the elliptic-motion formulae. But the advantages of the recurrence algorithm become particularly significant if the motion of the disturbing body cannot be approximated by elliptic motion. This is the case, for example, for artificial satellites of the Moon and distant satellites of the Earth, when one should use the analytical lunar theory. In such cases the coordinates x', y', z' and the coefficients of recurrences (5.2.11) and (5.2.14) should be expressed with the aid of the corresponding analytical theory. Then the functions U'_{kj} and V'_{kj} may be constructed by a PS processor in terms of the parameters of the theory of motion of the disturbing body.

The expressions for the general terms of the expansions of the third-body disturbing function follow directly from (5.3.12) and (5.3.16). Hence,

$$X_{kj} = r^k \sum_{l=0}^{k} F_{kjl}(i) \exp \mathrm{i}\left[(k-2l)(v+\omega) + j\Omega - \nabla_{kj}\frac{\pi}{2}\right] \qquad (5.3.22)$$

and

$$Y'_{kj} = r'^{-k-1} \sum_{l'=0}^{k} F_{kjl'}(i') \exp \mathrm{i} \left[(k-2l')(v'+\omega') + j\Omega' - \nabla_{kj} \frac{\pi}{2} \right]. \quad (5.3.23)$$

Therefore, by virtue of (5.1.1) and (5.1.12) one obtains the well-known expansion of the third-body disturbing function first given by Kaula (1962)

$$R' = Gm' \sum_{k=2}^{\infty} \frac{r^k}{r'^{k+1}} \sum_{l=0}^{k} \sum_{l'=0}^{k} \sum_{j=0}^{k} (2-\delta_{j0}) \frac{(1)_{k-j}}{(1)_{k+j}} F_{kjl}(i) F_{kjl'}(i') \times$$
$$\times \cos\left[(k-2l)(v+\omega) - (k-2l')(v'+\omega') + j(\Omega - \Omega')\right]. \quad (5.3.24)$$

Replacing now the indices k, l and l' by m, s and s' so that

$$k = \left| \frac{s-s'}{2} \right| + \left| \frac{s+s'}{2} \right| + 2m, \quad l = \frac{k-s}{2}, \quad l' = \frac{k+s'}{2} \quad (5.3.25)$$

(s and s' are numbers of the same parity) one may transform (5.3.24) to the three-argument trigonometric expansion

$$R' = \sum_{j=0}^{\infty} \sum_{s=-\infty}^{\infty} \sum_{s'=-\infty}^{\infty} C_{ss'j} \cos\left[s(v+\omega) + s'(v'+\omega') + j(\Omega - \Omega')\right] \quad (5.3.26)$$

with

$$C_{ss'j} = Gm'(2-\delta_{j0}) \sum_{m=m^*}^{\infty} \frac{r^k}{r'^{k+1}} \frac{(1)_{k-j}}{(1)_{k+j}} F_{kjl}(i) F_{kjl'}(i'), \quad (5.3.27)$$

where k, l and l' have values (5.3.25) and the summation starts with the value

$$m^* = \max\left\{0, \mathrm{E}\left[\frac{1}{2}\left(\max\{2, j\} - \left|\frac{s-s'}{2}\right| - \left|\frac{s+s'}{2}\right| + 1\right)\right]\right\}. \quad (5.3.28)$$

With respect to the inclination angles the general term (5.3.27) has the order

$$O\left(\left(\sin\frac{i}{2}\right)^{|s-j|} \left(\cos\frac{i}{2}\right)^{|s+j|} \left(\sin\frac{i'}{2}\right)^{|s'+j|} \left(\cos\frac{i'}{2}\right)^{|s'-j|}\right).$$

Again, the functions of r, v, r' and v' may be expanded in (e, M), (e', M') or (q, w), (q', w') series by applying (2.2.6) or (2.5.22), respectively.

In the case of non-Keplerian motion of the disturbing body it is appropriate to use only expansion (5.3.22), which is related to the coordinates of the satellite, and to retain Y'_{kj} in an explicit form. This corresponds to the representation of R' as follows:

$$R' = Gm' \sum_{k=2}^{\infty} \sum_{j=0}^{k} \sum_{l=0}^{k} (2 - \delta_{j0}) \frac{(1)_{k-j}}{(1)_{k+j}} \frac{r^k}{r'^{k+1}} P_{kj}(\sin\beta') F_{kjl}(i) \times$$

$$\times \cos\left[(k-2l)(v+\omega) + j(\Omega - \lambda') - \nabla_{kj}\frac{\pi}{2}\right] \tag{5.3.29}$$

with further replacement of the coordinates r', λ', β' of the disturbing body by means of the relevant theory of motion.

Let us note that in using (q, w) expansions for R_0 and R' it is no problem to find the derivatives of R with respect to a, i, Ω and ω but it is easier to find the derivatives with respect to e and M by differentiating (5.3.20) and (5.3.29) as functions of r and v and then applying (q, w) expansions (2.5.22) to these derivatives themselves.

5.4 Equations of Motion in Rotating Systems

The equations of satellite motion (5.3.1) have been written in an arbitrary inertial coordinate system. Sometimes it may be useful to study the motion of a satellite in some rotating coordinate system X, Y, Z. Let $\mathbf{r} = (X, Y, Z)$ be the coordinates of the satellite in this system and let $\boldsymbol{\omega} = (\omega_X\ \omega_Y,\ \omega_Z)$ be the angular rotational velocity of this system. In the Lagrange form the equations of motion referred to the system (X, Y, Z) will be

$$\frac{d}{dt}\frac{\partial L}{\partial \dot{\mathbf{r}}} - \frac{\partial L}{\partial \mathbf{r}} = 0 \tag{5.4.1}$$

with the Lagrangian

$$L = \frac{1}{2}\dot{\mathbf{r}}^2 + \frac{Gm_0}{r} + R + \Phi, \tag{5.4.2}$$

R being the disturbing function due to the non-sphericity of the primary body and the third-body attraction. This function should now be expressed in terms of X, Y, Z and time t. The rotation of the coordinate system leads to the additive Φ-term

$$\Phi = \tfrac{1}{2}\omega^2\mathbf{r}^2 - \tfrac{1}{2}(\boldsymbol{\omega}\mathbf{r})^2 + \boldsymbol{\omega}(\mathbf{r} \times \dot{\mathbf{r}}). \tag{5.4.3}$$

The function Φ depends on $\dot{\mathbf{r}}$ and for this reason one cannot directly use the Lagrange equations for the osculating elements. Of course, it is always possible to use the Gauss equations for osculating elements without taking advantage of the representation of the perturbing forces by means of the disturbing function. But it is still possible to use the Lagrange equations for the so-called contact elements related to the coordinates and impulses of the satellite just by the same formulae that relate the osculating elements to the satellite coordinates and velocities. Indeed, equations (5.4.1) may be written in the canonical form

$$\dot{\mathbf{r}} = \frac{\partial V}{\partial \mathbf{p}}, \qquad \dot{\mathbf{p}} = -\frac{\partial V}{\partial \mathbf{r}} \tag{5.4.4}$$

with the Hamiltonian

$$V = \frac{1}{2}\mathbf{p}^2 - \frac{Gm_0}{r} - R - \Psi, \tag{5.4.5}$$

where

$$\Psi = \Phi + \frac{1}{2}\left(\frac{\partial \Phi}{\partial \dot{\mathbf{r}}}\right)^2 = \boldsymbol{\omega}(\mathbf{r} \times \mathbf{p}). \tag{5.4.6}$$

The satellite velocity $\dot{\mathbf{r}}$ is related to the impulse $\mathbf{p}$ by

$$\dot{\mathbf{r}} = \mathbf{p} + (\mathbf{r} \times \boldsymbol{\omega}). \tag{5.4.7}$$

It is seen that the function

$$W = R + \Psi \tag{5.4.8}$$

may be regarded as the disturbing function for problem (5.4.4) provided that the angular velocity $\boldsymbol{\omega}$ is sufficiently small (the typical procedure in celestial mechanics is to take into account the big "regular" part of the rotation by means of the appropriate direct transformation and to treat the small "irregular" part of the rotation as a perturbation). Now, if $\mathbf{r}$ and $\mathbf{p}$ are expressed by means of the formulae for the two-body problem in terms of t and six (contact) Keplerian elements then in perturbed motion these contact elements will satisfy the usual Lagrange perturbing equations with the disturbing function W. One may use therewith any set of the Keplerian elements including the possible replacement of e by the Jacobi nome q for highly eccentric orbits. The influence of the rotation will be completely described by function (5.4.6), which in terms of the contact elements takes the form

$$\Psi = na^2(1 - e^2)^{\frac{1}{2}}\boldsymbol{\omega}\mathbf{k}, \tag{5.4.9}$$

$\mathbf{k}$ being the unit vector of two-body-problem solution (2.1.1) and (2.1.2).

More details about the contact elements may be found in the book by Brumberg (1991).

6 The Planetary Disturbing Function

6.1 General Structure

The main difficulty in expanding the disturbing function in planetary problems is that the ratio of the semi-major axes of the planetary orbits may be rather large (of the order of 0.7, for example, for the pair Venus–Earth). Therefore, efficient expansions in powers of this ratio in satellite problems cannot be directly used here. They are given below to characterize the general structure of the expansions.

Since the expansion of the indirect term of the disturbing function causes no difficulties we consider here the reciprocal of the mutual distance Δ between the planets. Some methods of planetary perturbation theory demand the expantion of not only Δ^{-1} but Δ^{-3}, Δ^{-5}, ... as well. Taking this into account we shall treat the problem of expanding Δ^{-n} for an arbitrary natural integer n.

In dealing with the function

$$\Delta = (r^2 + r'^2 - 2rr' \cos H)^{\frac{1}{2}} \tag{6.1.1}$$

we shall choose the case $r < r'$. Then the standard expansion in Gegenbauer polynomials results in

$$\Delta^{-n} = r'^{-n} \sum_{k=0}^{\infty} \left(\frac{r}{r'}\right)^k C_k^{n/2}(\cos H) . \tag{6.1.2}$$

The expansion of the Gegenbauer polynomials in terms of true anomalies v, v', arguments of pericentres ω, ω', longitudes of nodes Ω, Ω' and inclinations i, i' may be expressed in the form

$$C_k^{n/2}(\cos H) = \sum_{l=0}^{k} \sum_{l'=0}^{k} \sum_{j=0}^{k} (2 - \delta_{j0}) C_{k-l-l',l'-l,j}^{(k,n/2)}(i,i') \times$$
$$\times \cos\left[(k-2l)(v+\omega) - (k-2l')(v'+\omega') + j(\Omega - \Omega')\right] . \tag{6.1.3}$$

The coefficients $C_{pqj}^{(k,n/2)}(i,i')$ admit a simple representation in the case of Legendre polynomials for $n = 1$:

$$C_{pqj}^{(k,1/2)}(i,i') = \frac{(1)_{k-j}}{(1)_{k+j}} F_{k,j,\frac{k-p-q}{2}}(i) F_{k,j,\frac{k-p+q}{2}}(i') , \tag{6.1.4}$$

$F_{kjl}(i)$ being the Kaula inclination function (5.3.10). The representation of the Gegenbauer polynomial as the linear combination of a set of Legendre polynomials (Sack, 1964)

$$C_k^{n/2}(x) = \sum_{r=0}^{\mathrm{E}(\frac{k}{2})} (2k-4r+1) \frac{\left(\frac{n}{2}\right)_{k-r} \left(\frac{n-1}{2}\right)_r}{\left(\frac{3}{2}\right)_{k-r} (1)_r} C_{k-2r}^{1/2}(x) \tag{6.1.5}$$

enables us to express the coefficients of expansion (6.1.3) as the sum of the products of the Kaula inclination functions:

$$C_{pqj}^{(k,n/2)}(i,i') = \sum_{r=0}^{r^*} (2k-4r+1) \frac{\left(\frac{n}{2}\right)_{k-r} \left(\frac{n-1}{2}\right)_r}{\left(\frac{3}{2}\right)_{k-r} (1)_r} \times$$

$$\times \frac{(1)_{k-2r-j}}{(1)_{k-2r+j}} F_{k-2r,j,\frac{k-2r-p-q}{2}}(i) F_{k-2r,j,\frac{k-2r-p+q}{2}}(i') . \tag{6.1.6}$$

The summation is performed here up to the first vanishing term. Therefore, the limiting value $r = r^*$ is determined by

$$r^* = \min\left\{ \mathrm{E}\left(\frac{k-j}{2}\right),\ \frac{k-|p|-|q|}{2} \right\} . \tag{6.1.7}$$

From (6.1.3) it is evident that $p+q$ and k in (6.1.4), (6.1.6) and (6.1.7) are integers of the same parity. Formula (6.1.6) may be transformed to reveal the dependence on i and i' more explicitly. Indeed, in accordance with (5.3.9) and (5.3.10) one may put (Brumberg, 1967)

$$F_{kjl}(i) = \lambda_{kjl} \left(\sin\frac{i}{2}\right)^{|k-j-2l|} \left(\cos\frac{i}{2}\right)^{|k+j-2l|} \times$$

$$\times F\Bigg(\tfrac{1}{2}|k-j-2l| + \tfrac{1}{2}|k+j-2l| - k,$$

$$1 + k + \tfrac{1}{2}|k-j-2l| + \tfrac{1}{2}|k+j-2l|,$$

$$1 + |k-j-2l|,\ \sin^2\frac{i}{2}\Bigg) \tag{6.1.8}$$

with

$$\lambda_{kjl} = (-1)^{\mathrm{E}(\frac{k-j}{2}) + \max\{0,-k+j+2l\}} \frac{(1)_{k+j}}{2^k (1)_l (1)_{k-l}} \times$$

$$\times \frac{(1+k+j)_{\max\{0,k-j-2l\}} (1+k-j)_{\max\{0,-k+j+2l\}}}{(1)_{|k-j-2l|}} . \tag{6.1.9}$$

Substituting (6.1.8) into (6.1.6) one gets

$$
\begin{aligned}
C_{pqj}^{(k,n/2)}(i,i') = & \left(\sin\frac{i}{2}\right)^{|p+q-j|}\left(\cos\frac{i}{2}\right)^{|p+q+j|}\left(\sin\frac{i'}{2}\right)^{|p-q-j|}\times\\
& \times\left(\cos\frac{i'}{2}\right)^{|p-q+j|}\sum_{r=0}^{r^*} d_r(n,k,p,q,j)\times\\
& \times F\Big(\tfrac{1}{2}|p+q-j|+\tfrac{1}{2}|p+q+j|-k+2r,\\
& \qquad 1+k-2r+\tfrac{1}{2}|p+q-j|+\tfrac{1}{2}|p+q+j|,\\
& \qquad 1+|p+q-j|,\ \sin^2\frac{i}{2}\Big)\times\\
& \times F\Big(\tfrac{1}{2}|p-q-j|+\tfrac{1}{2}|p-q+j|-k+2r,\\
& \qquad 1+k-2r+\tfrac{1}{2}|p-q-j|+\tfrac{1}{2}|p-q+j|,\\
& \qquad 1+|p-q-j|,\ \sin^2\frac{i'}{2}\Big)
\end{aligned}
\tag{6.1.10}
$$

with numerical coefficients

$$
\begin{aligned}
d_r(n,k,p,q,j) = & (2k-4r+1)\frac{\left(\frac{n}{2}\right)_{k-r}\left(\frac{n-1}{2}\right)_r}{\left(\frac{3}{2}\right)_{k-r}(1)_r}\times\\
& \times\frac{(1)_{k-2r-j}}{(1)_{k-2r+j}}\lambda_{k-2r,j,\frac{k-2r-p-q}{2}}\lambda_{k-2r,j,\frac{k-2r-p+q}{2}}\ .
\end{aligned}
\tag{6.1.11}
$$

Hence, computation of coefficients (6.1.6) is reduced to the summation over the products of two hypergeometric polynomials multiplied by a numerical factor. The summation of quantities of the same order is a certain disadvantage of (6.1.10). Instead of (6.1.10) it is possible to give another representation of coefficients (6.1.6) involving the hypergeometric functions of two variables (Brumberg, 1966) but this is not essential for the algorithms considered below.

Substituting (6.1.3) into (6.1.2) one gets

$$
\begin{aligned}
\Delta^{-n} = & r'^{-n}\sum_{k=0}^{\infty}\left(\frac{r}{r'}\right)^k\sum_{l=0}^{k}\sum_{l'=0}^{k}\sum_{j=0}^{k}(2-\delta_{j0})C_{k-l-l',l'-l,j}^{(k,n/2)}(i,i')\times\\
& \times\cos\left[(k-2l)(v+\omega)-(k-2l')(v'+\omega')+j(\Omega-\Omega')\right]
\end{aligned}
\tag{6.1.12}
$$

or after changing the order of summation

$$
\begin{aligned}
\Delta^{-n} = & \sum_{j=0}^{\infty}\sum_{s=-\infty}^{\infty}\sum_{s'=-\infty}^{\infty}(\Delta^{-n})_{ss'j}\times\\
& \times\cos\left[s(v+\omega)+s'(v'+\omega')+j(\Omega-\Omega')\right]
\end{aligned}
\tag{6.1.13}
$$

with the coefficients

$$\left(\Delta^{-n}\right)_{ss'j} = r'^{-n}(2-\delta_{j0}) \sum_{m=m^*}^{\infty} \left(\frac{r}{r'}\right)^k C^{(k,n/2)}_{\frac{s-s'}{2},\frac{s+s'}{2},j}(i,i') \tag{6.1.14}$$

and the previous relation (5.3.25) for k. The only difference is in replacing $\max\{2,j\}$ by j in (5.3.28) so that

$$m^* = \max\left\{0,\ \mathrm{E}\left[\frac{1}{2}\left(j - \left|\frac{s-s'}{2}\right| - \left|\frac{s+s'}{2}\right| + 1\right)\right]\right\}. \tag{6.1.15}$$

Needless to say that s and s' in (6.1.13) again have the same parity.

The functions of r, v, r' and v' occurring in (6.1.12) may be expanded by (2.2.6) in a classical (e, M) series, resulting in the most general five-argument trigonometric expansion of Δ^{-n}:

$$\Delta^{-n} = \sum_{j=0}^{\infty} \sum_{p=-\infty}^{\infty} \sum_{p'=-\infty}^{\infty} \sum_{s=-\infty}^{\infty} \sum_{s'=-\infty}^{\infty} (\Delta^{-n})_{pp'ss'j} \times$$
$$\times \cos\left[pM + p'M' + s\omega + s'\omega' + j(\Omega - \Omega')\right] \tag{6.1.16}$$

with the coefficients

$$(\Delta^{-n})_{pp'ss'j} = a'^{-n}(2-\delta_{j0}) \sum_{m=m^*}^{\infty} \left(\frac{a}{a'}\right)^k X_p^{k,s}(e) X_{p'}^{-n-k,s'}(e') \times$$
$$\times C^{(k,n/2)}_{\frac{s-s'}{2},\frac{s+s'}{2},j}(i,i') \tag{6.1.17}$$

and the values of k and m^* from (5.3.25) and (6.1.15), respectively. Again, if the eccentricities e and e' are large it is reasonable to use (2.5.22), replacing the (e, M) and (e', M') series by (q, w) and (q', w') series, respectively. The Hansen coefficients in (6.1.17) are thus replaced by elliptic Hansen coefficients. Hence, the problem of large eccentricities (in studying the motion of some minor planets, for example) may now be treated efficiently. The problem of large inclinations (also in the case of some minor planets) presents no difficulties at all because coefficients (6.1.6) may be expressed in closed form with respect to the trigonometric functions of $i/2$ and $i'/2$. The main difficulty in dealing with (6.1.17) for planetary problems is related to the slow convergence of these series with respect to the ratio $\alpha = a/a'$ of the semi-major axes. At present, one uses two main techniques to overcome this difficulty. The first technique is to put the integral functions of k outside the summation sign in (6.1.17), replacing k by the operator $D = \alpha(d/d\alpha)$. This technique is based on the theorem of symbolic expansions (Brown and Shook, 1933) as follows.

If a function $f(x)$ can be presented by a series

$$f(x) = \sum_{s=-\infty}^{\infty} b_s x^s$$

and if p^D is expanded into the Taylor series

$$p^D = \sum_{m=0}^{\infty} a_m D^m$$

then

$$f(px) = p^D f(x)$$

with

$$D = x\frac{d}{dx}.$$

Considering that the Hansen coefficients $X_p^{k,s}(e)$, $X_{p'}^{-n-k,s'}(e')$ and the hypergeometric polynomials in (6.1.10) are integral functions of k and changing the order of summation

$$\sum_{m=m^*}^{\infty} \sum_{r=0}^{r^*} = \sum_{r=0}^{\infty} \sum_{m=m^*+r}^{\infty}$$

one can write (6.1.17) in the form

$$\begin{aligned}
(\Delta^{-n})_{pp'ss'j} = {} & a'^{-n}(2-\delta_{j0}) X_p^{D,s}(e) X_{p'}^{-n-D,s'}(e') \times \\
& \times \left(\sin\frac{i}{2}\right)^{|s-j|} \left(\cos\frac{i}{2}\right)^{|s+j|} \left(\sin\frac{i'}{2}\right)^{|s'+j|} \left(\cos\frac{i'}{2}\right)^{|s'-j|} \times \\
& \times \sum_{r=0}^{\infty} F\Big(\tfrac{1}{2}|s-j| + \tfrac{1}{2}|s+j| + 2r - D, \\
& \qquad 1 + \tfrac{1}{2}|s-j| + \tfrac{1}{2}|s+j| - 2r + D, \\
& \qquad 1 + |s-j|,\ \sin^2\frac{i}{2}\Big) \times \\
& \times F\Big(\tfrac{1}{2}|s'-j| + \tfrac{1}{2}|s'+j| + 2r - D, \\
& \qquad 1 + \tfrac{1}{2}|s'-j| + \tfrac{1}{2}|s'+j| - 2r + D, \\
& \qquad 1 + |s'+j|,\ \sin^2\frac{i'}{2}\Big) \times \\
& \times \sum_{m=m^*+r}^{\infty} d_r\left(n, k, \frac{s-s'}{2}, \frac{s+s'}{2}, j\right) \left(\frac{a}{a'}\right)^k
\end{aligned} \tag{6.1.18}$$

with values (5.3.25) and (6.1.15) for k and m^*, respectively. The operator $D = \alpha(d/d\alpha)$ is acting here on the functions of $\alpha = a/a'$. These functions represent the generalized hypergeometric functions of α replacing the Laplace coefficients of the classical expansions. Such a technique, which is particularly convenient for the case $n = 1$ (when only one term with $r = 0$ survives in the summation over r), has been developed by G. A. Krasinsky (1973a,b) for planetary problems (with small inclinations) and for close Earth satellite problems (with arbitrary inclinations).

Another technique for overcoming the difficulty related to the slow convergence of the series in powers of α has been suggested by Chapront (1970) in constructing purely analytical planetary theories. First, coefficients (6.1.17) are represented by the four-argument power series

$$(\Delta^{-n})_{pp'ss'j} = a'^{-n} \sum F^{(n)}_{k_1 k_2 k_3 k_4}(\alpha) e^{k_1} e'^{k_2} \times \\ \times \left(\sin \frac{i}{2}\right)^{k_3} \left(\sin \frac{i'}{2}\right)^{k_4}, \tag{6.1.19}$$

the coefficient $F(\alpha)$ being the series in powers of α^2 starting with terms of some definite degree with respect to α (the explicit dependence of these coefficients on the indices p, p', s, s' and j is omitted here). This series admits the representation

$$F^{(n)}_{k_1 k_2 k_3 k_4}(\alpha) = \frac{1}{(1-\alpha^2)^{n-1+k_1+k_2+k_3+k_4}} G^{(n)}_{k_1 k_2 k_3 k_4}(\alpha) . \tag{6.1.20}$$

The series $G(\alpha)$ is of the same form as $F(\alpha)$ but converges much faster.

Having got the general form of expansion (6.1.16) we turn to some practical algorithms for constructing such an expansion by computer.

6.2 Expansion Algorithms

The simplest algorithm for obtaining expansion (6.1.16) without any preliminary analytical operations is to construct with the aid of a Keplerian processor a Poisson series for

$$\frac{\Delta^2}{a'^2} = \frac{r'^2}{a'^2} - 2\alpha \frac{r}{a} \frac{r'}{a'} \cos H + \alpha^2 \frac{r^2}{a^2} \tag{6.2.1}$$

and then to raise the series obtained to the power $(-n/2)$ by means of a PS processor. As the initial approximation for the function $(a'/\Delta)^n$ one may take the value 1 or the series for $(a'/r')^n$ or the expansion of $(a'/\Delta)^n$ for circular orbits. This algorithm has been applied, for example, by Broucke and Smith (1971).

The same representation of $(a'/\Delta)^n$ in the form of a Poisson series is obtained by proceeding from the initial formulae (6.1.2). The Gegenbauer polynomials $C_k^{n/2}(\cos H)$ are represented by Poisson series with the aid of a Keplerian processor and are then multiplied by series for the functions $(r/a)^k$ and $(a'/r')^{n+k}$. In both cases one obtains expansion (6.1.16) and (6.1.17) with very slow convergence with respect to α.

Taking as the initial expansion the Fourier series

$$\Delta^{-n} = r'^{-n} \sum_{k=0}^{\infty} \left(1 - \tfrac{1}{2}\delta_{k0}\right) b_n^{(k)} \left(\frac{r}{r'}\right) \cos kH \tag{6.2.2}$$

with the Laplace coefficients

$$\frac{1}{2} b_n^{(k)}(x) = \frac{\left(\frac{n}{2}\right)_k}{(1)_k} x^k F\left(\frac{n}{2}, \frac{n}{2} + k, 1 + k, x^2\right) \tag{6.2.3}$$

one may construct expansion (6.1.16) in the form of an echeloned Poisson series separating the poles $(1 - \alpha^2)^{-1}$ in the coefficients. For this purpose Abu-El-Ata and Chapront (1975) suggested performing the transformation in (6.2.2) and (6.2.3)

$$x^2 = \alpha^2(1 + y) \tag{6.2.4}$$

with

$$x = \frac{r}{r'}, \qquad \alpha = \frac{a}{a'}, \qquad y = \left(\frac{r}{a}\frac{a'}{r'}\right)^2 - 1. \tag{6.2.5}$$

It is evident that the function y is of the order of e and e'. The substitution of (6.2.4) into (6.2.3) involves the Taylor series expansion of the hypergeometric function

$$F\left(\frac{n}{2}, \frac{n}{2} + k, 1 + k, x^2\right) = \sum_{m=0}^{\infty} \frac{\left(\frac{n}{2}\right)_m \left(\frac{n}{2} + k\right)_m}{(1 + k)_m (1)_m} y^m \times$$
$$\times \alpha^{2m} F\left(\frac{n}{2} + m, \frac{n}{2} + k + m, 1 + k + m, \alpha^2\right).$$

Hence, transformation (6.2.4) followed by Euler transformation (2.3.6) results in

$$\Delta^{-n} = a'^{-n} \sum_{k=0}^{\infty} (2 - \delta_{k0}) \left(\frac{r}{a}\right)^k \left(\frac{a'}{r'}\right)^{n+k} \cos kH \times$$
$$\times \sum_{m=0}^{\infty} \frac{\left(\frac{n}{2}\right)_m \left(\frac{n}{2}\right)_{k+m}}{(1)_m (1)_{k+m}} y^m \frac{\alpha^{k+2m}}{(1 - \alpha^2)^{n-1+m}} \times$$
$$\times F\left(1 - \frac{n}{2}, 1 + k - \frac{n}{2}, 1 + k + m, \alpha^2\right). \tag{6.2.6}$$

By expanding the functions $(r/a)^k$, $(a'/r')^{n+k}$, $\cos kH$ and y^m with the aid of a Keplerian processor we get an echeloned series with the poles corresponding to the power n and the minimum order of a given coefficient with respect to the eccentricities.

It is also possible to apply the technique of (6.1.18) to (6.2.2). From the representation

$$C_k^{n/2}(\cos H) = \sum_{s=0}^{k} \frac{\left(\frac{n}{2}\right)_s \left(\frac{n}{2}\right)_{k-s}}{(1)_s (1)_{k-s}} \cos(k-2s)H$$

it is easy to derive the well-known relation

$$C_k^1(\cos H) = \frac{\sin(k+1)H}{\sin H},$$

and then

$$\cos kH = \tfrac{1}{2}\left[C_k^1(\cos H) - C_{k-2}^1(\cos H)\right].$$

Therefore,

$$\Delta^{-n} = \tfrac{1}{2} r'^{-n} \sum_{k=0}^{\infty} \left[b_n^{(k)}\left(\frac{r}{r'}\right) - b_n^{(k+2)}\left(\frac{r}{r'}\right)\right] C_k^1(\cos H). \tag{6.2.7}$$

Substituting (6.1.3) and changing the order of summation one again obtains (6.1.13) with the coefficients

$$\left(\Delta^{-n}\right)_{ss'j} = r'^{-n}\left(1 - \tfrac{1}{2}\delta_{j0}\right) \times$$
$$\times \sum_{m=m^*}^{\infty} \left[b_n^{(k)}\left(\frac{r}{r'}\right) - b_n^{(k+2)}\left(\frac{r}{r'}\right)\right] C^{(k,\,1)}_{\frac{s-s'}{2},\,\frac{s+s'}{2},\,j}(i,\,i') \tag{6.2.8}$$

with values (5.3.25) and (6.1.15) for k and m^*, respectively. The Laplace coefficients may be expanded symbolically as

$$b_n^{(k)}\left(\frac{r}{r'}\right) = \left(\frac{r}{a}\right)^D \left(\frac{a'}{r'}\right)^D b_n^{(k)}(\alpha), \quad \alpha = \frac{a}{a'} \tag{6.2.9}$$

with

$$D = \alpha \frac{d}{d\alpha}.$$

Again using (6.1.10) and applying the same technique as in (6.1.18) we get (6.1.16) with the coefficients

$$\begin{aligned}
\left(\Delta^{-n}\right)_{pp'ss'j} = & a'^{-n}\left(1-\tfrac{1}{2}\delta_{j0}\right) X_p^{D,s}(e) X_{p'}^{-n-D,s'}(e')\times \\
& \times\left(\sin\frac{i}{2}\right)^{|s-j|}\left(\cos\frac{i}{2}\right)^{|s+j|}\left(\sin\frac{i'}{2}\right)^{|s'+j|}\left(\cos\frac{i'}{2}\right)^{|s'-j|}\times \\
& \times\sum_{r=0}^{\infty} F\Big(\tfrac{1}{2}|s-j|+\tfrac{1}{2}|s+j|+2r-D, \\
& \qquad 1+\tfrac{1}{2}|s-j|+\tfrac{1}{2}|s+j|-2r+D, \\
& \qquad 1+|s-j|, \sin^2\frac{i}{2}\Big)\times \\
& \quad\times F\Big(\tfrac{1}{2}|s'-j|+\tfrac{1}{2}|s'+j|+2r-D, \\
& \qquad 1+\tfrac{1}{2}|s'-j|+\tfrac{1}{2}|s'+j|-2r+D, \\
& \qquad 1+|s'+j|, \sin^2\frac{i'}{2}\Big)\times \\
& \times\sum_{m=m^*+r}^{\infty} d_r\left(2, k, \frac{s-s'}{2}, \frac{s+s'}{2}, j\right)\left[b_n^{(k)}(\alpha)-b_n^{(k+2)}(\alpha)\right].
\end{aligned} \tag{6.2.10}$$

Hence, this expansion technique is reduced to the classic problem of determing the Laplace coefficients and their derivatives (Brown and Shook, 1933). But both expansions (6.1.18) and (6.2.10) nowadays look a little bit out-of-date.

Some modern methods, such as that of Petrovskaya (1970), do not assume the permanent fulfillment of the condition $r < r'$ and are valid even for the case of nearly intersecting orbits (the Neptune–Pluto system). An interesting algorithm of this kind generalizing the algorithm by Brown and Shook (1933) has been elaborated by Yuasa and Hori (1979, 1984). For orbits intersecting in projection it is important to take into account the actual values of the inclinations. Hence, it is appropriate to introduce the function

$$\Delta_0^2 = r^2 + r'^2 - 2rr'(cc' - ss')^2\cos(W - W') \tag{6.2.11}$$

with $c = \cos(i/2)$, $s = \sin(i/2)$ and the true longitude in orbit $W = \Omega+\omega+v$. The primed quantities are referred to the second planet. Then the square of the mutual distance between the planets is

$$\begin{aligned}
\Delta^2 = & \Delta_0^2 - 2rr'\Big\{c^2 s'^2\cos(W + W' - 2\Omega')+ \\
& + s^2 c'^2\cos(W + W' - 2\Omega) + s^2 s'^2\cos(W - W' - 2\Omega + 2\Omega')+ \\
& + 2csc's'\left[\cos(W - W' - \Omega + \Omega') - \cos(W + W' - \Omega - \Omega')\right]+ \\
& + 2(csc's' - s^2 s'^2)\cos(W - W')\Big\}.
\end{aligned} \tag{6.2.12}$$

The absolute magnitude of the ratio $(\Delta^2 - \Delta_0^2)/\Delta_0^2$ is always less than 1 (with the only exception being for the case of actually intersecting orbits). Therefore the function Δ^{-n} may be expanded in powers of this ratio:

$$\Delta^{-n} = \Delta_0^{-n} \sum_{k=0}^{\infty} \frac{(-1)^k \left(\frac{n}{2}\right)_k}{(1)_k} \left(\frac{\Delta^2 - \Delta_0^2}{\Delta_0^2}\right)^k . \tag{6.2.13}$$

The function $(\Delta^2 - \Delta_0^2)$ is expanded in a Poisson series with the aid of a Keplerian processor. It remains to expand the function Δ_0^{-n}. To this end, Yuasa and Hori (1979) start from the representation

$$\Delta_0^{-n} = \frac{\left[q - \cos(W - W')\right]^{-\frac{n}{2}}}{\left(\sqrt{2rr'}|cc' - ss'|\right)^n} \tag{6.2.14}$$

with

$$q = \frac{r^2 + r'^2}{2rr'(cc' - ss')^2} > 1 . \tag{6.2.15}$$

Then the trigonometric expansion

$$(q - \cos\psi)^{-\frac{n}{2}} = \frac{\sqrt{2}}{\left(\frac{1}{2}\right)_{\frac{n-1}{2}} \pi} \exp\left(-\mathrm{i}\frac{n-1}{2}\pi\right) \left(q^{-1}\right)^{\frac{n-1}{2}} \left(1 - q^{-2}\right)^{-\frac{n-1}{4}} \times$$
$$\times \sum_{j=0}^{\infty} (2 - \delta_{j0})\, Q_{j-\frac{1}{2}}^{\frac{n-1}{2}}(q) \cos j\psi \tag{6.2.16}$$

involves the associated Legendre functions of the second kind

$$Q_{j-\frac{1}{2}}^{\frac{n-1}{2}} = \frac{\left(\frac{1}{2}\right)_{\frac{n-1}{2}} \left(\frac{n}{2}\right)_j \pi}{2^{j+\frac{1}{2}} (1)_j} \exp\left(\mathrm{i}\frac{n-1}{2}\pi\right) \left(q^{-1}\right)^{j+\frac{1}{2}} \left(1 - q^{-2}\right)^{\frac{n-1}{4}} \times$$
$$\times F\left(\frac{j}{2} + \frac{n}{4}, \frac{j+1}{2} + \frac{n}{4}, j+1, q^{-2}\right) . \tag{6.2.17}$$

Of course, for even n the generalized factorial occurring in (6.2.16) and (6.2.17) should be expressed in terms of the gamma-function

$$\left(\tfrac{1}{2}\right)_{\frac{n-1}{2}} = \frac{1}{\sqrt{\pi}} \Gamma\left(\frac{n}{2}\right) , \tag{6.2.18}$$

but it is possible also to substitute (6.2.17) into (6.2.16) and to use the expansion

$$(q - \cos\psi)^{-\frac{n}{2}} = \left(q^{-1}\right)^{\frac{n}{2}} \sum_{j=0}^{\infty} (2 - \delta_{j0}) \frac{\left(\frac{n}{2}\right)_j}{2^j (1)_j} \left(q^{-1}\right)^j \times$$
$$\times F\left(\frac{j}{2} + \frac{n}{4}, \frac{j+1}{2} + \frac{n}{4}, j+1, q^{-2}\right) \cos j\psi \tag{6.2.19}$$

directly. Expansion (6.2.14) and (6.2.19) with the power variable q^{-1} can be obtained fairly easily without using (6.2.16) and (6.2.17). Indeed, generalizing the approach of Brown and Shook (1933) one has

$$\Delta_0^2 = \frac{4rr'\mu}{k^2}\delta^2 \tag{6.2.20}$$

with

$$\mu = (cc' - ss')^2 , \tag{6.2.21}$$

$$k^2 = \frac{4rr'\mu}{r^2 + r'^2 + 2rr'\mu} , \tag{6.2.22}$$

$$W - W' = \pi - 2\varphi \tag{6.2.23}$$

and

$$\delta = \left(1 - k^2 \sin^2 \varphi\right)^{\frac{1}{2}} . \tag{6.2.24}$$

On the other hand, introducing

$$\alpha = \frac{k^2}{(1+k')^2} , \qquad k' = (1-k^2)^{\frac{1}{2}} , \qquad k^2 = \frac{4\alpha}{(1+\alpha)^2} \tag{6.2.25}$$

and

$$\zeta = -\exp(-\mathrm{i}\,2\varphi) = -\cos 2\varphi + \mathrm{i}\sin 2\varphi \tag{6.2.26}$$

one may rewrite (6.2.24) in the form

$$\delta = (1+\alpha)^{-1}(1-\alpha\zeta^{-1})^{\frac{1}{2}}(1-\alpha\zeta)^{\frac{1}{2}} . \tag{6.2.27}$$

Hence, in terms of function (2.3.1) one has

$$\delta^{-n} = (1+\alpha)^n \gamma\left(0, -\frac{n}{2}, -\frac{n}{2}, 0, \alpha, \zeta\right) \tag{6.2.28}$$

with the expansion

$$\delta^{-n} = (1+\alpha)^n \sum_{j=-\infty}^{\infty} \gamma_j\left(0, -\frac{n}{2}, -\frac{n}{2}, 0, \alpha\right)\zeta^j . \tag{6.2.29}$$

Considering that the coefficients in ζ^j and ζ^{-j} are the same and returning by (6.2.23) and (6.2.26) to the old trigonometric argument we get

$$\delta^{-n} = (1+\alpha)^n \sum_{j=0}^{\infty} (2-\delta_{j0})\,\gamma_j\left(0, -\frac{n}{2}, -\frac{n}{2}, 0, \alpha\right)\cos j(W - W') . \tag{6.2.30}$$

Therefore, in virtue of (6.2.20) the coefficients of the expansion

$$\Delta_0^{-n} = \sum_{j=0}^{\infty} \left(\Delta_0^{-n}\right)_j \cos j(W - W') \tag{6.2.31}$$

will be

$$\left(\Delta_0^{-n}\right)_j = \frac{(\alpha)^{\frac{n}{2}}}{(rr'\mu)^{\frac{n}{2}}} (2 - \delta_{j0})\, \gamma_j \left(0, -\frac{n}{2}, -\frac{n}{2}, 0, \alpha\right). \tag{6.2.32}$$

By (2.3.5) coefficients (6.2.32) are expressed in terms of α as

$$\gamma_j \left(0, -\frac{n}{2}, -\frac{n}{2}, 0, \alpha\right) = \frac{\left(\frac{n}{2}\right)_j}{(1)_j} \alpha^j F\left(\frac{n}{2}, \frac{n}{2} + j, 1 + j, \alpha^2\right). \tag{6.2.33}$$

The quadratic transformation for hypergeometric functions

$$F\left(a, b, a - b + 1, \alpha^2\right) = (1 + \alpha)^{-2a} F\left(a, a - b + \tfrac{1}{2}, 2a - 2b + 1, k^2\right) \tag{6.2.34}$$

with (6.2.25) enables one to express (6.2.33) in terms of k:

$$\gamma_j \left(0, -\frac{n}{2}, -\frac{n}{2}, 0, \alpha\right) = \frac{\left(\frac{n}{2}\right)_j}{(1)_j} \left(\frac{1 + k'}{2}\right)^n \left(\frac{k^2}{4}\right)^j \times$$
$$\times F\left(\frac{n}{2} + j, j + \tfrac{1}{2}, 1 + 2j, k^2\right). \tag{6.2.35}$$

Another quadratic transformation

$$F\left(a, b, 2b, k^2\right) = \left(1 - \tfrac{1}{2}k^2\right)^{-a} F\left(\frac{a}{2}, \frac{a + 1}{2}, b + \tfrac{1}{2}, \beta^2\right) \tag{6.2.36}$$

with

$$\beta = \frac{k^2}{2 - k^2} = \frac{2\alpha}{1 + \alpha^2}, \qquad k^2 = \frac{2\beta}{1 + \beta} \tag{6.2.37}$$

results in

$$\gamma_j \left(0, -\frac{n}{2}, -\frac{n}{2}, 0, \alpha\right) = \frac{\left(\frac{n}{2}\right)_j}{2^j (1)_j} \left(1 + \alpha^2\right)^{-\frac{n}{2}} \beta^j \times$$
$$\times F\left(\frac{n}{4} + \frac{j}{2}, \frac{n}{4} + \frac{j + 1}{2}, 1 + j, \beta^2\right). \tag{6.2.38}$$

Alternatively, substituting expressions (6.2.33), (6.2.35) and (6.2.38) into (6.2.32) one gets three forms representing the coefficients of expansion (6.2.31):

$$\left(\Delta_0^{-n}\right)_j = \frac{2-\delta_{j0}}{(rr'\mu)^{\frac{n}{2}}} \frac{\left(\frac{n}{2}\right)_j}{(1)_j} \alpha^{j+\frac{n}{2}} F\left(\frac{n}{2}, \frac{n}{2}+j, 1+j, \alpha^2\right) = \tag{6.2.39a}$$

$$= \frac{2-\delta_{j0}}{(rr'\mu)^{\frac{n}{2}}} \frac{\left(\frac{n}{2}\right)_j}{(1)_j} \left(\frac{k}{2}\right)^{2j+n} F\left(\frac{n}{2}+j, j+\tfrac{1}{2}, 1+2j, k^2\right) = \tag{6.2.39b}$$

$$= \frac{2-\delta_{j0}}{(rr'\mu)^{\frac{n}{2}}} \frac{\left(\frac{n}{2}\right)_j}{(1)_j} \left(\frac{\beta}{2}\right)^{j+\frac{n}{2}} F\left(\frac{n}{4}+\frac{j}{2}, \frac{n}{4}+\frac{j+1}{2}, 1+j, \beta^2\right). \tag{6.2.39c}$$

From (6.2.21), (6.2.22) and (6.2.37) it is evident that

$$\beta = \frac{2rr'(cc'-ss')^2}{r^2+r'^2} = q^{-1} \tag{6.2.40}$$

so that expression (6.2.39c) again results in expansion (6.2.14) and (6.2.19).

In a similar way the function Δ^{-n} may be expressed starting from (6.2.13) in terms of r, r', W, W' and μ. Further expansion in powers of eccentricities (or corresponding Jacobi nomes) is produced by applying the symbolic expansion technique of Brown and Shook (1933):

$$f(r, r') = \left(\frac{r}{a}\right)^{\mathcal{D}} \left(\frac{r'}{a'}\right)^{\mathcal{D}'} f(a, a') \tag{6.2.41}$$

with

$$\mathcal{D} = a\frac{\partial}{\partial a}, \qquad \mathcal{D}' = a'\frac{\partial}{\partial a'}. \tag{6.2.42}$$

All the algorithms considered above enable one, if necessary, to construct expansions in closed form with respect to the orbital inclinations. This is evident for expansion (6.2.31) and (6.2.39). For other expansions it is sufficient to apply the subroutine COSINE of a Keplerian processor for $\cos H$ in version (2.6.1) (d), rigorously taking into account the trigonometric functions of half arguments i and i'. However, in planetary problems with small eccentricities and inclinations it may be reasonable to use expansions in powers of these parameters from the very beginning. Amongst the algorithms intended for this case note the algorithm of Abu-El-Ata and Chapront (1975), which results in coefficients of type (6.1.19) and (6.1.20). Here we give another algorithm (Brumberg, 1970) that leads to the same kind of expansion. This algorithm might be more time-consuming but its universal form enables one to use different systems of eccentric and oblique variables.

Instead of the rectangular coordinates x, y and z one introduces the dimensionless variables p, q and w, representing deviations from planar circular motion with semi-major axis a and mean longitude λ:

$$\begin{aligned} x + \mathrm{i}y &= a(1-p)\exp \mathrm{i}\lambda\,, \\ x - \mathrm{i}y &= a(1-q)\exp(-\mathrm{i}\lambda)\,, \\ z &= aw\,. \end{aligned} \tag{6.2.43}$$

The complex conjugate variables p and q are of the order of the eccentricity e. The real variable w is of the order of the inclination i. The analogous primed variables will be referred to the second planet.

In terms of the new variables the radius-vector r is determined by

$$r^2 = a^2\left[(1-p)(1-q) + w^2\right]. \tag{6.2.44}$$

By successive application of the binomial expansion one gets

$$\left(\frac{a}{r}\right)^n = \sum_{k=0}^{\infty}\sum_{l=0}^{\infty}\sum_{m=0}^{\infty} \alpha_{klm}^{(n)} p^k q^l w^{2m} \tag{6.2.45}$$

with

$$\alpha_{klm}^{(n)} = (-1)^m \frac{\left(\frac{n}{2}\right)_m \left(\frac{n}{2}+m\right)_k \left(\frac{n}{2}+m\right)_l}{(1)_m (1)_k (1)_l}. \tag{6.2.46}$$

The mutual distance Δ between two planets is determined by

$$\begin{aligned}\Delta^2 = &\left[a(1-p) - a'(1-p')\zeta^{-1}\right]\left[a(1-q) - a'(1-q')\zeta\right] + \\ &+ (aw - a'w')^2\end{aligned} \tag{6.2.47}$$

with

$$\zeta = \exp \mathrm{i}(\lambda - \lambda'). \tag{6.2.48}$$

Application of the binomial expansion yields

$$\begin{aligned}\left(\frac{a}{\Delta}\right)^n = &\sum_{m=0}^{\infty} \frac{(-1)^m \left(\frac{n}{2}\right)_m}{(1)_m} \left(1 - \frac{a'}{a}\zeta^{-1} - p + \frac{a'}{a}p'\zeta^{-1}\right)^{-\frac{n}{2}-m} \times \\ &\times \left(1 - \frac{a'}{a}\zeta - q + \frac{a'}{a}q'\zeta\right)^{-\frac{n}{2}-m} \left(w - \frac{a'}{a}w'\right)^{2m}.\end{aligned} \tag{6.2.49}$$

The first binomial raising is performed as follows:

$$\begin{aligned}&\left(1 - \frac{a'}{a}\zeta^{-1} - p + \frac{a'}{a}p'\zeta^{-1}\right)^{-\frac{n}{2}-m} = \\ &= \sum_{k=0}^{\infty} \frac{(-1)^k \left(\frac{n}{2}+m\right)_k}{(1)_k} \left(1 - \frac{a'}{a}\zeta^{-1}\right)^{-\frac{n}{2}-m-k} \left(\frac{a'}{a}p'\zeta^{-1} - p\right)^k = \\ &= \sum_{k=0}^{\infty}\sum_{r=0}^{k} \frac{\left(\frac{n}{2}+m\right)_k (-k)_r}{(1)_k (1)_r} \left(1 - \frac{a'}{a}\zeta^{-1}\right)^{-\frac{n}{2}-m-k} \left(\frac{a'}{a}\right)^r p^{k-r} p'^r \zeta^{-r} = \\ &= \sum_{k=0}^{\infty}\sum_{r=0}^{\infty} \frac{(-1)^r \left(\frac{n}{2}+m\right)_{k+r} (1+r)_k}{(1)_{k+r} (1)_k} \left(1 - \frac{a'}{a}\zeta^{-1}\right)^{-\frac{n}{2}-m-k-r} \left(\frac{a'}{a}\right)^r \times \\ &\quad \times p^k p'^r \zeta^{-r}.\end{aligned}$$

This final expression is obtained from the preceding one by changing the order of summation, replacing k by $k+r$ and transforming the factorial coefficient. The last multiplier in (6.2.49) involves no difficulties:

$$\left(w - \frac{a'}{a}w'\right)^{2m} = \sum_{t=0}^{2m} \frac{(-2m)_t}{(1)_t} \left(\frac{a'}{a}\right)^t w^{2m-t} w'^t =$$
$$= \sum_{t=-m}^{m} (-1)^{m+t} \frac{(1+m-t)_{m+t}}{(1)_{m+t}} \left(\frac{a'}{a}\right)^{m+t} w^{m-t} w'^{m+t} .$$

Introducing function (2.3.1) and substituting into (6.2.49) the intermediate results one gets

$$\left(\frac{a}{\Delta}\right)^n = \sum_{k=0}^{\infty}\sum_{l=0}^{\infty}\sum_{r=0}^{\infty}\sum_{s=0}^{\infty}\sum_{m=0}^{\infty}\sum_{t=-m}^{m} \beta_{klrsmt}^{(n)}\left(\frac{a'}{a}, \zeta\right) p^k q^l p'^r q'^s w^{m-t} w'^{m+t} \tag{6.2.50}$$

with

$$\beta_{klrsmt}^{(n)}(\alpha, \zeta) =$$
$$= \frac{(-1)^t \left(\frac{n}{2}\right)_m \left(\frac{n}{2}+m\right)_{k+r} \left(\frac{n}{2}+m\right)_{l+s} (1+r)_k (1+s)_l (1+m-t)_{m+t}}{(1)_m (1)_{k+r} (1)_{l+s} (1)_k (1)_l (1)_{m+t}} \times$$
$$\times \gamma\left(r+s+m+t, -\frac{n}{2}-m-k-r, -\frac{n}{2}-m-l-s, -r+s, \alpha, \zeta\right) . \tag{6.2.51}$$

This formula is valid both for $\alpha < 1$ and $\alpha > 1$ but in the latter case one should perform transformation (2.3.2). Application of (2.3.4) enables one to represent coefficients (6.2.51) by series in positive and negative powers of ζ involving coefficients (2.3.5) or (2.3.7). Irrespective of the cases $a' < a$ or $a < a'$ the power of the pole in (2.3.7) is equal to

$$-x - y - 1 = n - 1 + k + r + l + s + 2m . \tag{6.2.52}$$

This value is consistent with representation (6.1.20).

These results may be applied to expand the force function for the problem of the heliocentric motion of N planets. In the heliocentric coordinates x_i, y_i, z_i $(i = 1, 2, \ldots, N)$ one has

$$U_i = \frac{G(m_0 + m_i)}{r_i} + \sum_{j=1}^{N}{}^{(i)} G m_j \left(\frac{1}{\Delta_{ij}} - \frac{x_i x_j + y_i y_j + z_i z_j}{r_j^3}\right) \tag{6.2.53}$$

or in terms of the new variables p_i, q_i w_i

$$U_i = n_i^2 a_i^2 \left\{ \frac{a_i}{r_i} + \mu \sum_{j=1}^{N} {}^{(i)}\kappa_{ij} \left[\frac{a_i}{\Delta_{ij}} - \right.\right.$$
$$\left.\left. - \frac{1}{2} a_i^2 a_j \frac{(1-p_i)(1-q_j)\zeta_{ij} + (1-q_i)(1-p_j)\zeta_{ij}^{-1} + 2w_i w_j}{r_j^3} \right] \right\}. \quad (6.2.54)$$

Here the mean motions n_i and the semi-major axes a_i are related by Kepler's third law

$$n_i^2 a_i^3 = G(m_0 + m_i)\,, \quad (6.2.55)$$

the small parameter μ and the mass coefficients κ_{ij} are introduced by

$$\mu\kappa_{ij} = \frac{m_j}{m_0 + m_i}\,, \qquad \mu = 10^{-3} \quad (6.2.56)$$

and the exponential arguments ζ_{ij} are the functions of the difference of the mean longitudes λ_i and λ_j:

$$\zeta_{ij} = \exp \mathrm{i}(\lambda_i - \lambda_j)\,. \quad (6.2.57)$$

Substitution of (6.2.45) and (6.2.50) into (6.2.54) results in the expansion

$$U_i = n_i^2 a_i^2 \left\{ \sum_{k=0}^{\infty}\sum_{l=0}^{\infty}\sum_{m=0}^{\infty} A_{klm} p_i^k q_i^l w_i^{2m} + \right.$$
$$\left. + \mu \sum_{j=1}^{N} {}^{(i)}\kappa_{ij} \sum_{k=0}^{\infty}\sum_{l=0}^{\infty}\sum_{r=0}^{\infty}\sum_{s=0}^{\infty}\sum_{m=0}^{\infty}\sum_{t=-m}^{m} B_{klrsmt}^{(ij)} p_i^k q_i^l p_j^r q_j^s w_i^{m-t} w_j^{m+t} \right\}. \quad (6.2.58)$$

The numerical coefficients A_{klm} are determined by

$$A_{klm} = \alpha_{klm}^{(1)}\,. \quad (6.2.59)$$

The coefficients $B_{klrsmt}^{(ij)}$ are functions of the semi-major axes and the differences of the mean longitudes

$$B_{klrsmt}^{(ij)} = \beta_{klrsmt}^{(1)}\left(\frac{a_j}{a_i}, \zeta_{ij}\right) + \left(\frac{a_i}{a_j}\right)^2 \alpha_{rsm}^{(1)} \left[2m\delta_{k0}\delta_{l0}\delta_{t,m-1} + \right.$$
$$+ \left(m + r + \tfrac{1}{2}\right)(\delta_{k1} - \delta_{k0})\delta_{l0}\delta_{tm}\zeta_{ij} +$$
$$\left. + \left(m + s + \tfrac{1}{2}\right)(\delta_{l1} - \delta_{l0})\,\delta_{k0}\delta_{tm}\zeta_{ij}^{-1} \right]. \quad (6.2.60)$$

In deriving the expression for the indirect term the relation

$$(r+1)\alpha_{r+1,s,m}^{(1)} = \left(m + r + \tfrac{1}{2}\right)\alpha_{rsm}^{(1)}$$

has been used. With the aid of PS and Keplerian processors it is easy to obtain the modern expansions in different systems of elements from (6.2.58), in particular, the expansion by Duriez (1977) used in constructing the general planetary theory in elliptic variables.

6.3 Expansion with the Aid of Elliptic Functions

In Section 2.5 elliptic functions (in particular, their transformation laws and trigonometric expansions) were applied to overcome the difficulties related to large eccentricities. In typical planetary problems the main difficulty is caused by the large values of the ratios of the semi-major axes. The key problem is to express function (2.3.1) in closed form, admitting the integration in terms of elliptic functions or in the form of a fast converging series replacing ordinary expansion (2.3.4). Proceeding from the work of Richardson (1982), Williams et al. (1987) and Brumberg (1992, 1994) one can start with the representation of the Laplace coefficients in terms of the complete elliptic integrals. The negative powers of the mutual distance between two planets moving on coplanar circular orbits are expanded in series:

$$\begin{aligned}\Delta_0^{-n} &= [(a - a'\zeta^{-1})(a - a'\zeta)]^{-\frac{n}{2}} = \\ &= \left(\max\{a,\, a'\}\right)^{-n} \gamma\left(0,\, -\frac{n}{2},\, -\frac{n}{2},\, 0,\, \alpha,\, \zeta\right) = \\ &= \frac{1}{2} \sum_{\sigma=-\infty}^{\infty} c_n^{(\sigma)}(a,\, a')\zeta^{\sigma}\,, \qquad c_n^{(-\sigma)} = c_n^{(\sigma)}\end{aligned} \tag{6.3.1}$$

with

$$\alpha = \frac{\min\{a,\, a'\}}{\max\{a,\, a'\}}\,, \qquad \zeta = \exp \mathrm{i}(\lambda - \lambda')\,, \tag{6.3.2}$$

λ and λ' being the mean longitudes of the planets. The coefficients $c_n^{(\sigma)}(a,\, a')$, which are symmetrical with respect to their arguments, are called Laplace symmetrical coefficients. They are related to the ordinary Laplace coefficients (6.2.3) by means of

$$c_n^{(\sigma)}(a,\, a') = \left(\frac{\alpha}{aa'}\right)^{\frac{n}{2}} b_n^{(\sigma)}(\alpha)\,. \tag{6.3.3}$$

Putting $\mu = 1$, $r = a$ and $r' = a'$ in (6.2.20) and (6.2.22) one obtains from (6.2.39a,b,c) three expressions for these coefficients ($\sigma \geq 0$):

$$\frac{1}{2} c_n^{(\sigma)}(a,\, a') = \frac{\left(\frac{n}{2}\right)_\sigma}{(1)_\sigma (aa')^{\frac{n}{2}}} \alpha^{\sigma+\frac{n}{2}} F\left(\frac{n}{2},\, \frac{n}{2} + \sigma,\, 1 + \sigma,\, \alpha^2\right) = \tag{6.3.4a}$$

$$= \frac{\left(\frac{n}{2}\right)_\sigma}{(1)_\sigma (aa')^{\frac{n}{2}}} \left(\frac{k}{2}\right)^{2\sigma+n} F\left(\frac{n}{2} + \sigma,\, \sigma + \tfrac{1}{2},\, 1 + 2\sigma,\, k^2\right) = \tag{6.3.4b}$$

$$= \frac{\left(\frac{n}{2}\right)_\sigma}{(1)_\sigma (aa')^{\frac{n}{2}}} \left(\frac{\beta}{2}\right)^{\sigma+\frac{n}{2}} F\left(\frac{n}{4} + \frac{\sigma}{2},\, \frac{n}{4} + \frac{\sigma+1}{2},\, 1 + \sigma,\, \beta^2\right) \tag{6.3.4c}$$

with previous relations (6.2.25) and (6.2.37) between k, α and β. On the one hand,

$$F\left(\tfrac{1}{2}, \tfrac{1}{2}, 1, z^2\right) = \frac{2}{\pi} K(z), \qquad F\left(-\tfrac{1}{2}, \tfrac{1}{2}, 1, z^2\right) = \frac{2}{\pi} E(z). \tag{6.3.5}$$

On the other hand, $F(a+k, b+m, c+n, z)$ with integer k, m and n may be reduced by the Gauss relations to $F(a, b, c, z)$ and one of its contiguous functions (Erdélyi, 1953). Taking this into account we see that it is evident that forms (6.3.4a) and (6.3.4b) are reduced to complete elliptic integrals with modulus α and k, respectively. Indeed, using (6.3.5) and the Gauss relations between the contiguous hypergeometric functions one obtains from (6.3.4a) and (6.3.4b) the classical results

$$c_1^{(0)}(a, a') = \frac{4}{\pi} \left(\frac{\alpha}{aa'}\right)^{\frac{1}{2}} K(\alpha) = \frac{4}{\pi} \frac{K(k)}{(a+a')}, \tag{6.3.6}$$

$$\begin{aligned} c_1^{(1)}(a, a') &= \frac{4}{\pi} \left(\frac{\alpha}{aa'}\right)^{\frac{1}{2}} \frac{1}{\alpha} [K(\alpha) - E(\alpha)] = \\ &= \frac{4}{\pi} \frac{1}{(a+a')k^2} \left[(2-k^2)K(k) - 2E(k)\right], \end{aligned} \tag{6.3.7}$$

$$\begin{aligned} c_3^{(0)}(a, a') &= \frac{4}{\pi} \left(\frac{\alpha}{aa'}\right)^{\frac{3}{2}} \frac{1}{(1-\alpha^2)^2} [-(1-\alpha^2)K(\alpha) + 2E(\alpha)] = \\ &= \frac{4}{\pi} \frac{E(k)}{(a+a')^3(1-k^2)}, \end{aligned} \tag{6.3.8}$$

$$\begin{aligned} c_3^{(1)}(a, a') &= \frac{4}{\pi} \left(\frac{\alpha}{aa'}\right)^{\frac{3}{2}} \frac{1}{\alpha(1-\alpha^2)^2} [-(1-\alpha^2)K(\alpha) + (1+\alpha^2)E(\alpha)] = \\ &= \frac{4}{\pi} \frac{1}{(a+a')^3 k^2} \left[-2K(k) + \frac{2-k^2}{1-k^2} E(k)\right], \end{aligned} \tag{6.3.9}$$

$$\begin{aligned} c_3^{(2)}(a, a') &= \frac{4}{\pi} \left(\frac{\alpha}{aa'}\right)^{\frac{3}{2}} \frac{1}{\alpha^2(1-\alpha^2)^2} [-(1-\alpha^2)(2-\alpha^2)K(\alpha) + \\ &\qquad + 2(1-\alpha^2+\alpha^4)E(\alpha)] = \\ &= \frac{4}{\pi} \frac{1}{(a+a')^3 k^4} \left[-8(2-k^2)K(k) + \frac{16-16k^2+k^4}{1-k^2} E(k)\right], \end{aligned} \tag{6.3.10}$$

$$\begin{aligned} c_5^{(0)}(a, a') &= \frac{4}{3\pi} \left(\frac{\alpha}{aa'}\right)^{\frac{5}{2}} \frac{1}{(1-\alpha^2)^4} [-(1-\alpha^2)(5+3\alpha^2)K(\alpha) + \\ &\qquad + 8(1+\alpha^2)E(\alpha)] = \\ &= \frac{4}{3\pi} \frac{1}{(a+a')^5(1-k^2)} \left[-K(k) + \frac{2(2-k^2)}{1-k^2} E(k)\right]. \end{aligned} \tag{6.3.11}$$

To extend this list of $c_n^{(\sigma)}$ one may also use the classical recurrence relations

$$2n(1-k^2)(a+a')^2 c_{n+2}^{(\sigma)} = (n+2\sigma)(2-k^2)c_n^{(\sigma)} + (n-2\sigma-2)k^2 c_n^{(\sigma+1)} \tag{6.3.12}$$

and

$$(2\sigma - n + 2)k^2 c_n^{(\sigma+1)} = 4\sigma(2-k^2)c_n^{(\sigma)} - (2\sigma+n-2)k^2 c_n^{(\sigma-1)} \tag{6.3.13}$$

or their equivalents in terms of α or β.

As can be seen from (6.2.51) and (6.2.60), in the expansion of the planetary perturbing function each of the parameters x and y occurring in function (2.3.1) is equal to one half a negative odd integer. Hence, the coefficients (2.3.5) of expansion (2.3.4) may always be reduced to the combination of complete elliptic integrals with modulus α or k. This is not of much importance because formulae (2.3.5) and (2.3.7) are good enough for calculating the coefficients of (2.3.4). Our aim is to replace expansion (2.3.4) itself by the closed-form representation to be integrated with the aid of the elliptic functions or to be expanded into the fast converging trigonometric series stemming from the theory of elliptic functions. Putting

$$\lambda - \lambda' = \pi - 2\varphi \tag{6.3.14}$$

and regarding K and L as positive odd integers one may represent function (2.3.1) in the form

$$\gamma\left(n, -\frac{K}{2}, -\frac{L}{2}, \nu, \alpha, \zeta\right) = \frac{\alpha^n}{(1+\alpha)^{\max\{K,L\}}} \frac{(-\zeta)^\nu}{\delta^{\max\{K,L\}}} \times \\ \times \left(1 - \alpha\zeta^{\operatorname{sgn}(K-L)}\right)^{\frac{|K-L|}{2}} \tag{6.3.15}$$

with previous relations (6.2.24)–(6.2.26). Evidently,

$$\Delta_0 = (a+a')\delta\,. \tag{6.3.16}$$

Introducing now the incomplete elliptic integrals of the first and second kinds,

$$u = F(\varphi, k) = \int_0^\varphi \frac{d\varphi}{\delta}\,, \qquad E(\varphi, k) = \int_0^\varphi \delta\, d\varphi\,, \tag{6.3.17}$$

one may deal in natural way with the Jacobi elliptic functions

$$\varphi = \operatorname{am} u\,, \qquad \sin\varphi = \operatorname{sn} u\,, \qquad \cos\varphi = \operatorname{cn} u\,, \qquad \delta = \operatorname{dn} u\,. \tag{6.3.18}$$

On the other hand, from (6.3.3) and (6.3.14) one returns again to (6.2.26), which results in

$$\zeta^\sigma = (-1)^\sigma(\cos 2\sigma\varphi - \mathrm{i}\sin 2\sigma\varphi) \tag{6.3.19}$$

with

$$\cos 2\sigma\varphi = F\left(-\sigma, \sigma, \tfrac{1}{2}, \sin^2\varphi\right) \tag{6.3.20}$$

and

$$\sin 2\sigma\varphi = \sigma \sin 2\varphi\, F\left(-\sigma+1,\, \sigma+1,\, \tfrac{3}{2},\, \sin^2\varphi\right) . \tag{6.3.21}$$

$|K-L|/2$ being an integer, it is evident that function (6.3.15) is reduced to a finite polynomial in powers of $\mathrm{sn}^2\, u$, $\mathrm{sn}\, u\, \mathrm{cn}\, u$ and $1/\,\mathrm{dn}\, u$. Therefore, the planetary disturbing function may be expanded in powers of the eccentricity and inclination variables, the coefficients being representable in closed form with respect to the mean longitudes and the semi-major axes by means of the elliptic functions of the argument u and modulus k. In the problem of N planets as in (6.2.58) one should use for each pair of the planets

$$k_{ij}^2 = \frac{4a_i a_j}{(a_i+a_j)^2}\,, \qquad \lambda_i - \lambda_j = \pi - 2\varphi_{ij}\,, \qquad \varphi_{ij} = \mathrm{am}(u_{ij},\, k_{ij})\,,$$

$$\zeta_{ij} = \exp \mathrm{i}(\lambda_i - \lambda_j)\,. \tag{6.3.22}$$

One then has

$$k_{ji}^2 = k_{ij}^2\,, \quad \varphi_{ji} = \pi - \varphi_{ij}\,, \quad \zeta_{ji} = \zeta_{ij}^{-1}\,, \quad u_{ji} = 2K(k_{ij}) - u_{ij}\,. \tag{6.3.23}$$

The closed-form integration of function (6.3.15) involves no major difficulties. Such an integration may be facilitated by applying an elliptic-function processor as described, for instance, by Abad and Deprit (1991). Hence, the first-order planetary theory may in principle also be constructed by means of a power series in the eccentricity and inclination variables with closed-form coefficients dependent on the mean longitudes and the semi-major axes. An example of such a theory is considered in Chapter 10.

To ensure the relationship of the argument u with the physical time t or to advance to the second-order theory one has to abandon the closed-form representation and make use of the trigonometric expansions of elliptic functions. Introducing

$$\tau = \exp \mathrm{i}\frac{\pi u}{K(k)} \tag{6.3.24}$$

we see from (2.5.69), (2.5.70) and (2.5.74) that finite polynomial (6.3.15) in $\mathrm{sn}^2\, u$, $\mathrm{sn}\, u\, \mathrm{cn}\, u$ and $1/\,\mathrm{dn}\, u$ may be expanded into a series in positive and negative powers of τ with rational coefficients with respect to the Jacobi nome defined by (2.5.12) for any given k. Using relation (6.3.19) for $\sigma = 1$ and expansions (2.5.69) and (2.5.70) one finds the relationship between ζ and τ in the form

$$\zeta = \frac{1}{k^2 K}\left[(2-k^2)K - 2E\right] - \frac{4\pi^2}{k^2 K^2} \sum_{m=-\infty}^{\infty} |m| \frac{q^{(2+\mathrm{sgn}\, m)|m|}}{1-q^{4|m|}} \tau^m\,. \tag{6.3.25}$$

In the case of N planets coefficients (6.2.60) of the disturbing function may be represented by exponential series in

$$\tau_{ij} = \exp \mathrm{i} \frac{\pi u_{ij}}{K(k_{ij})} \tag{6.3.26}$$

with rational coefficients in

$$q_{ij} = \exp \left[-\frac{\pi K'(k_{ij})}{K(k_{ij})} \right]. \tag{6.3.27}$$

In addition to (6.3.23) one therefore has

$$q_{ji} = q_{ij}, \qquad \tau_{ji} = \tau_{ij}^{-1}. \tag{6.3.28}$$

In the theory of motion of eight major planets for the most unfavourable case $i = 2$, $j = 3$ one has

$$\alpha = 0.723, \qquad k = 0.987, \qquad q = 0.215 \tag{6.3.29}$$

so that q is comparatively small. Hence, the τ-series stemming from the theory of elliptic functions seem to be quite effective in planetary problems.

The differential relationships between the arguments introduced so far are of the form

$$(n - n')dt = -2\,d\varphi = -2\,\mathrm{dn}\,u\,du = -\mathrm{i}\frac{d\zeta}{\zeta} = \mathrm{i}\frac{2K}{\pi}\,\mathrm{dn}\,u\frac{d\tau}{\tau}. \tag{6.3.30}$$

From definitions (6.3.14) and (6.3.18) it follows that

$$2\,\mathrm{am}\,u = \pi - (\lambda - \lambda'). \tag{6.3.31}$$

Putting

$$w = \frac{\pi u}{K}, \qquad M = \pi - (\lambda - \lambda') \tag{6.3.32}$$

and using expansion (2.5.27) for $\mathrm{am}\,u$ one may rewrite this equation in the form

$$w + \sum_{m=1}^{\infty} d_m(q) \sin mw = M \tag{6.3.33}$$

with the coefficients

$$d_m(q) = \frac{4}{m} \frac{q^m}{1 + q^{2m}}. \tag{6.3.34}$$

Hence, we again obtain equation (2.5.32) with solution (2.5.34). Therefore, the problem of relating u or τ to the time t presents no difficulties.

It goes without saying that the algorithms discussed above do not pretend to cover the entire variety of methods for expanding the planetary disturbing function.

7 Iteration Techniques of Perturbation Theory

7.1 Iteration Versions of the Classical Methods

This section deals with iteration versions of the classical methods for determining perturbations in rectangular coordinates elaborated by Broucke (1969). Without pretending to construct long-term theories of motion, these methods do not involve the separation of short- and long-period perturbations and removal of secular terms. However, such methods, most particularly in the iteration version, are very convenient for fast constructing short-term theories aimed at representing motion over comparatively short intervals of time (say, several periods of revolution around the central body) but with a high level of accuracy.

In relative rectangular coordinates the equations of the perturbed two-body motion are written in the form

$$\ddot{\mathbf{r}} = -Gm\frac{\mathbf{r}}{r^3} + \mathbf{X}. \tag{7.1.1}$$

Let $\boldsymbol{\rho}$ be the position vector in the undisturbed Keplerian motion described by the equations

$$\ddot{\boldsymbol{\rho}} = -Gm\frac{\boldsymbol{\rho}}{\rho^3}. \tag{7.1.2}$$

Here we denote

$$r = |\mathbf{r}|, \qquad \rho = |\boldsymbol{\rho}|,$$

m being the sum of the masses of two bodies. Putting

$$\mathbf{r} = \boldsymbol{\rho} + \mathbf{s} \tag{7.1.3}$$

one obtains for the vector of corrections $\mathbf{s}$ the following equation:

$$\ddot{\mathbf{s}} = -Gm\left(\frac{\mathbf{r}}{r^3} - \frac{\boldsymbol{\rho}}{\rho^3}\right) + \mathbf{X}. \tag{7.1.4}$$

This equation may be replaced by an equivalent equation

$$\ddot{\mathbf{s}} + S\mathbf{s} = -Gm\left(\frac{\mathbf{r}}{r^3} - \frac{\boldsymbol{\rho}}{\rho^3}\right) + S(\mathbf{r} - \boldsymbol{\rho}) + \mathbf{X}, \tag{7.1.5}$$

$S = S(t)$ being a square matrix to be determined below.

Equation (7.1.5) is solved by Picard iterations. If some approximate solution $\mathbf{r} = \mathbf{r}(t)$ is known, then the right-hand side of (7.1.5) becomes a given function of time $\mathbf{Q} = \mathbf{Q}(t)$. The vectorial function $\mathbf{s}$ results from solving the linear inhomogeneous equation

$$\ddot{\mathbf{s}} + S\mathbf{s} = \mathbf{Q}(t)\,. \tag{7.1.6}$$

Then (7.1.3) gives a new, more accurate value of $\mathbf{r}$. As a first approximation one usually adopts the Keplerian motion $\mathbf{r}^{(1)} = \boldsymbol{\rho}$. But it is also possible to take as $\mathbf{r}^{(1)}$ the series of a long term analytical theory of the moving body. In this case the iteration techniques considered below serve to improve the analytical theories over some finite interval of time.

Convergence of the Picard iterations for equation (7.1.5) is usually ensured in typical celestial mechanics problems. But the rate of convergence depends essentially on the magnitude of the disturbing acceleration $\mathbf{X}$ and the choice of the matrix S. Broucke discusses four, most natural techniques, dependent on the choice of S . In three versions the matrix S is chosen in such a way that the solution of equation (7.1.6) without the right-hand side may be presented in the form

$$\mathbf{s} = \mathbf{K}_1 q_1 + \mathbf{K}_2 q_2\,. \tag{7.1.7}$$

Here $\mathbf{K}_1$ and $\mathbf{K}_2$ are vectorial constants whereas q_1 and q_2 are the linearly independent solutions of the homogeneous equation satisfying the condition

$$q_1\dot{q}_2 - q_2\dot{q}_1 = 1\,. \tag{7.1.8}$$

In accordance with the method of the variation of arbitrary constants the solution of inhomogeneous equation (7.1.6) may also be presented in form (7.1.7), provided that $\mathbf{K}_1$ and $\mathbf{K}_2$ are functions of time satisfying the equations

$$\dot{\mathbf{K}}_1 = -q_2\mathbf{Q}\,, \qquad \dot{\mathbf{K}}_2 = q_1\mathbf{Q}\,. \tag{7.1.9}$$

Therefore, denoting by $\mathbf{Q}^{(m)}$ the result of the substitution of the approximate solution $\mathbf{r}^{(m)}(t)$ into the right-hand side of (7.1.5) one may obtain the general solution of equation (7.1.6) at step m of the iteration process in the form

$$\mathbf{s}^{(m)} = \mathbf{C}_1^{(m)} q_1 + \mathbf{C}_2^{(m)} q_2 - q_1 \int_0^t q_2\mathbf{Q}^{(m)}dt + q_2 \int_0^t q_1\mathbf{Q}^{(m)}dt\,, \tag{7.1.10}$$

$\mathbf{C}_1^{(m)}$ and $\mathbf{C}_2^{(m)}$ being arbitrary constants determined by the specific conditions of the problem under consideration. In virtue of (7.1.3) the more accurate approximation will be

$$\mathbf{r}^{(m+1)} = \boldsymbol{\rho} + \mathbf{s}^{(m)}\,. \tag{7.1.11}$$

In the first method one chooses

$$S = 0 \,, \tag{I}$$

involving the values

$$q_1 = 1 \,, \qquad q_2 = t \,. \tag{7.1.12}$$

The two integrals in (7.1.10) may be combined therewith in a double integral

$$-\int_0^t t\mathbf{Q}^{(m)} dt + t \int_0^t \mathbf{Q}^{(m)} dt = \int_0^t \int_0^t \mathbf{Q}^{(m)} dt \, dt \,. \tag{7.1.13}$$

This method is the simplest of all the techniques considered here. In the second method

$$S = n^2 E \,. \tag{II}$$

Here E stands for the unit matrix and n is the mean motion of the moving body. Therefore,

$$q_1 = \frac{1}{\sqrt{n}} \cos \lambda \,, \qquad q_2 = \frac{1}{\sqrt{n}} \sin \lambda \tag{7.1.14}$$

with $\lambda = nt + \varepsilon$ being the mean longitude. In the third method the matrix S is again of diagonal form but represents a function of time:

$$S = \frac{Gm}{\rho^3} E \,. \tag{III}$$

In this case

$$q_1 = \frac{1}{\sqrt{n}} (\cos g - e) \,, \qquad q_2 = \frac{1}{\sqrt{n}} \sin g \tag{7.1.15}$$

with the eccentricity e and the eccentric anomaly g satisfying the equation

$$\dot{g} = \frac{n}{1 - e \cos g} \,. \tag{7.1.16}$$

The Keplerian solution $\boldsymbol{\rho}$ is thus also expressed in terms of q_1 and q_2:

$$\boldsymbol{\rho} = \sqrt{n} a \left(\mathbf{A}_1 q_1 + \sqrt{1 - e^2} \mathbf{A}_2 q_2 \right) \tag{7.1.17}$$

with semi-major axis a and unit orthogonal vectors $\mathbf{A}_1$ and $\mathbf{A}_2$. Evidently, method (III) represents an iteration version of the classical methods of Encke and Hill.

In versions (II) and (III) two integrals in (7.1.10) may be combined by the well-known Hansen device

$$-q_1 \int_0^t q_2 \mathbf{Q}^{(m)} dt + q_2 \int_0^t q_1 \mathbf{Q}^{(m)} dt = \int_0^t (\tilde{q}_2 q_1 - \tilde{q}_1 q_2) \, \mathbf{Q}^{(m)} dt \,. \tag{7.1.18}$$

Here

$$\tilde{q}_2 q_1 - \tilde{q}_1 q_2 = \frac{1}{n} \sin(\tilde{\lambda} - \lambda)$$

for version (II) and

$$\tilde{q}_1 q_2 - \tilde{q}_1 q_2 = \frac{1}{n} \left[\sin(\tilde{g} - g) - e(\sin\tilde{g} - \sin g)\right]$$

for version (III). A quantity marked by a tilde is treated as constant under integration and regains its functional dependence on time after integration.

In the fourth technique by Broucke, which originates from Brouwer's method, the matrix S is defined by

$$S = Gm \operatorname{grad}\left(\frac{\boldsymbol{\rho}}{\rho^3}\right) \tag{IV}$$

or for the individual elements of $S = ||S_{ij}||$ and $\boldsymbol{\rho} = (\xi_1, \xi_2, \xi_3)$

$$S_{ij} = \frac{Gm}{\rho^3}\left(\delta_{ij} - \frac{3}{\rho^2}\xi_i\xi_j\right).$$

With such a choice of S homogeneous equation (7.1.6) corresponds to the equations in variations for system (7.1.2). Hence, if the general solution of (7.1.2) is given by the expressions

$$\boldsymbol{\rho} = \boldsymbol{\rho}(t, c_1, ..., c_6), \qquad \dot{\boldsymbol{\rho}} = \dot{\boldsymbol{\rho}}(t, c_1, ..., c_6) \tag{7.1.19}$$

with arbitrary constants $c_1, ..., c_6$ then the general solution of the equations in variations will be

$$\mathbf{s} = \sum_{i=1}^{6} K_i \frac{\partial \boldsymbol{\rho}}{\partial c_i}. \tag{7.1.20}$$

This expression results in the solution of inhomogeneous equation (7.1.6) provided that the functions $K_i = K_i(t)$ satisfy the conditions

$$\sum_{i=1}^{6} \dot{K}_i \frac{\partial \boldsymbol{\rho}}{\partial c_i} = 0, \qquad \sum_{i=1}^{6} \dot{K}_i \frac{\partial \dot{\boldsymbol{\rho}}}{\partial c_i} = \mathbf{Q}, \tag{7.1.21}$$

involving

$$\dot{K}_i = \frac{\partial c_i}{\partial \dot{\boldsymbol{\rho}}} \mathbf{Q}. \tag{7.1.22}$$

Hence, iteration formula (7.1.10) is replaced in this method by the following one:

$$\mathbf{s}^{(m)} = \sum_{i=1}^{6} \frac{\partial \boldsymbol{\rho}}{\partial c_i}\left(C_i^{(m)} + \int_0^t \frac{\partial c_i}{\partial \dot{\boldsymbol{\rho}}} \mathbf{Q}^{(m)} dt\right). \tag{7.1.23}$$

In specific implementations of the Brouwer method this formula occurs in various forms, depending on the choice of the parameters c_i (particular sets of elements or initial coordinates and velocities), the independent variable (time or different anomalies), the arbitrary constants $C_i^{(m)}$ and the representation of the derivatives $\partial c_i/\partial \dot{p}$ from equations (7.1.19). The rate of convergence of iterations in version (IV) is the best, compared to versions (I)–(III). But the execution time for one iteration is a maximum for version (IV) so that finally all the versions (I)–(IV) turn out to be almost equivalent.

The actual application of techniques (I)–(IV) is greatly facilitated by using some CAS for manipulating series of a specific kind. Besides this, for versions (III) and (IV) it is appropriate to use a Keplerian processor. Short-term theories of motion are usually constructed in terms of series of the following type:

1. series of Chebyshev polynomials in t;
2. polynomial–trigonometric series in t;
3. polynomial–trigonometric series in different anomalies.

The most compact theories are obtained with the aid of Chebyshev polynomial series. Nowadays traditional astronomical ephemerides are being replaced more and more often by such series. The associated problems and techniques for their solution are described by Deprit et al. (1975, 1978). The Chebyshev polynomial series for problem (7.1.1) can be constructed by iterations most simply with the aid of version (I). The related operations with Chebyshev polynomials involve no difficulties (Broucke, 1973; Stumpff, 1974). An analogous algorithm has been developed by Chapront (1977) for the problem of representing the motion of the major planets in Keplerian elements. In this method the equations of planetary motion in elements in the vectorial form

$$\dot{X} = F(X)$$

are considered over some finite interval $t_0 \le t \le t_1$. The transformation of the independent argument

$$t = \frac{t_1 - t_0}{2} x + \frac{t_1 + t_0}{2}$$

reduces this interval to $-1 \le x \le 1$. Then the right-hand sides are expanded in series of Chebyshev polynomials. The integration of these series gives the desired function X. The iterations are repeated until the demanded accuracy is attained. The arbitrary constants arising in integrating the Chebyshev polynomial series are determined from the initial data for the vector X.

Note that Chebyshev polynomial series may be constructed without iterations by reducing the equations of planetary motion to polynomial form and constructing the solution by Taylor expansions with their subsequent conversion into a Chebyshev polynomial series (Section 4.1). The efficient evaluation of functions represented by Chebyshev polynomial series is performed by the well-known algorithm of Clenshaw (Press et al., 1992). The same algorithm may be applied to calculate the derivatives or the integrals of such functions.

Along with Chebyshev polynomial series one also often uses polynomial–trigonometric series in t for problem (7.1.1). The simplest way to construct such a solution is to apply techniques (II) or (III) with the aid of a Keplerian processor.

The use of some anomaly (true, eccentric or elliptic) as an independent argument in methods (III) or (IV) instead of time as a rule results in diminishing a number of terms in the final series. But this makes the corresponding operations more complicated. One of the most important operations is related to an integration of the type

$$I = \int F(\lambda) G(\lambda')\, dt\,, \tag{7.1.24}$$

where F and G are implicit functions of the mean longitudes $\lambda(= nt+\varepsilon)$ and $\lambda'(= n't+\varepsilon')$ of the disturbed and disturbing bodies, respectively. In satellite problems where the mean motions n and n' satisfy the inequality $n' \ll n$ it is appropriate to use integration by parts:

$$I = \frac{1}{n} F^{(-1)}(\lambda) G(\lambda') - \frac{n'}{n} \int F^{(-1)}(\lambda) G^{(1)}(\lambda')\, dt \tag{7.1.25}$$

with

$$F^{(-1)}(\lambda) = \int F(\lambda)\, d\lambda\,, \qquad G^{(1)}(\lambda') = \frac{dG(\lambda')}{d\lambda'}\,.$$

By repeating this process one has

$$I = \frac{1}{n} \sum_{k=1}^{m} (-1)^{k-1} \left(\frac{n'}{n}\right)^{k-1} F^{(-k)}(\lambda) G^{(k-1)}(\lambda') +$$

$$+ (-1)^m \left(\frac{n'}{n}\right)^m \int F^{(-m)}(\lambda) G^{(m)}(\lambda')\, dt \tag{7.1.26}$$

with

$$F^{(-k)}(\lambda) = \int F^{(-k+1)}(\lambda)\, d\lambda\,, \qquad G^{(k)}(\lambda') = \frac{dG^{(k-1)}(\lambda')}{d\lambda'}\,,$$

for $k = 1, 2, \ldots, m$. For sufficiently large m the last term in (7.1.26) gives a negligibly small contribution so that the integral I will be presented in closed form. To evaluate the integrals $F^{(-k)}(\lambda)$ one needs a closed-form Keplerian processor because the function $F(\lambda)$ is usually given in terms of some anomaly. Algorithm (7.1.24)–(7.1.26) has been successfully applied by A. V. Egorova (1960) in the problem of determining luni-solar perturbations in the motion of Earth's artificial satellites.

In planetary problems when the ratio n'/n is not small one has to use another technique. Consider the integral

$$I = \int f(s,\, s')\, ds\,, \tag{7.1.27}$$

where s and s' are some anomalies (true, eccentric or elliptic) of the disturbed and disturbing bodies, respectively. The integrand function $f(s,\, s')$ usually represents a Fourier series in s and s' but this does not facilitate the integration because s' is related to s in a rather complicated way. This difficulty may be overcome by a skillful device due to Hansen. Let M and M' be linear functions of time with frequencies n and n', respectively (mean anomalies, for example):

$$M = nt + M_0\,, \qquad M' = n't + M_0'\,, \tag{7.1.28}$$

M_0 and M_0' being the initial values. It is thus assumed that the one-to-one correspondences $s \leftrightarrow M$ and $s' \leftrightarrow M'$ are known. In virtue of (7.1.28) one has the identity

$$M' = \frac{n'}{n}s - \frac{n'}{n}(s - M) + c \tag{7.1.29}$$

with the constant

$$c = M_0' - \frac{n'}{n}M_0\,. \tag{7.1.30}$$

Introducing a new variable

$$s_1 = M' + \frac{n'}{n}(s - M) \tag{7.1.31}$$

one has from (7.1.29)

$$s_1 = \frac{n'}{n}s + c\,. \tag{7.1.32}$$

For a given s and s' the function s_1 is determined in a unique way by (7.1.31). Therefore, the function $f(s,\, s')$ may be regarded as some function $g(s,\, s_1)$ representable as a Fourier series in s and s_1. But in virtue of (7.1.32) s_1 is a linear function of s so that the integration is now performed without difficulty. The resulting series may again be converted into a Fourier series in s and s'.

To illustrate this technique consider the case when s and s' are the elliptic anomalies w and w' of the disturbed and disturbing bodies, respectively (Brumberg E., 1992). In this case the relationship between w and M has form (2.5.32). An analogous relationship exists for w' and M'. Then equation (7.1.31) takes the form

$$s_1 = w' + \sum_{m=1}^{\infty} d'_m \sin mw' - \frac{n'}{n} \sum_{m=1}^{\infty} d_m \sin mw \tag{7.1.33}$$

with the coefficients $d_m = d_m(q)$ and $d'_m = d_m(q')$ determined by (2.5.33). This equation is the Lagrange implicit equation with respect to w'

$$w' + \sum_{m=1}^{\infty} d'_m \sin mw' = x \tag{7.1.34}$$

with the known right-hand side

$$x = s_1 + \frac{n'}{n} \sum_{m=1}^{\infty} d_m \sin mw \,. \tag{7.1.35}$$

Equation (7.1.34) may be solved as in Section 2.5, resulting in the representation of any specific function of w' in terms of some function $\psi(x)$ developable in multiples of x. This step is not necessary if s' represents the function M' itself (tesseral harmonic perturbations in the Earth's satellite problem or third-body perturbations caused by a body in circular orbit). In this case equations (7.1.33) and (7.1.34) are replaced by

$$s_1 = M' - \frac{n'}{n} \sum_{m=1}^{\infty} d_m \sin mw \tag{7.1.36}$$

and

$$M' = x \,. \tag{7.1.37}$$

It remains to transform the trigonometric expansion $\psi(x)$ into a Fourier series in w and s_1. This may be done either numerically by harmonic analysis techniques or analytically by using some CAS or employing with the aid of some CAS the standard expansion with Bessel coefficients

$$\exp \mathrm{i}(A \sin \lambda + B) = \sum_{k=-\infty}^{\infty} J_k(A) \exp \mathrm{i}(k\lambda + B) \tag{7.1.38}$$

with

$$J_{-k}(A) = (-1)^k J_k(A) \,.$$

Hence, for any integer p

$$\exp \mathrm{i} px = \sum_{k=-\infty}^{\infty} A_k(p) \exp \mathrm{i}(ps_1 + kw) \tag{7.1.39}$$

with

$$A_k(p) = \sum_{k_1,\ldots,k_{m^*}} J_{k_1}\left(\frac{n'}{n}pd_1\right)\ldots J_{k_{m^*}}\left(\frac{n'}{n}pd_{m^*}\right), \tag{7.1.40}$$

where the summation is taken over all values of $k_1, k_2, \ldots, k_{m^*}$ such that

$$k_1 + 2k_2 + \ldots + m^* k_{m^*} = k \tag{7.1.41}$$

and

$$|k_1| + 2|k_2| + \ldots + m^*|k_{m^*}| \leq m^* . \tag{7.1.42}$$

Here m^* denotes the maximum value of m actually taken into account in (7.1.35). Evidently, $A_{-k}(p) = A_k(-p)$.

In such a manner one may find the trigonometric expansion of the function $\psi(x)$ in multiples of w and s_1, enabling one to find the expression for integral (7.1.27) in terms of w and s_1. To restore the original form in terms of w and s' it is necessary to substitute expression (7.1.33) or (7.1.36) for s_1 and by the same techniques to get the expansion in terms of w and s'.

The Hansen device is particularly beneficial in combination with expansions in multiples of the elliptic anomaly. In using the true or eccentric anomalies one has the eccentricity e as a small parameter and may apply this technique only for low-eccentricity orbits. The Hansen device in combination with $(q,\, w)$ expansions seems to be a fairly efficient tool for analytically solving, for example, the problem of the tesseral harmonic perturbations of the motion of a satellite of the Earth. Recall that the relegation of the nodes technique proposed for the same problem (see Section 2.1) might be efficient only for distant Earth satellites with a large value of r_0/r, r_0 being the mean equatorial radius of the Earth.

7.2 Intermediate Orbits in the N-Planet Problem

To illustrate the application of the iteration technique we consider the problem of the motion of N major planets. This problem is solved here by a technique of type (II) but in rotating coordinates. The initial equations in the rectangular heliocentric coordinates $\mathbf{r}_i = (x_i,\, y_i,\, z_i)$ $(i = 1, \ldots, N)$ are of the form

$$\ddot{\mathbf{r}}_i = \frac{\partial U_i}{\partial \mathbf{r}_i} \tag{7.2.1}$$

with force function (6.2.53). As in (6.2.43) we introduce instead of x_i, y_i and z_i the dimensionless complex variables p_i, $q_i = \bar{p}_i$ and real variables w_i characterizing the deviation of the actual planetary orbits from planar circular orbits with semi-major axes a_i and mean motions n_i. Then

$$x_i + \mathrm{i}\, y_i = a_i(1 - p_i) \exp \mathrm{i}\lambda_i , \tag{7.2.2}$$

$$z_i = a_i w_i \tag{7.2.3}$$

with mean longitudes

$$\lambda_i = n_i t + \varepsilon_i \,. \tag{7.2.4}$$

n_i and a_i are related by Kepler's third law (6.2.55). In terms of the new variables equations (7.2.1) may be reduced to the form

$$\ddot{p}_i + 2\,\mathrm{i}\, n_i \dot{p}_i - \tfrac{3}{2} n_i^2 (p_i + q_i) = n_i^2 P_i \,, \tag{7.2.5}$$

$$\ddot{w}_i + n_i^2 w_i = n_i^2 W_i \tag{7.2.6}$$

with right-hand sides

$$P_i = -1 - \frac{1}{2} p_i - \frac{3}{2} q_i + \frac{2}{n_i^2 a_i^2} \frac{\partial U_i}{\partial q_i} \tag{7.2.7}$$

and

$$W_i = w_i + \frac{1}{n_i^2 a_i^2} \frac{\partial U_i}{\partial w_i} \,. \tag{7.2.8}$$

As can be seen from (6.2.58) right-hand members (7.2.7) and (7.2.8) are small quantities of a higher order than p_i and q_i. Indeed, their undisturbed parts start with terms of the second degree with respect to p_i and w_i whereas the perturbing terms are proportional to the small parameter μ, which of the order of the ratio of the planetary masses to the mass of the Sun ($\mu = 10^{-3}$). Therefore, system (7.2.5) and (7.2.6) may be solved by Picard iteration techniques.

Putting $P_i = W_i = 0$ and omitting for a moment the index i one obtains from (7.2.5) and (7.2.6) a homogeneous system with the general solution

$$p = A \exp \mathrm{i}\lambda - 3\bar{A} \exp(-\mathrm{i}\lambda) + \mathrm{i}\, n(B + 3Ct) - 2C \,, \tag{7.2.9}$$

$$w = D \exp \mathrm{i}\lambda + \bar{D} \exp(-\mathrm{i}\lambda) \,, \tag{7.2.10}$$

$$\dot{p} = \mathrm{i}\, n\big[A \exp \mathrm{i}\lambda + 3\bar{A} \exp(-\mathrm{i}\lambda) + 3C\big] \tag{7.2.11}$$

and

$$\dot{w} = \mathrm{i}\, n\big[D \exp \mathrm{i}\lambda - \bar{D} \exp(-\mathrm{i}\lambda)\big] \,. \tag{7.2.12}$$

Here B and C are real constants. A and D are complex constants. By the technique of the variation of arbitrary constants a particular solution of inhomogeneous system (7.2.5) and (7.2.6) results from (7.2.9)–(7.2.12) provided that A, B, C and D are functions of time as follows:

$$A = \frac{1}{4} \mathrm{i}\, n \int (3\bar{P} - P) \exp(-\mathrm{i}\lambda)\, dt \,, \tag{7.2.13}$$

$$B = -\int \big[P + \bar{P} + \tfrac{3}{2} \mathrm{i}\, n(P - \bar{P})t\big]\, dt \,, \tag{7.2.14}$$

$$C = \frac{1}{2}\,\mathrm{i}\,n\int (P - \bar{P})\,dt \tag{7.2.15}$$

and

$$D = -\frac{1}{2}\,\mathrm{i}\,n\int W\exp(-\mathrm{i}\lambda)\,dt\,. \tag{7.2.16}$$

In the general case the solution of equations (7.2.5) and (7.2.6) for p_i, w_i has the form of the polynomial–exponential series of the type

$$S = \sum_k S_k(t)\exp\mathrm{i}(k\lambda)\,, \tag{7.2.17}$$

where k is a multi-index $k = (k_1, \ldots, k_N)$, λ is a vector with components $\lambda_1, \ldots, \lambda_N$ and $(k\lambda) = k_1\lambda_1 + \ldots + k_N\lambda_N$. $S_k(t)$ represents polynomials of t with complex coefficients. The polynomial structure of these coefficients is caused by the secular motions of the perihelia and nodes of the planetary orbits. In taking quadrature (7.2.15) one should add to C a constant equal to $2P^*/3$, P^* being a constant (independent of t) part of P. Such a choice removes the appearance in (7.2.9) of fictitious secular terms having no physical meaning. This choice is possible provided that P^* is a real constant. Such a condition is fulfilled by itself in celestial-mechanics problems.

Restoring now the index i and putting

$$P_i = P_i^* + P_i^+ \tag{7.2.18}$$

one may write the iteration algorithm for determining p_i and w_i in the form

$$\begin{aligned} p_i = \mathrm{i}\,n_i\Bigg[&\frac{3}{4}\exp(-\mathrm{i}\lambda_i)\int (3P_i^+ - \bar{P}_i^+)\exp\mathrm{i}\lambda_i\,dt + \\ &+ \frac{1}{4}\exp\mathrm{i}\lambda_i\int (3\bar{P}_i^+ - P_i^+)\exp(-\mathrm{i}\lambda_i)\,dt - 2\int P_i^+\,dt\Bigg] - \\ &- \frac{3}{2}n_i^2\int\!\!\int (P_i^+ - \bar{P}_i^+)\,dt\,dt - \frac{1}{3}P_i^* + \tilde{p}_i \end{aligned} \tag{7.2.19}$$

and

$$w_i = \frac{1}{2}\,\mathrm{i}\,n_i\left[\exp(-\mathrm{i}\lambda_i)\int W_i\exp\mathrm{i}\lambda_i\,dt - \exp\mathrm{i}\lambda_i\int W_i\exp(-\mathrm{i}\lambda_i)\,dt\right] + \tilde{w}_i\,. \tag{7.2.20}$$

The derivatives $\dot{p}_i$ and $\dot{w}_i$ may be found, if necessary, with the formulae

$$\begin{aligned} \dot{p}_i = n_i^2\Bigg[&\frac{3}{4}\exp(-\mathrm{i}\lambda_i)\int (3P_i^+ - \bar{P}_i^+)\exp\mathrm{i}\lambda_i\,dt - \\ &- \frac{1}{4}\exp\mathrm{i}\lambda_i\int (3\bar{P}_i^+ - P_i^+)\exp(-\mathrm{i}\lambda_i)\,dt - \frac{3}{2}\int (P_i^+ - \bar{P}_i^+)\,dt\Bigg] + \dot{\tilde{p}}_i \end{aligned} \tag{7.2.21}$$

and

$$\dot{w}_i = \frac{1}{2} n_i^2 \left[\exp(-\mathrm{i}\lambda_i) \int W_i \exp \mathrm{i}\lambda_i \, dt + \exp \mathrm{i}\lambda_i \int W_i \exp(-\mathrm{i}\lambda_i) \, dt \right] + \dot{\tilde{w}}_i \,. \tag{7.2.22}$$

The condition

$$P_i^* = \bar{P}_i^*$$

may serve as a checking relation at each step of the iteration process.

The functions $\tilde{p}_i$, $\tilde{w}_i$, $\dot{\tilde{p}}_i$ and $\dot{\tilde{w}}_i$ in the right-hand members (7.2.19)–(7.2.22) represent the contributions due to the general solution (7.2.9)–(7.2.12) of the homogeneous system. In taking into account these contributions one should annul the constants C_i to avoid the appearance of the fictitious secular terms. The constants B_i may also be omitted. Then the constants A_i and D_i together with n_i and ε_i give the complete set of $6N$ real constants characterizing the initial conditions in the N-planet problem. Comparison with the solution of the two-body problem in Keplerian elements clearly shows that to a first approximation one can estimate the constants A_i and D_i by means of the relations

$$A = -\tfrac{1}{2} e \exp(-\mathrm{i}\pi) \,, \qquad D = -\tfrac{1}{2} \mathrm{i} \sin i \exp(-\mathrm{i}\Omega) \,, \tag{7.2.23}$$

e, i, π and Ω being the eccentricity, inclination, longitude of the pericentre and longitude of the node, respectively.

If a software package to manipulate series (7.2.17) is available then the iterations by formulae (7.2.19) and (7.2.20) involve no difficulties. But, as in any iteration technique, particular attention should be given to the most accurate computation of the right-hand sides. It is therefore appropriate to consider this question in more detail.

In virtue of (7.2.7), (7.2.8) and (6.2.54) one has

$$P_i = -1 - \frac{1}{2} p_i - \frac{3}{2} q_i + (1 - p_i) \left(\frac{a_i}{r_i} \right)^3 + \mu \sum_{j=1}^{N} {}^{(i)} P_{ij} \tag{7.2.24}$$

and

$$W_i = w_i - w_i \left(\frac{a_i}{r_i} \right)^3 + \mu \sum_{j=1}^{N} {}^{(i)} W_{ij} \tag{7.2.25}$$

with

$$P_{ij} = \kappa_{ij} \left\{ \left[1 - p_i - \frac{a_j}{a_i} (1 - p_j) \zeta_{ij}^{-1} \right] \left(\frac{a_i}{\Delta_{ij}} \right)^3 + \right. \\ \left. + \zeta_{ij}^{-1} \left(\frac{a_i}{a_j} \right)^2 (1 - p_j) \left(\frac{a_j}{r_j} \right)^3 \right\} , \tag{7.2.26}$$

$$W_{ij} = -\kappa_{ij} \left[\left(w_i - \frac{a_j}{a_i} w_j \right) \left(\frac{a_i}{\Delta_{ij}} \right)^3 + \left(\frac{a_i}{a_j} \right)^2 w_j \left(\frac{a_j}{r_j} \right)^3 \right] \tag{7.2.27}$$

and previous designations (6.2.55)–(6.2.57).

To avoid a loss of accuracy in computing (7.2.24) and (7.2.25) it is necessary, first of all, to transform the Keplerian terms. Rewriting relation (6.2.44) in the form

$$\left(\frac{r_i}{a_i}\right)^2 = 1 - S_i\,, \tag{7.2.28}$$

$$S_i = p_i + q_i - (p_i q_i + w_i^2) \tag{7.2.29}$$

one has for any natural number n

$$\left(\frac{a_i}{r_i}\right)^n = 1 + V_i^{(n)} \tag{7.2.30}$$

with

$$V_i^{(n)} = \frac{n}{2} S_i + T_i^{(n)} \tag{7.2.31}$$

and

$$T_i^{(n)} = (1 - S_i)^{-\frac{n}{2}} - 1 - \frac{n}{2} S_i = \sum_{k=2}^{\infty} \frac{\left(\frac{n}{2}\right)_k}{(1)_k} S^k\,. \tag{7.2.32}$$

Hence, the formulae for the actual computation of the right-hand sides P_i and W_i take the form

$$P_i = -\frac{3}{2}(p_i q_i + w_i^2) - p_i V_i^{(3)} + T_i^{(3)} + \mu \sum_{j=1}^{N}{}^{(i)} P_{ij} \tag{7.2.33}$$

and

$$W_i = -w_i V_i^{(3)} + \mu \sum_{j=1}^{N}{}^{(i)} W_{ij}\,. \tag{7.2.34}$$

In computing P_{ij} and W_{ij} the main difficulty is to deal with the quantities $(a_i/\Delta_{ij})^3$. Writing (6.2.47) in the form

$$\left(\frac{\Delta_{ij}}{a_i}\right)^2 = \left(\frac{\Delta_{ij}^{(0)}}{a_i}\right)^2 (1 - U_{ij}) \tag{7.2.35}$$

one finds for any natural number n that

$$\left(\frac{a_i}{\Delta_{ij}}\right)^n = \left(\frac{a_i}{\Delta_{ij}^{(0)}}\right)^n \left(1 + V_{ij}^{(n)}\right), \tag{7.2.36}$$

where

$$V_{ij}^{(n)} = (1 - U_{ij})^{-\frac{n}{2}} - 1 = \sum_{k=1}^{\infty} \frac{\left(\frac{n}{2}\right)_k}{(1)_k} (U_{ij})^k . \tag{7.2.37}$$

The initial approximation $\Delta_{ij}^{(0)}$ should be chosen so that $|U_{ij}| \ll 1$, and the expansions of negative powers of $\Delta_{ij}^{(0)}$ are constructed in an elementary way. In the typical planetary case both conditions are satisfied by choosing $\Delta_{ij}^{(0)}$ for planar circular orbits as in (6.3.1). Such a choice involves the expression of U_{ij} as follows:

$$\begin{aligned} U_{ij} = &\left(\frac{a_i}{\Delta_{ij}^{(0)}}\right)^2 \left[\left(1 - \frac{a_j}{a_i}\zeta_{ij}\right)\left(p_i - \frac{a_j}{a_i}p_j\zeta_{ij}^{-1}\right) + \right. \\ &+ \left(1 - \frac{a_j}{a_i}\zeta_{ij}^{-1}\right)\left(q_i - \frac{a_j}{a_i}q_j\zeta_{ij}\right) - \left(p_i - \frac{a_j}{a_i}p_j\zeta_{ij}^{-1}\right)\left(q_i - \frac{a_j}{a_i}q_j\zeta_{ij}\right) - \\ &\left. - \left(w_i - \frac{a_j}{a_i}w_j\right)^2\right] . \end{aligned} \tag{7.2.38}$$

Hence, by applying (7.2.30)–(7.2.32) and (7.2.36)–(7.2.38) one may compute the right-hand sides P_i and W_i without any difficulty.

In the particular case when $r_i \ll r_j$ (to be more specific when $r_i/r_j <$ $< 1/10$ in the problem of the motion of the major planets) it is appropriate to use in (7.2.26) and (7.2.27) an Encke-type transformation. In this case it is convenient to choose

$$\Delta_{ij}^{(0)} = r_j . \tag{7.2.39}$$

Then

$$\begin{aligned} U_{ij} = \frac{a_i}{a_j}&\left[(1 - p_i)(1 - q_j)\zeta_{ij} + (1 - p_j)(1 - q_i)\zeta_{ij}^{-1} + 2w_iw_j - \right. \\ &\left. - \frac{a_i}{a_j}\left(\frac{r_i}{a_i}\right)^2\right]\left(\frac{a_j}{r_j}\right)^2 . \end{aligned} \tag{7.2.40}$$

Expressions (7.2.26) and (7.2.27) are thus replaced by

$$P_{ij} = \kappa_{ij}\left(\frac{a_i}{a_j}\right)^2\left(\frac{a_j}{r_j}\right)^3\left\{\frac{a_i}{a_j}(1 - p_i) + \left[\frac{a_i}{a_j}(1 - p_i) - (1 - p_j)\zeta_{ij}^{-1}\right]V_{ij}^{(3)}\right\} \tag{7.2.41}$$

and

$$W_{ij} = -\kappa_{ij}\left(\frac{a_i}{a_j}\right)^2\left(\frac{a_j}{r_j}\right)^3\left[\frac{a_i}{a_j}w_i + \left(\frac{a_i}{a_j}w_i - w_j\right)V_{ij}^{(3)}\right] . \tag{7.2.42}$$

Hence, the right-hand sides P_i and W_i may again be computed without any loss of accuracy.

Let us mention now two applications of iteration process (7.2.19) and (7.2.20).

If one starts the iterations with the values $p_i^{(0)} = \tilde{p}_i = 0$, $w_i^{(0)} = \tilde{w}_i = 0$ then after completing the process one will still have $w_i = 0$. Variables p_i will be represented by purely exponential series with respect to the differences of the mean longitudes of the planets (series (7.2.17) with numerical real coefficients S_k and the condition $k_1 + \ldots + k_N = 0$ for each term). This particular solution, which corresponds to the planar quasi-periodic motions in the N-planet problem ($N = 8$), was constructed by Brumberg (1970) and Brumberg et al. (1975) using two different techniques distinguished by the manner of computing the right-hand sides (power-series expansions and the iteration technique). In both cases the resulting series for p_i were constructed term by term by applying to equations (7.2.5) the method of indefinite coefficients. The use of iteration process (7.2.19) excludes operations with particular terms of the series and increases the efficiency of iterations. In Chapter 10 we shall consider the possibility of constructing such a solution with the aid of elliptic functions and their expansions.

Iterations by means of (7.2.19) and (7.2.20) enable one to improve the semi-analytical series of the general planetary theory (Brumberg and Chapront, 1973; Brumberg et al., 1978). These series of form (7.2.17) result from the substitution of the numerical values for the orbital elements into the literal Poisson series of the general planetary theory. The inevitable neglect of higher-order terms with respect to the planetary masses and higher-degree terms in eccentricities and inclinations as well as inevitable limitations on the admissible values of the exponential indices introduce systematic errors into the final series. Iterations by means of (7.2.19) and (7.2.20) lead to more accurate estimations of the coefficients of these series.

7.3 Iterations in the Two-Body Problem

Iteration algorithm (7.2.19) and (7.2.20) is applied here to construct literal expansions for the coordinates in the two-body problem in powers of Laplace-type variables. Similar expansions (2.2.18) and (2.2.19) have been derived in Section 2.2 from trigonometric series (2.2.13) and (2.2.14). Omitting now the unnecessary index i one may write the required expansions in the form

$$p = \sum p_{pqrs} a^p \bar{a}^q b^r \bar{b}^s \,, \tag{7.3.1}$$

$$w = \sum w_{pqrs} a^p \bar{a}^q b^r \bar{b}^s \tag{7.3.2}$$

with

$$a = \alpha \exp i\lambda \,, \qquad b = \beta \exp i\lambda \,, \tag{7.3.3}$$

$$\lambda = nt + \varepsilon \,, \tag{7.3.4}$$

α and β being elements of type (2.2.20) of the order of eccentricity and inclination, respectively (the designation of the semi-major axis and one of the Laplace-type power variables by one and the same letter should not lead to misunderstanding). The summation in (7.3.1) and (7.3.2) is extended over all non-negative values of indices p, q, r and s satisfying the condition $p+q+$ $+r+s \geq 1$. The sum $r+s$ is even in (7.3.1) and odd in (7.3.2). In constructing series (7.3.1) and (7.3.2) by means of iterations (7.2.19) and (7.2.20) one needs only very simple integration formulae:

$$\int a^p\bar{a}^q b^r\bar{b}^s \, dt = \frac{a^p\bar{a}^q b^r\bar{b}^s}{in(p-q+r-s)}, \quad p-q+r-s \neq 0 \tag{7.3.5}$$

$$\int\int a^p\bar{a}^q b^r\bar{b}^s \, dt\, dt = -\frac{a^p\bar{a}^q b^r\bar{b}^s}{n^2(p-q+r-s)^2}, \quad p-q+r-s \neq 0 \tag{7.3.6}$$

and

$$\int a^p\bar{a}^q b^r\bar{b}^s \exp i\lambda \, dt = \frac{a^p\bar{a}^q b^r\bar{b}^s \exp i\lambda}{in(p-q+r-s+1)}, \quad p-q+r-s\neq-1. \tag{7.3.7}$$

The first steps of this algorithm are listed below. With initial values

$$p^{(1)} = \tilde{p} = -\tfrac{1}{2}a + \tfrac{3}{2}\bar{a}, \tag{7.3.8}$$

and

$$w^{(1)} = \tilde{w} = b + \bar{b} \tag{7.3.9}$$

one finds by means of (7.2.28)–(7.2.34) (with $\mu = 0$) the second-degree terms on the right-hand sides P and W:

$$P^{(2)} = \frac{15}{4}a^2 - \frac{3}{2}a\bar{a} + \frac{3}{4}\bar{a}^2 - \frac{3}{2}(b^2 + 2b\bar{b} + \bar{b}^2)$$

and

$$W^{(2)} = -\frac{3}{2}(ab + \bar{a}b + a\bar{b} + \bar{a}\bar{b}).$$

After integration with the aid of (7.3.5)–(7.3.7) one obtains p and w accurate up to the second-degree terms inclusive. These second-degree terms are

$$p^{(2)} = -\frac{3}{8}a^2 + \frac{1}{2}a\bar{a} - \frac{1}{8}\bar{a}^2 + b\bar{b} + \bar{b}^2 \tag{7.3.10}$$

and

$$w^{(2)} = \frac{1}{2}ab - \frac{3}{2}\bar{a}b - \frac{3}{2}a\bar{b} + \frac{1}{2}\bar{a}\bar{b}. \tag{7.3.11}$$

Continuing this process one gets

$$p^{(3)} = -\frac{1}{3}a^3 + \frac{9}{8}a\bar{a}^2 - \frac{1}{24}\bar{a}^3 - \frac{3}{2}a\bar{b}^2 + \frac{1}{2}\bar{a}\bar{b}^2 , \tag{7.3.12}$$

$$w^{(3)} = \frac{3}{8}a^2 b + \frac{3}{8}\bar{a}^2\bar{b} , \tag{7.3.13}$$

$$\begin{aligned} p^{(4)} = &-\frac{125}{384}a^4 - \frac{3}{16}a^3\bar{a} + \frac{49}{64}a^2\bar{a}^2 - \frac{11}{48}a\bar{a}^3 - \frac{3}{128}\bar{a}^4 - \\ &-\frac{3}{8}a^2 b\bar{b} - \frac{1}{16}a^2\bar{b}^2 + \frac{3}{2}a\bar{a}b\bar{b} + \frac{1}{2}a\bar{a}\bar{b}^2 - \frac{1}{16}\bar{a}^2 b^2 - \\ &-\frac{3}{8}\bar{a}^2 b\bar{b} + \frac{3}{8}\bar{a}^2\bar{b}^2 + b^2\bar{b}^2 + b\bar{b}^3 , \end{aligned} \tag{7.3.14}$$

$$\begin{aligned} w^{(4)} = &\frac{1}{3}a^3 b - \frac{1}{48}a^3\bar{b} + \frac{1}{4}a^2\bar{a}b - \frac{27}{16}a^2\bar{a}\bar{b} - \frac{27}{16}a\bar{a}^2 b + \\ &+\frac{1}{4}a\bar{a}^2\bar{b} - \frac{1}{48}\bar{a}^3 b + \frac{1}{3}\bar{a}^3\bar{b} + \frac{1}{2}ab^2\bar{b} - \frac{3}{2}ab\bar{b}^2 - \\ &-\frac{3}{2}\bar{a}b^2\bar{b} + \frac{1}{2}\bar{a}b\bar{b}^2 , \end{aligned} \tag{7.3.15}$$

$$\begin{aligned} p^{(5)} = &-\frac{27}{80}a^5 - \frac{1}{3}a^4\bar{a} + \frac{19}{8}a^2\bar{a}^3 - \frac{5}{48}a\bar{a}^4 - \frac{1}{60}\bar{a}^5 - \frac{2}{3}a^3 b\bar{b} - \\ &-\frac{39}{16}a^2\bar{a}\bar{b}^2 + \frac{21}{8}a\bar{a}^2 b\bar{b} + \frac{1}{2}a\bar{a}^2\bar{b}^2 - \frac{1}{16}\bar{a}^3 b^2 - \frac{5}{24}\bar{a}^3 b\bar{b} + \\ &+\frac{1}{3}\bar{a}^3\bar{b}^2 - 3ab\bar{b}^3 + \bar{a}b\bar{b}^3 , \end{aligned} \tag{7.3.16}$$

$$\begin{aligned} w^{(5)} = &\frac{125}{384}a^4 b - \frac{3}{128}a^4\bar{b} + \frac{3}{8}a^3\bar{a}b + \frac{3}{8}a\bar{a}^3\bar{b} - \frac{3}{128}\bar{a}^4 b + \\ &+\frac{125}{384}\bar{a}^4\bar{b} + \frac{3}{4}a^2 b^2\bar{b} + \frac{3}{4}\bar{a}^2 b\bar{b}^2 , \end{aligned} \tag{7.3.17}$$

and so on. Again, this algorithm is simpler compared with the term by term determination of series (7.3.1) and (7.3.2), due to Brumberg and Chapront (1973). The relationship of α, β and ε with the Keplerian elements may be easily found by comparing (7.3.1) and (7.3.2) with the classical expansions using any PS processor (Brumberg et al., 1978). Indeed, on the one hand one has series (7.3.1) and (7.3.2) representable in the form

$$p = -\tfrac{1}{2}a + \tfrac{3}{2}\bar{a} + \delta p(a,\, \bar{a},\, b,\, \bar{b}) \tag{7.3.18}$$

and

$$w = b + \bar{b} + \delta w(a,\, \bar{a},\, b,\, \bar{b}) , \tag{7.3.19}$$

where δp and δw are the power series in a, $\bar{a}$, b and $\bar{b}$ starting with the second-degree terms. On the other hand, one has expansions (2.2.18) and (2.2.19), which may be written as

$$\frac{1}{a}(x + iy)\exp(-i\Lambda) = 1 + \frac{1}{2}K - \frac{3}{2}\bar{K} + S(K,\, \bar{K},\, L,\, \bar{L}) \tag{7.3.20}$$

and

$$2\mathrm{i}\frac{z}{a} = L - \bar{L} + 2\mathrm{i}T(K,\, \bar{K},\, L,\, \bar{L})\,, \tag{7.3.21}$$

where S and T are power series in K, $\bar{K}$, L and $\bar{L}$ starting with the second-degree terms. Up to the terms of the fifth-degree inclusive one has

$$\begin{aligned} S = {} & \frac{3}{8}K^2 - \frac{1}{2}K\bar{K} + \frac{1}{8}\bar{K}^2 - \frac{1}{4}L\bar{L} + \frac{1}{4}\bar{L}^2 + \frac{1}{3}K^3 - \frac{3}{8}K^2\bar{K} + \frac{1}{24}\bar{K}^3 - \\ & - \frac{1}{8}KL\bar{L} - \frac{3}{8}K\bar{L}^2 + \frac{3}{8}\bar{K}L\bar{L} + \frac{1}{8}\bar{K}\bar{L}^2 + \frac{125}{384}K^4 - \frac{3}{8}K^3\bar{K} - \\ & - \frac{1}{64}K^2\bar{K}^2 + \frac{1}{24}K\bar{K}^3 + \frac{3}{128}\bar{K}^4 - \frac{3}{32}K^2L\bar{L} + \frac{1}{32}K^2\bar{L}^2 + \frac{1}{8}K\bar{K}L\bar{L} - \\ & - \frac{1}{8}K\bar{K}\bar{L}^2 - \frac{1}{32}\bar{K}^2L\bar{L} + \frac{3}{32}\bar{K}^2\bar{L}^2 - \frac{1}{16}L^2\bar{L}^2 + \frac{1}{16}L\bar{L}^3 + \frac{27}{80}K^5 - \\ & - \frac{5}{12}K^4\bar{K} + \frac{5}{96}K^3\bar{K}^2 + \frac{1}{96}K\bar{K}^4 + \frac{1}{60}\bar{K}^5 - \frac{1}{12}K^3L\bar{L} + \frac{1}{96}K^3\bar{L}^2 + \\ & + \frac{3}{32}K^2\bar{K}L\bar{L} - \frac{3}{32}K\bar{K}^2\bar{L}^2 - \frac{1}{96}\bar{K}^3L\bar{L} + \frac{1}{12}\bar{K}^3\bar{L}^2 - \frac{1}{32}KL^2\bar{L}^2 - \\ & - \frac{3}{32}KL\bar{L}^3 + \frac{3}{32}\bar{K}L^2\bar{L}^2 + \frac{1}{32}\bar{K}L\bar{L}^3 + \ldots \end{aligned} \tag{7.3.22}$$

and

$$\begin{aligned} 2\mathrm{i}T = {} & \frac{1}{2}KL + \frac{3}{2}K\bar{L} - \frac{3}{2}\bar{K}L - \frac{1}{2}\bar{K}\bar{L} + \frac{3}{8}K^2L - \frac{1}{8}K^2\bar{L} - \frac{1}{2}K\bar{K}L + \\ & + \frac{1}{2}K\bar{K}\bar{L} + \frac{1}{8}\bar{K}^2L - \frac{3}{8}\bar{K}^2\bar{L} + \frac{1}{3}K^3L - \frac{1}{24}K^3\bar{L} - \frac{3}{8}K^2\bar{K}L + \\ & + \frac{3}{8}K\bar{K}^2\bar{L} + \frac{1}{24}\bar{K}^3L - \frac{1}{3}\bar{K}^3\bar{L} + \frac{125}{384}K^4L - \frac{3}{128}K^4\bar{L} - \\ & - \frac{3}{8}K^3\bar{K}L - \frac{1}{24}K^3\bar{K}\bar{L} - \frac{1}{64}K^2\bar{K}^2L + \frac{1}{64}K^2\bar{K}^2\bar{L} + \frac{1}{24}K\bar{K}^3L + \\ & + \frac{3}{8}K\bar{K}^3\bar{L} + \frac{3}{128}\bar{K}^4L - \frac{125}{384}\bar{K}^4\bar{L} + \ldots\,. \end{aligned} \tag{7.3.23}$$

The variables K and L are determined by (2.2.15). The Keplerian mean longitude Λ is different from λ by its initial value

$$\Lambda = nt + \mathcal{E}\,, \tag{7.3.24}$$

where $\mathcal{E}$ is the Keplerian element. The relationship of a, b and ε with K, L and $\mathcal{E}$ has the form

$$a = K + F(K,\, \bar{K},\, L,\, \bar{L})\,, \tag{7.3.25}$$

$$b = -\tfrac{1}{2}\mathrm{i}\,L + G(K,\, \bar{K},\, L,\, \bar{L}) \tag{7.3.26}$$

and

$$\exp\mathrm{i}(\varepsilon - \mathcal{E}) = 1 + H(K,\, \bar{K},\, L,\, \bar{L})\,. \tag{7.3.27}$$

F, G and H are power series in K, $\bar{K}$, L and $\bar{L}$. F and G contain only odd-degree terms starting with the third degree. H contains only even-degree terms starting with the fourth degree. Moreover, with p, q, r and s denoting the powers of K, $\bar{K}$, L and $\bar{L}$, respectively, the combination $p-q+r-s$ is always unity for each term of F and G and zero for each term of H. Substituting (7.3.18)–(7.3.21) and (7.3.25)–(7.3.27) into (7.2.2) and (7.2.3) one obtains the relations

$$\tfrac{1}{2}F - \tfrac{3}{2}\bar{F} + H = \Phi\,, \tag{7.3.28}$$

$$G + \bar{G} = \Psi \tag{7.3.29}$$

with the right-hand sides

$$\Phi = S + (1+H)\delta p + \left(-\tfrac{1}{2}K + \tfrac{3}{2}\bar{K} - \tfrac{1}{2}F + \tfrac{3}{2}\bar{F}\right) H \tag{7.3.30}$$

and

$$\Psi = T - \delta w\,. \tag{7.3.31}$$

If F, G and H are known up to terms of some definite degree then substituting (7.3.25) and (7.3.26) into (7.3.18) and (7.3.19) one may determine the right–hand sides (7.3.30) and (7.3.31) more accurately resulting in new, more accurate expressions for F, G and H. If necessary, it is easy with the aid of a PS processor to find the inverse transformation for (7.3.25) and (7.3.26). Up to the fifth-degree terms relations (7.3.25)–(7.3.27) are as follows:

$$\begin{aligned} a = K &- \frac{3}{4}K^2\bar{K} - \frac{1}{4}KL\bar{L} + \frac{5}{48}K^3\bar{K}^2 + \frac{1}{192}K^3\bar{L}^2 + \frac{3}{16}K^2\bar{K}L\bar{L} + \\ &+ \frac{1}{64}K\bar{K}^2L^2 - \frac{1}{16}KL^2\bar{L}^2 + \ldots\,, \end{aligned} \tag{7.3.32}$$

$$b = \frac{1}{2\mathrm{i}}\left(L - \frac{1}{8}K^2\bar{L} - \frac{1}{2}K\bar{K}L - \frac{1}{24}K^3\bar{K}\bar{L} - \frac{1}{64}K^2\bar{K}^2L + \ldots\right), \tag{7.3.33}$$

and

$$\mathrm{i}(\varepsilon - \mathcal{E}) = \frac{1}{64}K^2\bar{L}^2 - \frac{1}{64}\bar{K}^2L^2 + \ldots\,. \tag{7.3.34}$$

If a and b are replaced by (7.3.3) and K and L are replaced by

$$K = k\exp\mathrm{i}\Lambda\,, \qquad L = l\exp\mathrm{i}\Lambda \tag{7.3.35}$$

with original meaning (2.2.20) then

$$\begin{aligned} \alpha = k &- \frac{3}{4}k^2\bar{k} - \frac{1}{4}kl\bar{l} + \frac{5}{48}k^3\bar{k}^2 - \frac{1}{96}k^3\bar{l}^2 + \frac{3}{16}k^2\bar{k}l\bar{l} + \\ &+ \frac{1}{32}k\bar{k}^2l^2 - \frac{1}{16}kl^2\bar{l}^2 + \ldots\,, \end{aligned} \tag{7.3.36}$$

$$\beta = \frac{1}{2\mathrm{i}}\Big(l - \frac{1}{8}k^2\bar{l} - \frac{1}{2}k\bar{k}l - \frac{1}{24}k^3\bar{k}\bar{l} - \frac{1}{64}k^2\bar{k}^2 l - \frac{1}{64}k^2 l\bar{l}^2 + \\ + \frac{1}{64}\bar{k}^2 l^3 + \ldots\Big) \tag{7.3.37}$$

and

$$\mathrm{i}(\varepsilon - \mathcal{E}) = \frac{1}{64}k^2\bar{l}^2 - \frac{1}{64}\bar{k}^2 l^2 + \ldots . \tag{7.3.38}$$

This section demonstrates one more version of a Keplerian processor based entirely on operations with power series.

8 Separation of Variables in Elements

8.1 The Krylov–Bogoljubov Method

In this and the following chapters we present some techniques for constructing long-term theories of motion for celestial bodies. In the present chapter we consider methods applied to equations in elements whereas further on we shall deal with methods applied to equations in rectangular coordinates. But the differences between these two groups of methods are not of major significance. Their common features are of greater importance. In essence, all contemporary methods for constructing long-term theories of motion are based on a unified fundamental principle of separation of short-period and long-period terms. Following this idea one develops a transformation of the variables, reducing the original system of the equations of motion to a system that does not contain fast-changing variables. This final system, comprising secular and long-period terms (as well as resonance terms if the original system belongs to the class of resonance systems), is usually of polynomial form and may be treated by simpler analytical techniques than the original system (Taylor expansions, Birkhoff normalization, etc.). In addition, it is always possible to integrate the resulting system numerically with a large integration step.

We can begin our discussion of the various techniques for constructing long-term theories in elements with the Krylov–Bogoljubov method in the form developed by Musen (1965a). This version may be easily implemented on computer for a wide range of problems.

Let the initial system be of the form

$$\dot{x} = X(x,\, y,\, z)\,, \tag{8.1.1}$$

$$\dot{y} = \lambda(x) + Y(x,\, y,\, z)\,, \tag{8.1.2}$$

$$\dot{z} = Z(x,\, y,\, z)\,, \tag{8.1.3}$$

each symbol generally denoting some vectorial quantity. The right-hand sides are Poisson series with polynomial variables x and trigonometric variables y and z. In addition, the functions X, Y and Z are proportional to some small parameter. Hence, x is the vector of action variables, y stands for the vector of fast angular variables and z is the vector of slow angular variables. The problem is to find a transformation

$$x = x^* + a(x^*, y^*, z^*), \tag{8.1.4}$$

$$y = y^* + b(x^*, y^*, z^*), \tag{8.1.5}$$

$$z = z^* + c(x^*, y^*, z^*) \tag{8.1.6}$$

such that the equations in new variables x^*, y^* and z^* do not involve fast angular variables

$$\dot{x}^* = X^*(x^*, -, z^*), \tag{8.1.7}$$
$$\dot{y}^* = \lambda(x^*) + Y^*(x^*, -, z^*), \tag{8.1.8}$$
$$\dot{z}^* = Z^*(x^*, -, z^*). \tag{8.1.9}$$

One thus assumes that there are no resonance relations between the components of the vector $\lambda(x)$ of the original system. Otherwise the right-hand sides of the resulting system (8.1.7)–(8.1.9) would contain the resonance combinations of the variables y^*.

In virtue of equations (8.1.7)–(8.1.9) the operator of differentiation with respect to time is equal to

$$\frac{d}{dt} = \lambda(x^*)\frac{\partial}{\partial y^*} + D \tag{8.1.10}$$

with

$$D = X^*\frac{\partial}{\partial x^*} + Y^*\frac{\partial}{\partial y^*} + Z^*\frac{\partial}{\partial z^*}. \tag{8.1.11}$$

By differentiating (8.1.4)–(8.1.6) and using equations (8.1.1)–(8.1.3) and (8.1.7)–(8.1.9) one gets

$$\lambda(x^*)\frac{\partial a}{\partial y^*} = X(x^* + a, y^* + b, z^* + c) - X^* - Da, \tag{8.1.12}$$

$$\lambda(x^*)\frac{\partial b}{\partial y^*} = Y(x^* + a, y^* + b, z^* + c) - Y^* + \lambda(x^* + a) - \lambda(x^*) - Db, \tag{8.1.13}$$

$$\lambda(x^*)\frac{\partial c}{\partial y^*} = Z(x^* + a, y^* + b, z^* + c) - Z^* - Dc. \tag{8.1.14}$$

These equations serve to determine both the right-hand sides X^*, Y^*, Z^* of the final system and the functions a, b and c of the basic transformation. If a, b and c are known to some approximation then X^*, Y^* and Z^* are determined by

$$X^* = MX(x^* + a, y^* + b, z^* + c), \tag{8.1.15}$$

$$Y^* = \lambda(x^* + a) - \lambda(x^*) + MY(x^* + a, y^* + b, z^* + c), \tag{8.1.16}$$

$$Z^* = MZ(x^* + a, y^* + b, z^* + c), \tag{8.1.17}$$

where M stands for the averaging operator with respect to the fast variables y^*. This operator separates in the trigonometric expansions in y^* and z^* the terms not containing y^*. With values (8.1.15)–(8.1.17) the right-hand sides (8.1.12)–(8.1.14) are expanded into trigonometric series in y^* and z^* with, necessarily, the appearance of y^* in each term. Therefore, equations (8.1.12)–(8.1.14) are reduced to the set of scalar equations, which are independent of one another,

$$\lambda(x^*)\frac{\partial \psi}{\partial y^*} = \Psi(x^*, y^*, z^*) \tag{8.1.18}$$

with the unknown scalar function ψ. The right-hand side of this equation is expanded into the series

$$\Psi = \sum_{m \neq 0} \left[C_{mn}(x^*)\cos(my^* + nz^*) + S_{mn}(x^*)\sin(my^* + nz^*)\right], \tag{8.1.19}$$

where summation is performed with the aid of the multi-indices m and n. Integrating (8.1.19) one gets

$$\psi = \sum_{m \neq 0} \frac{1}{(m\lambda)}\left[C_{mn}(x^*)\sin(my^* + nz^*) - S_{mn}(x^*)\cos(my^* + nz^*)\right] \tag{8.1.20}$$

with $(m\lambda)$ denoting the scalar product of the multi-index m and vector $\lambda(x^*)$. In such a form one finds the functions a, b and c. The values of a, b and c obtained are used to obtain the right-hand sides of (8.1.12)–(8.1.14) more accurately. Iterations are repeated so long as the necessary accuracy is achieved. Initial values for this iteration process are given by $a = b = c = 0$. The basic operation of the process reduces to the expansion of X, Y and Z resulting from substituting (8.1.4)–(8.1.6). This expansion can be produced by the subroutine TAYLOR of a PS processor.

In the original version (Musen, 1965a) this method was related to developing a solution in the form of series in powers of a small parameter. One may determine these series term by term by the recurrence formulae. The substitution of (8.1.4)–(8.1.6) is performed with the aid of the operator representation

$$F(x^* + a, y^* + b, z^* + c) = (1 + T)F(x^*, y^*, z^*). \tag{8.1.21}$$

Operator T defined by the relation

$$1 + T(x^*, y^*, z^*) = \exp\left(a\frac{\partial}{\partial x^*} + b\frac{\partial}{\partial y^*} + c\frac{\partial}{\partial z^*}\right) \tag{8.1.22}$$

may also be expanded in powers of a small parameter. If one has a processor to manipulate series like (8.1.20), all these details of the analytical operations are unnecessary and it is more convenient to implement the iteration process in the general form given above.

Let us consider now the form of series (8.1.20) in more detail. If the components of the vector λ are constants then these series are simply Poisson series. They also retain the form of Poisson series in the case where there is only one fast variable and the scalar quantity λ is the power function of only one of the action variables. In the general case series (8.1.20) are echeloned Poisson series, each coefficient being the ratio of two polynomials with respect to the polynomial variables. If a processor to deal with such series is available then the problem considered in this section may be implemented without any difficulties. It is sufficient to describe the right-hand sides of (8.1.1)–(8.1.3) as the echeloned series and to indicate the desired accuracy of obtaining transformation (8.1.4)–(8.1.6) and resulting system (8.1.7)–(8.1.9).

8.2 The von Zeipel Method and Its Modifications

One would not be exaggerating to say that the von Zeipel method is at present one of the most widespread analytical techniques of celestial mechanics. Like the Krylov–Bogoljubov method, it is based on the principle of separating fast- and slow-changing variables but it is designated for the canonical systems of differential equations. The essence of this method is to find a canonical transformation of variables excluding fast variables from the right-hand sides of new equations. The Von Zeipel method reduces to the determination of only two scalar functions, the generating function of the canonical transformation and the new Hamiltonian. In this respect, the von Zeipel method is more compact compared to the Krylov–Bogoljubov method.

The von Zeipel method in its classical version has a disadvantage caused by the dependence of the generating function of the canonical transformation both on new and old variables (usually new action variables and old angular variables). Modern versions of this method originated in the papers by Hori (1966) and Deprit (1969) and are free from this defect. In these versions the canonical transformation is constructed explicitly in terms only of new (or old) variables. Nowadays there are a lot of different modifications of the Hori–Deprit technique, partly presented in the works by Giacaglia (1972) and Stumpff (1974) and in great detail by Schneider (1993). Some versions have been realized in the form of specialized software (for example, by Palacián, 1992). But if one can use an advanced PS processor then the problem of finding the explicit form of the basic transformation involves no difficulties. Therefore, one should not regard the von Zeipel method as being out of date. The principle of the separation of variables may be presented very concisely in just its classical form, as given, for example, by Musen (1965a). But as with the Krylov–Bogoljubov method an iteration version is preferred below over the power-series expansions with respect to a small parameter.

Let us consider a canonical system

$$\begin{aligned} \dot{x} &= \frac{\partial F}{\partial y}, & \dot{y} &= -\frac{\partial F}{\partial x}, \\ \dot{u} &= \frac{\partial F}{\partial z}, & \dot{z} &= -\frac{\partial F}{\partial u} \end{aligned} \tag{8.2.1}$$

with the Hamiltonian

$$F(x,\, u,\, y,\, z) = F_0(x) + \tilde{F}(x,\, u,\, y,\, z)\,. \tag{8.2.2}$$

This Hamiltonian represents a Poisson series with power variables x and u and trigonometric variables y and z. The function $\tilde{F}$ is assumed to be small compared with the main term $F_0(x)$. The vectors x and y have one and the same dimension. The vectors u and z also have one and the same dimension. Evidently, the vector y corresponds to fast variables. The vector z describes slow variables. The problem is to find the generating function

$$S = S(x^*,\, u^*,\, y,\, z) \tag{8.2.3}$$

of the canonical transformation

$$\begin{aligned} x &= x^* + \frac{\partial S}{\partial y}, & y^* &= y + \frac{\partial S}{\partial x^*}, \\ u &= u^* + \frac{\partial S}{\partial z}, & z^* &= z + \frac{\partial S}{\partial u^*} \end{aligned} \tag{8.2.4}$$

so that the Hamiltonian F^* of the new equations

$$\begin{aligned} \dot{x}^* &= \frac{\partial F^*}{\partial y^*}, & \dot{y}^* &= -\frac{\partial F^*}{\partial x^*}, \\ \dot{u}^* &= \frac{\partial F^*}{\partial z^*}, & \dot{z}^* &= -\frac{\partial F^*}{\partial u^*} \end{aligned} \tag{8.2.5}$$

does not contain fast variables y^*.

Owing to the canonicity of transformation (8.2.4) one has

$$F(x,\, u,\, y,\, z) = F^*(x^*,\, u^*,\, -,\, z^*)\,, \tag{8.2.6}$$

which leads to the identity with respect to x^*, u^*, y and z

$$F\left(x^* + \frac{\partial S}{\partial y},\, u^* + \frac{\partial S}{\partial z},\, y,\, z\right) = F^*\left(x^*,\, u^*,\, -,\, z + \frac{\partial S}{\partial u^*}\right). \tag{8.2.7}$$

This identity enables one to determine S and F^*. Writing similarly to (8.2.2)

$$F^*(x^*,\, u^*,\, -,\, z^*) = F_0^*(x^*) + \tilde{F}^*(x^*,\, u^*,\, -,\, z^*) \tag{8.2.8}$$

one may rewrite (8.2.7) in the form

$$\begin{aligned} F_0\left(x^* + \frac{\partial S}{\partial y}\right) - F_0^*(x^*) + \tilde{F}\left(x^* + \frac{\partial S}{\partial y},\, u^* + \frac{\partial S}{\partial z},\, y,\, z\right) = \\ = \tilde{F}^*\left(x^*,\, u^*,\, -,\, z + \frac{\partial S}{\partial u^*}\right). \end{aligned} \tag{8.2.9}$$

S and $\tilde{F}^*$ being as small as $\tilde{F}$, one finds, at once, that

$$F_0^*(x^*) = F_0(x^*) \,. \tag{8.2.10}$$

Functions F_0 and $\tilde{F}^*$ are expanded in variations of their arguments as follows:

$$F_0\left(x^* + \frac{\partial S}{\partial y}\right) = F_0(x^*) - \lambda(x^*)\frac{\partial S}{\partial y} + T_2 F_0\left(x^* + \frac{\partial S}{\partial y}\right) \tag{8.2.11}$$

and

$$\tilde{F}^*\left(x^*,\, u^*,\, -,\, z + \frac{\partial S}{\partial u^*}\right) = \tilde{F}^*(x^*,\, u^*,\, -,\, z) + \\ + T_1\tilde{F}^*\left(x^*,\, u^*,\, -,\, z + \frac{\partial S}{\partial u^*}\right) \tag{8.2.12}$$

with T_k denoting the Taylor expansion starting with the k-degree term. Besides, the vector $\lambda(x^*)$ means

$$\lambda(x^*) = -\frac{\partial F_0(x^*)}{\partial x^*} \,. \tag{8.2.13}$$

Substituting (8.2.11) and (8.2.12) into (8.2.9) one obtains the basic equation

$$-\lambda(x^*)\frac{\partial S}{\partial y} + \Phi(x^*,\, u^*,\, y,\, z) = \tilde{F}^*(x^*,\, u^*,\, -,\, z) \tag{8.2.14}$$

with

$$\Phi(x^*,\, u^*,\, y,\, z) = T_2 F_0\left(x^* + \frac{\partial S}{\partial y}\right) + \tilde{F}\left(x^* + \frac{\partial S}{\partial y},\, u^* + \frac{\partial S}{\partial z},\, y,\, z\right) - \\ - T_1\tilde{F}^*\left(x^*,\, u^*,\, -,\, z + \frac{\partial S}{\partial u^*}\right) . \tag{8.2.15}$$

Equation (8.2.14) serves to determine the functions S and $\tilde{F}^*$ by iteration. At each step of the iteration the function Φ admits the expansion

$$\Phi = \sum\Big[C_{mn}(x^*,\, u^*)\cos(my + nz) + S_{mn}(x^*,\, u^*)\sin(my + nz)\Big] , \tag{8.2.16}$$

the summation being performed over integer values of the multi-indices m and n. The function $\tilde{F}^*$ results from (8.2.14) by applying to Φ an averaging operator M with respect to the fast variables so that

$$\tilde{F}^*(x^*,\, u^*,\, -,\, z) = \sum\Big[C_{0n}(x^*,\, u^*)\cos(nz) + S_{0n}(x^*,\, u^*)\sin(nz)\Big] . \tag{8.2.17}$$

Then S is determined by integrating the remaining part of Φ not containing the fast variables

$$S(x^*, u^*, y, z) = \sum_{m \neq 0} \frac{1}{(m\lambda)} \Big[C_{mn}(x^*, u^*) \sin(my + nz) - $$
$$ - S_{mn}(x^*, u^*) \cos(my + nz) \Big] . \qquad (8.2.18)$$

The iterations start with the value $S = 0$ and are repeated until the desired accuracy is attained.

In the original paper (Musen, 1965a) S and F^* are determined by series in powers of a small parameter. With this aim basic identity (8.2.7) is written in the form

$$(1 + T)F(x^*, u^*, y, z) = (1 + T^*)F^*(x^*, u^*, -, z) \qquad (8.2.19)$$

with the operators

$$1 + T(x^*, u^*, y, z) = \exp\left(\frac{\partial S}{\partial y}\frac{\partial}{\partial x^*} + \frac{\partial S}{\partial z}\frac{\partial}{\partial u^*}\right) \qquad (8.2.20)$$

and

$$1 + T^*(x^*, u^*, y, z) = \exp\left(\frac{\partial S}{\partial u^*}\frac{\partial}{\partial z}\right) . \qquad (8.2.21)$$

These operators are also expanded in powers of a small parameter.

In any case, having determined S and F^*, one is interested in the explicit transformation from the old variables to the new ones or vice versa. This may be easily performed with the aid of an adequate processor. Combining the action variables x and u and the angular variables y and z into vectors p and q, respectively, and taking into account that the generating function S depends on the new action variables p^* and the old angular variables q one may describe transformation (8.2.4) as

$$p = p^* + P(p^*, q) \qquad (8.2.22)$$

and

$$q^* = q + Q(p^*, q) . \qquad (8.2.23)$$

The inversion of these equations performed by a processor results in

$$p^* = p + U(p, q) \qquad (8.2.24)$$

and

$$q = q^* + V(p^*, q^*) . \qquad (8.2.25)$$

Substituting (8.2.25) into P one gets from (8.2.22) and (8.2.25) explicit expressions for p and q in terms of p^* and q^*. In just the same manner, substituting (8.2.24) into Q one gets from (8.2.23) and (8.2.24) explicit expressions for p^* and q^* in terms of p and q.

If the original Hamiltonian F is a Poisson series then in the general case expressions (8.2.17) and (8.2.18) represent echeloned series, the coefficients in trigonometric terms being rational functions of power variables.

As with the Krylov–Bogoljubov method, the von Zeipel method may be implemented in a general form. The particular case of resonance relations between the components of the frequency vector λ is not considered here.

System (8.2.5), in addition to the energy integral

$$F^*(x^*, u^*, -, z^*) = \text{const}, \tag{8.2.26}$$

has the integral

$$x^* = \text{const}. \tag{8.2.27}$$

Therefore, this system is reduced to the system for u^* and z^* with the Hamiltonian F^* where the components of x^* occur as parameters. If the secular part of F^* (not containing z^*) dominates over its periodic part then one may repeat iterations and obtain the solution of the problem in a purely trigonometric form.

8.3 The von Zeipel Method as a "Coordinate" Method

In classic works on celestial mechanics a distinction is made between methods to determine perturbations in elements or in coordinates. In fact, this distinction is not crucial, being related only to the manner in which the short-period perturbations are taken into account. The secular and long-period perturbations always imply utilization of mean elements or analogous quantities. The conditional character of this distinction has been underlined by Izsak (1963), who showed the possibility of considering the von Zeipel method as a method for determining perturbations in the coordinates rather than a perturbation-of-elements method. The analysis of Izsak is based on the canonical system deduced by Hill (1913). The original equations of disturbed motion

$$\begin{aligned}
\ddot{x} + \frac{\mu x}{r^3} &= \frac{\partial W}{\partial x}, \\
\ddot{y} + \frac{\mu y}{r^3} &= \frac{\partial W}{\partial y}, \\
\ddot{z} + \frac{\mu z}{r^3} &= \frac{\partial W}{\partial z}
\end{aligned} \tag{8.3.1}$$

(with designation $\mu = Gm$) are rewritten in the canonical form

$$\begin{aligned}
\dot{X} &= \frac{\partial F}{\partial x}, & \dot{x} &= -\frac{\partial F}{\partial X}, \\
\dot{Y} &= \frac{\partial F}{\partial y}, & \dot{y} &= -\frac{\partial F}{\partial Y}, \\
\dot{Z} &= \frac{\partial F}{\partial z}, & \dot{z} &= -\frac{\partial F}{\partial Z},
\end{aligned} \tag{8.3.2}$$

with the Hamiltonian

$$F = -\frac{1}{2}\left(X^2 + Y^2 + Z^2\right) + \frac{\mu}{r} + W(t,\, x,\, y,\, z)\,. \tag{8.3.3}$$

The canonical transformation by Hill

$$\begin{pmatrix} X & Y & Z \\ x & y & z \end{pmatrix} \rightarrow \begin{pmatrix} R & \Phi & H \\ r & \phi & h \end{pmatrix} \tag{8.3.4}$$

reduces (8.3.2) to the system

$$\begin{aligned} \dot{R} &= \frac{\partial F}{\partial r}\,, & \dot{r} &= -\frac{\partial F}{\partial R}\,, \\ \dot{\Phi} &= \frac{\partial F}{\partial \phi}\,, & \dot{\phi} &= -\frac{\partial F}{\partial \Phi}\,, \\ \dot{H} &= \frac{\partial F}{\partial h}\,, & \dot{h} &= -\frac{\partial F}{\partial H} \end{aligned} \tag{8.3.5}$$

with the Hamiltonian

$$F = -\frac{1}{2}\left(R^2 + \frac{\Phi^2}{r^2}\right) + \frac{\mu}{r} + W\,(t,\, r,\, \phi,\, h,\, \Phi,\, H)\,. \tag{8.3.6}$$

In explicit form transformation (8.3.4) is as follows:

$$\begin{aligned} x &= r\left(\cos\phi\cos h - \frac{H}{\Phi}\sin\phi\sin h\right), \\ y &= r\left(\cos\phi\sin h + \frac{H}{\Phi}\sin\phi\cos h\right), \\ z &= r\frac{\sqrt{\Phi^2 - H^2}}{\Phi}\sin\phi\,, \\ xY &- yX = H\,, \\ X^2 &+ Y^2 + Z^2 = R^2 + \frac{\Phi^2}{r^2}\,, \\ xX &+ yY + zZ = rR\,. \end{aligned} \tag{8.3.7}$$

Φ, H and h are slowly changing variables. In terms of the Keplerian elements they are

$$\Phi = \sqrt{\mu a(1 - e^2)}\,, \qquad H = \Phi\cos i\,, \qquad h = \Omega\,. \tag{8.3.8}$$

$R = \dot{r}$ and the radius-vector r and argument of the latitude $\phi\,(= v + \omega)$ represent the fast-changing variables.

The correctness of equations (8.3.5) may be most easily established by comparison with the well-known equations in Delaunay variables

$$\begin{aligned} \dot{L} &= \frac{\partial F}{\partial l}, & \dot{l} &= -\frac{\partial F}{\partial L}, \\ \dot{G} &= \frac{\partial F}{\partial g}, & \dot{g} &= -\frac{\partial F}{\partial G}, \\ \dot{H} &= \frac{\partial F}{\partial h}, & \dot{h} &= -\frac{\partial F}{\partial H} \end{aligned} \tag{8.3.9}$$

with the Hamiltonian

$$F = \frac{\mu^2}{2L^2} + W(t,\, l,\, g,\, h,\, L,\, G,\, H)\,. \tag{8.3.10}$$

l, $g\,(=\omega)$ and h are the mean anomaly, argument of the pericentre and longitude of the ascending node, respectively. The action variables are

$$L = \sqrt{\mu a}\,, \qquad G = L\sqrt{1-e^2}\,, \qquad H = G\cos i\,. \tag{8.3.11}$$

The transformation

$$\begin{pmatrix} L & G & H \\ l & g & h \end{pmatrix} \to \begin{pmatrix} R & \Phi & H \\ r & \phi & h \end{pmatrix} \tag{8.3.12}$$

is given by the formulae

$$\begin{aligned} \frac{\mu^2}{L^2} &= \frac{2\mu}{r} - R^2 - \frac{U^2}{r^2}\,, \\ G &= \Phi\,, \\ e\cos(l+f) &= \frac{\Phi^2}{\mu r} - 1\,, \\ e\sin(l+f) &= \frac{R\Phi}{\mu}\,, \\ l+f+g &= \phi \end{aligned} \tag{8.3.13}$$

with $f = f(r,\, R,\, \Phi)$ being the equation of the centre, i.e. the difference $v - l$ between the true and mean anomalies. By differentiating the trigonometric relations in (8.3.13) one finds that

$$dl + df = \frac{R\Phi^3}{\mu^2 e^2 r^2}\,dr - \frac{\Phi}{\mu e^2}\left(1 - \frac{\Phi^2}{\mu r}\right) dR - \frac{R}{\mu e^2}\left(1 + \frac{\Phi^2}{\mu r}\right) d\Phi \tag{8.3.14}$$

with

$$e^2 = 1 - \frac{\Phi^2}{L^2}\,.$$

On the other hand, f is determined by (2.1.28), or in terms of r, R and Φ

$$f = 2\arctan\frac{rLR}{\Phi L + \mu r} + \frac{rR}{L}\,, \tag{8.3.15}$$

where L is to be understood as a function of r, R and Φ determined by the first relation in (8.3.13). From (8.3.15) it follows that

$$df = \frac{R}{L}\left[1 + \frac{L^2\Phi^3}{\mu^2 r^2 (L+U)}\right] dr + \left[\frac{2r}{L} - \frac{L\Phi}{\mu(L+\Phi)}\left(1 - \frac{\Phi^2}{\mu r}\right)\right] dR - $$
$$- \frac{LR}{\mu(L+\Phi)}\left(1 + \frac{\Phi^2}{\mu r}\right) d\Phi\,. \tag{8.3.16}$$

With the aid of (8.3.14) and (8.3.16) it is easy to verify the identity with respect to R, Φ, r and ϕ

$$L\,dl + G\,dg - R\,dr - \Phi\,d\phi = (L-\Phi)(dl + df) - L\,df - R\,dr = -2d(rR)\,. \tag{8.3.17}$$

Since the right-hand side is the total differential, transformation (8.3.12), given explicitly by (8.3.13) and (8.3.15), is canonical, leading from (8.3.9) to (8.3.5).

Equations (8.3.5) are well adapted for the application of the von Zeipel method, especially when it is desirable to obtain the closed-form solution with no expansion in powers of the eccentricity. Eliminating the fast angular variable ϕ one finds in a straightforward manner the short-period perturbations in coordinate form. This technique has been demonstrated by Izsak in application to the main problem of the motion of a satellite of the Earth.

The variable R is of the order of the eccentricity. In the case of a small inclination it is appropriate to use this circumstance by means of one more canonical transformation

$$\begin{pmatrix} R & \Phi & H \\ r & \phi & h \end{pmatrix} \to \begin{pmatrix} R & \Phi & K \\ r & \psi & h \end{pmatrix}, \tag{8.3.18}$$

determined explicitly by

$$K = H - \Phi\,, \qquad \psi = \phi + h\,. \tag{8.3.19}$$

K is now of the order of the square of the inclination and ψ is the true longitude in the orbit.

The Hill variables were successfully used by Aksnes (1972) in artificial-satellite theory. Broucke (1978) has elaborated an elegant technique for deriving the equations in these and similar variables using the Pfaff formalism.

8.4 The Kolmogorov–Arnold Method Using Howland's Technique

From the theoretical point of view the importance of the Kolmogorov–Arnold method consists in constructing a converging iteration process to represent the general quasiperiodic solution of a dynamical system with a small parameter. Based on Newton-type iterations the process possesses quadratic convergence. But this convergence is ensured under two conditions:

1. at each step of the iterations one has to rarefy the domain of the admissible initial values in order to eliminate too small divisors occurring in the process of integration;
2. a perturbing parameter of the dynamical system should be sufficiently small.

In solving specific problems when the small parameters are fixed and the initial conditions are given only within some limits of accuracy the convergence of Kolmogorov–Arnold iterations may be problematic. But in practical applications this method may be regarded as a constructive technique enabling one to obtain a formal solution by a smaller number of iterations compared, for example, with the von Zeipel method based on linear iterations. Such an approach to the Kolmogorov–Arnold method as applied to problems of celestial mechanics was developed in detail by Howland (1977, 1979, 1986). The Kolmogorov–Arnold method was implemented by Howland in the framework of the formalism of Hori and Deprit avoiding the inversion of the series. However, as stated above, this procedure encounters no difficulties when modern PS processors are used. The original version of the Kolmogorov–Arnold method leads to a fairly simple algorithm.

The Kolmogorov method is applied to solve a non-degenerate dynamical system (Howland, 1977)

$$\dot{x} = \frac{\partial H}{\partial y}, \qquad \dot{y} = -\frac{\partial H}{\partial x} \tag{8.4.1}$$

with the Hamiltonian $H(x, y)$ holomorphic in x and periodic in trigonometric variables y:

$$\begin{aligned} H(x, y) = H_0 - \sum_i \lambda_i x_i + A(y) + \sum_i B_i(y)x + \\ + \sum_i \sum_j C_{ij}(y) x_i x_j + D(x, y) \, . \end{aligned} \tag{8.4.2}$$

Here H_0 is a constant and λ_i are constant incommensurable frequencies. The functions $A(y)$, $B_i(y)$, $C_{ij}(y) = C_{ji}(y)$ and $D(x, y)$ are expanded into trigonometric series in the components of the vector y. The function $D(x, y)$ is at least of the third degree with respect to the components of the vector x. The functions $A(y)$ and $B_i(y)$ are assumed to be proportional to some small parameter ε. In addition, the function $A(y)$ does not contain a constant term (otherwise such a term might be combined with H_0).

The von Zeipel method involves the sequence of the canonical transformations

$$(x,\, y) \to (X,\, Y) \tag{8.4.3}$$

leading in the limit to a Hamiltonian that does not depend on the angular variables Y. By the Kolmogorov method one finds a sequence (8.4.3) that eventually excludes the angular variables Y from the terms of zero and first degrees in X. Hence, the final aim of the canonical transformations of the Kolmogorov method is the Hamiltonian of type (8.4.2) with $A = B_i = 0$. The corresponding system admits the solution

$$X_i = 0\,, \qquad Y_i = \lambda_i t + Y_i^{(0)} \tag{8.4.4}$$

with the frequencies λ_i conserving their initial constant values.

Transformation (8.4.3) is constructed with the generating function $S = S(y,\, X)$:

$$x_i = X_i + \frac{\partial S}{\partial y_i}\,, \qquad Y_i = y_i + \frac{\partial S}{\partial X_i} \tag{8.4.5}$$

with the explicit expression

$$S(y,\, X) = \sum_i \zeta_i y_i + U(y) + \sum_i V_i(y) X_i\,. \tag{8.4.6}$$

$U(y)$ and $V_i(y)$ are unknown periodic functions of y with no constant terms. The function S also contains unknown constants ζ_i. From (8.4.5) and (8.4.6) we have

$$x_i = X_i + \zeta_i + \frac{\partial U}{\partial y_i} + \sum_j X_j \frac{\partial V_j}{\partial y_i}\,, \tag{8.4.7}$$

$$Y_i = y_i + V_i\,. \tag{8.4.8}$$

Substituting these relations into (8.4.2) one has

$$\begin{aligned} H = H_0^* - \sum_i \lambda_i X_i + A^{(1)}(y) + \sum_i B_i^{(1)}(y) X_i + A^*(Y) + \\ + \sum_i B_i^*(Y) X_i + \sum_i \sum_j C_{ij}^*(Y) X_i X_j + D^*(X,\, Y) \end{aligned} \tag{8.4.9}$$

with

$$H_0^* = H_0 - \sum_i \lambda_i \zeta_i\,, \tag{8.4.10}$$

$$A^{(1)}(y) = A - \sum_i \lambda_i \frac{\partial U}{\partial y_i}, \tag{8.4.11}$$

$$B_i^{(1)}(y) = B_i - \sum_j \lambda_j \frac{\partial V_i}{\partial y_j} + 2\sum_j C_{ij}\left(\zeta_j + \frac{\partial U}{\partial y_j}\right), \tag{8.4.12}$$

$$\begin{aligned} A^*(Y) = {} & \sum_i B_i\left(\zeta_i + \frac{\partial U}{\partial y_i}\right) + \\ & + \sum_i \sum_j C_{ij}\left(\zeta_i + \frac{\partial U}{\partial y_i}\right)\left(\zeta_j + \frac{\partial U}{\partial y_j}\right) + D, \end{aligned} \tag{8.4.13}$$

$$\begin{aligned} B_i^*(Y) = {} & \sum_j B_j \frac{\partial V_i}{\partial y_j} + 2\sum_k \sum_j C_{kj}\left(\zeta_k + \frac{\partial U}{\partial y_k}\right)\frac{\partial V_i}{\partial y_j} + \\ & + \sum_j \frac{\partial D}{\partial x_j}\left(\delta_{ij} + \frac{\partial V_i}{\partial y_j}\right) \end{aligned} \tag{8.4.14}$$

and

$$\begin{aligned} C_{ij}^*(Y) = {} & C_{ij} + 2\sum_k C_{kj}\frac{\partial V_i}{\partial y_k} + \sum_k \sum_l C_{kl}\frac{\partial V_i}{\partial y_k}\frac{\partial V_j}{\partial y_l} + \\ & + \frac{1}{2}\sum_k \sum_l \frac{\partial^2 D}{\partial x_k \partial x_l}\left(\delta_{ik} + \frac{\partial V_i}{\partial y_k}\right)\left(\delta_{jl} + \frac{\partial V_j}{\partial y_l}\right). \end{aligned} \tag{8.4.15}$$

Arguments x_i in the function D occurring in (8.4.13)–(8.4.15) should be taken as

$$x_i = \zeta_i + \frac{\partial U}{\partial y_i}. \tag{8.4.16}$$

The function $D^*(X, Y)$, which is holomorphic in X and starts with the third-degree terms in X, is obtained by expanding D for values (8.4.16) and omitting the terms of zero, first and second degrees in X.

The unknown functions $U(y)$, ζ_i and $V_i(y)$ are now determined by the equations

$$\sum_i \lambda_i \frac{\partial U}{\partial y_i} = A, \tag{8.4.17}$$

$$\sum_j \langle C_{ij}\rangle \zeta_j = -\frac{1}{2}\langle B_i\rangle - \sum_j \left\langle C_{ij}\frac{\partial U}{\partial y_j}\right\rangle \tag{8.4.18}$$

and

$$\sum_j \lambda_j \frac{\partial V_i}{\partial y_j} = B_i + 2\sum_j C_{ij}\left(\zeta_j + \frac{\partial U}{\partial y_j}\right). \tag{8.4.19}$$

Angle brackets here mean the constant part of the corresponding function.

In virtue of the assumption $\langle A \rangle = 0$ and the incommensurability of the frequencies λ_i equation (8.4.17) belongs to the type considered above and may always be solved in the purely trigonometric form. Relations (8.4.18) form a system of linear algebraic equations that are always solvable in the non-degenerate case. As consequence of (8.4.18) the constant terms on the right-hand sides of differential equations (8.4.19) vanish so that these equations also admit the purely trigonometric solution.

Hence, the quantities $U(y)$, ζ_i and $V_i(y)$ may also be determined since they have the same order of smallness as the original functions $A(y)$ and $B_i(y)$. Therefore, the terms with $A^{(1)}$ and $B_i^{(1)}$ in Hamiltonian (8.4.9) vanish and the zero- and first-degree terms in X are determined only by the functions A^* and B_i^*. From expressions (8.4.13) and (8.4.14) it is seen that these functions are at least of the second order of smallness compared with the functions A and B_i.

By repeating transformation (8.4.7), (8.4.8) with (8.4.17)–(8.4.19) at each step of the iteration, one constructs a process with quadratic convergence. Having attained the necessary accuracy one may neglect the functions A^* and B_i^* so that the resulting system has solution (8.4.4). The main technical difficulties of this process are related to the necessity of inverting the transformation (8.4.8) and of performing substitution (8.4.7) in the function D. But all these problems are now treated by means of universal and specialized CAS.

The Arnold method is applied to solve degenerate systems (8.4.1) with the Hamiltonian $H(x, y)$ of the type (Howland, 1977)

$$H(x, y) = H_0(x) + \varepsilon H_1(x) + \varepsilon^m H_m(x, y) \,. \tag{8.4.20}$$

$H_0(x)$ does not contain all components of the action variable vector x. Some of these components occur only in the term $H_1(x)$ accompanied by the small parameter ε. Therefore, the trigonometric variables y are separated into fast and slowly changing variables. In contrast with the standard von Zeipel technique, which involves the separate elimination of these variables, the Arnold method enables one to exclude them one at a time. This is possible if $m \geq 2$ in (8.4.20). The Arnold method is of Newton-type. Through eliminating the slowly changing variables the parameter ε occurs in denominators and so the order of accuracy at each step of the iteration is given by ε^{2m-1}. The transformation of type (8.4.5) results in the new Hamiltonian

$$\begin{aligned} H = {} & H_0(X) + \varepsilon H_1(X) + \sum_i \left(\frac{\partial H_0}{\partial x_i} + \varepsilon \frac{\partial H_1}{\partial x_i} \right) \frac{\partial S}{\partial y_i} + \\ & + \varepsilon^m H_m(X, y) + \frac{1}{2} \sum_i \sum_j \left(\frac{\partial^2 H_0}{\partial x_i \partial x_j} + \varepsilon \frac{\partial^2 H_1}{\partial x_i \partial x_j} \right) \frac{\partial S}{\partial y_i} \frac{\partial S}{\partial y_j} + \\ & + \varepsilon^m \sum_i \frac{\partial H_m}{\partial x_i} \frac{\partial S}{\partial y_i} + \ldots \,. \end{aligned} \tag{8.4.21}$$

All the derivatives of H_0, H_1 and H_m should be taken here for the values $x = X$. The generating function $S = S(y, X)$ is determined as a solution of the equation

$$\sum_i \left(\frac{\partial H_0}{\partial x_i} + \varepsilon \frac{\partial H_1}{\partial x_i} \right) \frac{\partial S}{\partial y_i} = -\varepsilon^m H_m(X, y) \,. \tag{8.4.22}$$

The constant term in the trigonometric expansion for H_m may be assumed to be zero (otherwise it may be combined with H_1). Therefore, this equation admits in the non-resonance case a purely trigonometric solution and the corresponding function S is of the order ε^{m-1}. Hamiltonian (8.4.21) retains original form (8.4.20) but Y-dependent terms are now of order ε^{2m-1}. This is evident for the terms with H_1 and H_m. As far as the term with the second derivatives of H_0 is concerned, one has from (8.4.22)

$$\sum_i \frac{\partial H_0}{\partial x_i} \frac{\partial S}{\partial y_i} \sim \varepsilon^m \,.$$

Therefore, the term in (8.4.21) containing the second derivatives of H_0 is also of order ε^{2m-1}. Again, the main technical difficulties are related to the inversion of (8.4.5) and the substitution of this transformation in the Hamiltonian H. These procedures are typical for PS processors.

9 Separation of Variables in Rectangular Coordinates

9.1 Reducibility of the Equations of Variations

In this chapter we reproduce with slight modifications a technique of Brumberg (1978) to compute the perturbations in rectangular coordinates. This method is similar to the von Zeipel technique of separation of short- and long-period perturbations in elements. The general solution of the homogeneous equations of variations for the two-body problem is presented here only on the basis of a transformation leading to a differential system with constant coefficients in Jordan form. Therefore, the equations of perturbed motion have linear left-hand sides in Jordan form whereas their right-hand sides are holomorphic functions with respect to the unknown variables. A further transformation excludes all short-period terms and leads to a polynomial system with slowly changing coefficients. This system determines the long-period terms. The elements of the intermediate orbit of the two-body problem may be taken either in analytical or numerical form. Moreover, the mean motion always preserves its value in the process of solution. This method extends the technique of the general planetary theory (Brumberg, 1970; Brumberg and Chapront, 1973; see also Chapter 10) to non-planar and non-circular intermediary and may be used for constructing theories of motion for the major planets, asteroids and satellites. For the efficient realization of the method on a computer it is helpful to use the facilities of the PS and Keplerian processors.

Consider the equations of variations

$$\delta\ddot{\mathbf{r}} + \frac{Gm}{\rho^3}\left[\delta\mathbf{r} - \frac{3}{\rho^2}(\boldsymbol{\rho}\delta\mathbf{r})\boldsymbol{\rho}\right] = 0 \tag{9.1.1}$$

resulting from the equations of the unperturbed two-body problem

$$\ddot{\boldsymbol{\rho}} + \frac{Gm}{\rho^3}\boldsymbol{\rho} = 0\,, \tag{9.1.2}$$

where m is again the sum of the masses of two bodies, G is the gravitational constant, $\boldsymbol{\rho}$ is the unperturbed position vector of the moving body and $\delta\mathbf{r}$ is the variation of this vector. To solve (9.1.1) one can apply the usual decomposition of $\delta\mathbf{r}$ along vectors $\boldsymbol{\rho}$, $\dot{\boldsymbol{\rho}}$ and $\mathbf{k}$, the latter vector of unit length being directed along the normal to the plane of the unperturbed orbit (see Section 2.1)

$$\delta\mathbf{r} = T\boldsymbol{\rho} + \frac{1}{n}S\dot{\boldsymbol{\rho}} + wa\mathbf{k}\,. \tag{9.1.3}$$

To deal with the dimensionless components S, T and w we have introduced here the semi-major axis a of the unperturbed orbit and the observed mean motion n related by Kepler's third law

$$n^2a^3 = Gm\,. \tag{9.1.4}$$

From (9.1.3) we have

$$\delta\dot{\mathbf{r}} = \left(\dot{T} - \frac{a^3}{\rho^3}nS\right)\boldsymbol{\rho} + \left(T + \frac{1}{n}\dot{S}\right)\dot{\boldsymbol{\rho}} + \dot{w}a\mathbf{k}\,. \tag{9.1.5}$$

Differentiating (9.1.5) and substituting into (9.1.1) with the aid of (9.1.2) and (9.1.3) one gets

$$\left(\ddot{T} - 2\frac{a^3}{\rho^3}n\dot{S} - 3\frac{a^3}{\rho^3}n^2T\right)\boldsymbol{\rho} + \left(\frac{1}{n}\ddot{S} + 2\dot{T}\right)\dot{\boldsymbol{\rho}} + \left(\ddot{w} + \frac{a^3}{\rho^3}n^2w\right)a\mathbf{k} = 0\,. \tag{9.1.6}$$

Hence, T, S and w are determined by the equations

$$\ddot{T} - 2\frac{a^3}{\rho^3}n\dot{S} - 3\frac{a^3}{\rho^3}n^2T = 0\,, \tag{9.1.7}$$

$$\ddot{S} + 2n\dot{T} = 0 \tag{9.1.8}$$

and

$$\ddot{w} + \frac{a^3}{\rho^3}n^2w = 0\,. \tag{9.1.9}$$

Instead of S and T we introduce new complex conjugate variables p and q:

$$S = \tfrac{1}{2}\mathrm{i}\,(p-q)\,, \qquad T = -\tfrac{1}{2}(p+q)\,. \tag{9.1.10}$$

Equations (9.1.7) and (9.1.8) may be replaced now by a single equation:

$$\ddot{p} + \mathrm{i}n\left(1 + \frac{a^3}{\rho^3}\right)\dot{p} + \mathrm{i}n\left(1 - \frac{a^3}{\rho^3}\right)\dot{q} - \frac{3}{2}n^2\frac{a^3}{\rho^3}(p+q) = 0\,. \tag{9.1.11}$$

Transformation (9.1.3) and (9.1.5) expressed in new variables now takes the form

$$\delta\mathbf{r} = \frac{1}{2}p\left(\frac{\mathrm{i}}{n}\dot{\boldsymbol{\rho}} - \boldsymbol{\rho}\right) - \frac{1}{2}q\left(\frac{\mathrm{i}}{n}\dot{\boldsymbol{\rho}} + \boldsymbol{\rho}\right) + wa\mathbf{k} \tag{9.1.12}$$

and

$$\delta\dot{\mathbf{r}} = -\frac{1}{2}p\left(\mathrm{i}n\frac{a^3}{\rho^3}\boldsymbol{\rho} + \dot{\boldsymbol{\rho}}\right) + \frac{1}{2}q\left(\mathrm{i}n\frac{a^3}{\rho^3}\boldsymbol{\rho} - \dot{\boldsymbol{\rho}}\right) + \\ + \frac{1}{2}\dot{p}\left(\frac{\mathrm{i}}{n}\dot{\boldsymbol{\rho}} - \boldsymbol{\rho}\right) - \frac{1}{2}\dot{q}\left(\frac{\mathrm{i}}{n}\dot{\boldsymbol{\rho}} + \boldsymbol{\rho}\right) + \dot{w}a\mathbf{k}\,. \tag{9.1.13}$$

There exists a transformation of variables (Brumberg, 1978)

$$(p,\, \dot{p},\, w,\, \dot{w}) \to (\xi,\, h,\, u,\, v) \tag{9.1.14}$$

such that equations of variations (9.1.11) and (9.1.9) expressed in new variables take the Jordan form

$$\begin{aligned} \dot{\xi} &= \mathrm{i}nh, \\ \dot{h} &= 0, \\ \dot{u} &= \mathrm{i}nu, \\ \dot{v} &= \mathrm{i}nv\,. \end{aligned} \tag{9.1.15}$$

Evidently, ξ is a purely imaginary variable, h is a real variable, and u and v are complex variables (v should not be confused with the true anomaly designated in Chapter 2 by the same letter). The explicit form of transformation (9.1.14) is as follows:

$$\begin{aligned} p &= \xi - \tfrac{2}{3}h + \lambda u + \mu\bar{u}\,, \\ \dot{p} &= \mathrm{i}n(h + \gamma u + \delta\bar{u})\,, \\ w &= \varsigma v + \bar{\varsigma}\bar{v}\,, \\ \dot{w} &= \mathrm{i}n(\nu v - \bar{\nu}\bar{v}) \end{aligned} \tag{9.1.16}$$

with

$$\lambda = \frac{\rho}{a}\left[\frac{e}{4\eta}\frac{\rho}{a} + (\eta - \tfrac{1}{2})\sigma - \frac{e}{4\eta}\frac{\rho}{a}\sigma^2\right]\tau^{-1} - e\tau^{-1}\,, \tag{9.1.17}$$

$$\mu = \frac{\rho}{a}\left[-\frac{e}{4\eta}\frac{\rho}{a} - (\eta + \tfrac{1}{2})\sigma^{-1} + \frac{e}{4\eta}\frac{\rho}{a}\sigma^{-2}\right]\tau - e\tau\,, \tag{9.1.18}$$

$$\gamma = \left[\left(\frac{\rho}{a} - \frac{1}{2\eta}\right)\sigma + \frac{1}{2}e\left(3 - \frac{1}{\eta}\right)\right]\tau^{-1}\,, \tag{9.1.19}$$

$$\delta = \left[\left(\frac{\rho}{a} + \frac{1}{2\eta}\right)\sigma^{-1} + \frac{1}{2}e\left(3 + \frac{1}{\eta}\right)\right]\tau\,, \tag{9.1.20}$$

$$\varsigma = \frac{\rho}{a}\sigma\tau^{-1} \tag{9.1.21}$$

and

$$\nu = \frac{1}{\eta}(e + \sigma)\tau^{-1}\,. \tag{9.1.22}$$

Here η, σ and τ have the same meaning as in (2.1.9) and (2.3.12), e being the eccentricity of the unperturbed orbit. The bar denotes a complex conjugate quantity. The general solution of (9.1.15) is

$$h = -B, \qquad \xi = -\mathrm{i}(A + BM), \qquad u = C\tau, \qquad v = D\tau \tag{9.1.23}$$

with arbitrary real constants A and B and complex constants C and D. We do not need such a solution but it may be directly checked that expressions (9.1.16)–(9.1.23) satisfy equations (9.1.11) and (9.1.9). Depending on specific applications coefficients (9.1.17)–(9.1.22) may be retained in closed form or alternatively they may be expanded by a Keplerian processor in multiples of mean or elliptic anomalies. The latter expansions are obtained from initial series (2.5.22), (2.5.32) and (2.5.36). For example, to expand τ^p into (q, w) series (the designation of elliptic anomaly and a new variable by the same letter should not lead to misunderstanding!) one may again use the techniques indicated at the end of Section 7.1. The expansion for τ^p results directly from (7.1.39)–(7.1.40) by putting $x = M$, $s' = w$ and $n' = n$.

9.2 The Perturbed Two-Body Problem

Consider now the equations of the perturbed motion

$$\ddot{\mathbf{r}} + \frac{Gm}{r^3}\mathbf{r} = \mathbf{G} \tag{9.2.1}$$

with a small perturbative force $\mathbf{G}$. Putting

$$\mathbf{r} = \boldsymbol{\rho} + \delta\mathbf{r} \tag{9.2.2}$$

and taking (9.1.2) into account one gets

$$\delta\ddot{\mathbf{r}} + \frac{Gm}{\rho^3}\left[\delta\mathbf{r} - \frac{3}{\rho^2}(\boldsymbol{\rho}\delta\mathbf{r})\boldsymbol{\rho}\right] = \mathbf{Q}\,, \tag{9.2.3}$$

where

$$\mathbf{Q} = \mathbf{G} - Gm\left(\frac{\mathbf{r}}{r^3} - \frac{\boldsymbol{\rho}}{\rho^3}\right) + \frac{Gm}{\rho^3}\left[\delta\mathbf{r} - \frac{3}{\rho^2}(\boldsymbol{\rho}\delta\mathbf{r})\boldsymbol{\rho}\right]. \tag{9.2.4}$$

Using the operator D (Musen and Carpenter, 1963; Musen, 1965b)

$$D = \nabla \exp(\delta\mathbf{r}\cdot\nabla) = \nabla \sum_{k=0}^{\infty} \frac{1}{(1)_k}(\delta\mathbf{r}\cdot\nabla)^k\,, \qquad \nabla = \operatorname{grad}\boldsymbol{\rho} \tag{9.2.5}$$

one may rewrite this expression as follows:

$$\mathbf{Q} = \mathbf{G} + Gm(D - \nabla - (\delta\mathbf{r}\cdot\nabla)\cdot\nabla)\frac{1}{\rho}\,. \tag{9.2.6}$$

Instead of (9.1.6) transformation (9.1.3)–(9.1.5) now leads to

$$\left(\ddot{T} - 2\frac{a^3}{\rho^3}n\dot{S} - 3\frac{a^3}{\rho^3}n^2T\right)\boldsymbol{\rho} + \left(\frac{1}{n}\ddot{S} + 2\dot{T}\right)\dot{\boldsymbol{\rho}} + \left(\ddot{w} + \frac{a^3}{\rho^3}n^2w\right)a\mathbf{k} = \mathbf{Q}\,. \tag{9.2.7}$$

Subsequently multiplying this equation by $(\dot{\boldsymbol{\rho}} \times \mathbf{k})$, $(\mathbf{k} \times \boldsymbol{\rho})$ and $\mathbf{k}$ and remembering that

$$\boldsymbol{\rho} \times \dot{\boldsymbol{\rho}} = na^2\eta\mathbf{k} \tag{9.2.8}$$

one obtains

$$\ddot{T} - 2\frac{a^3}{\rho^3}n\dot{S} - 3\frac{a^3}{\rho^3}n^2T = \frac{1}{na^2\eta}\mathbf{Q}(\dot{\boldsymbol{\rho}} \times \mathbf{k})\,, \tag{9.2.9}$$

$$\ddot{S} + 2n\dot{T} = \frac{1}{a^2\eta}\mathbf{Q}(\mathbf{k} \times \boldsymbol{\rho}) \tag{9.2.10}$$

and

$$\ddot{w} + \frac{a^3}{\rho^3}n^2w = \frac{1}{a}\mathbf{Q}\mathbf{k}\,. \tag{9.2.11}$$

Hence, the equations of the perturbed motion in terms of the variables p and w take the form

$$\ddot{p} + \mathrm{i}n\left(1 + \frac{a^3}{\rho^3}\right)\dot{p} + \mathrm{i}n\left(1 - \frac{a^3}{\rho^3}\right)\dot{q} - \tfrac{3}{2}n^2\frac{a^3}{\rho^3}(p+q) = n^2P \tag{9.2.12}$$

and

$$\ddot{w} + n^2\frac{a^3}{\rho^3}w = n^2W \tag{9.2.13}$$

with

$$P = \frac{\mathrm{i}}{n^2a^2\eta}\mathbf{Q}\left[\left(\boldsymbol{\rho} + \frac{\mathrm{i}}{n}\dot{\boldsymbol{\rho}}\right) \times \mathbf{k}\right] \tag{9.2.14}$$

and

$$W = \frac{1}{n^2a}\mathbf{Q}\mathbf{k}\,. \tag{9.2.15}$$

If the perturbative force $\mathbf{G}$ admits a potential $H = H(\mathbf{r},\, t)$

$$\mathbf{G} = \frac{\partial H}{\partial \mathbf{r}} \tag{9.2.16}$$

then it may be appropriate to transform these expressions. Using (9.2.8) and some other basic relations of the Keplerian motion it is easy to find that

$$\left(\rho+\frac{\mathrm{i}}{n}\dot{\rho}\right)\times\mathbf{k}=\frac{\mathrm{i}}{2\eta}\left\{\left(\rho+\frac{\mathrm{i}}{n}\dot{\rho}\right)\left(2\frac{a}{\rho}-1+\frac{\rho^2}{a^2}\right)+\right.$$
$$\left.+\left(\rho-\frac{\mathrm{i}}{n}\dot{\rho}\right)\left[2\frac{a}{\rho}-1-\frac{\rho^2}{a^2}-\frac{\rho}{a}\frac{e}{\eta}(\sigma-\sigma^{-1})\right]\right\}, \tag{9.2.17}$$

$$\rho\left[\left(\rho+\frac{\mathrm{i}}{n}\dot{\rho}\right)\times\mathbf{k}\right]=\mathrm{i}a^2\eta \tag{9.2.18}$$

and

$$\dot{\rho}\left[\left(\rho+\frac{\mathrm{i}}{n}\dot{\rho}\right)\times\mathbf{k}\right]=-na^2\eta\,. \tag{9.2.19}$$

On the other hand, introducing the force function

$$K=\frac{Gm}{r}+H \tag{9.2.20}$$

one has

$$\mathbf{Q}=n^2\frac{a^3}{\rho^3}\left[\rho+\delta\mathbf{r}-\frac{3}{\rho^2}(\rho\delta\mathbf{r})\rho\right]+\frac{\partial K}{\partial\mathbf{r}}\,. \tag{9.2.21}$$

Therefore, in virtue of (9.1.12) P and W are transformed as follows:

$$P=\frac{a^3}{\rho^3}\left\{-1+\tfrac{1}{2}\left[-1+\tfrac{3}{2}\frac{a}{\rho}\frac{e}{\eta}(\sigma-\sigma^{-1})\right]p-\tfrac{3}{2}\left[1+\tfrac{1}{2}\frac{a}{\rho}\frac{e}{\eta}(\sigma-\sigma^{-1})\right]q\right\}+$$
$$+\frac{1}{n^2a^2\eta^2}\left\{\left(-1+2\frac{a}{\rho}+\frac{\rho^2}{a^2}\right)\frac{\partial K}{\partial q}+\right.$$
$$\left.+\left[-1+2\frac{a}{\rho}-\frac{\rho^2}{a^2}-\frac{\rho}{a}\frac{e}{\eta}(\sigma-\sigma^{-1})\right]\frac{\partial K}{\partial p}\right\} \tag{9.2.22}$$

and

$$W=\frac{a^3}{\rho^3}w+\frac{1}{n^2a^2}\frac{\partial K}{\partial w}, \tag{9.2.23}$$

where the force function K is supposed to be expressed by means of (9.2.2) and (9.1.12) in terms of p, q and w. This function may be expanded in powers of p, q and w. As a result, the zero- and first-degree terms not containing the small parameter cancel out in the expressions for P and W. Hence, P and W are holomorphic functions of p, q and w with the zero- and first-degree terms proportional to the disturbing small parameter.

Equations (9.2.12) and (9.2.13) may now be put into Jordan form. Introducing the vector V of the unknown variables and the vector B of the right-hand members

$$V = \begin{pmatrix} p \\ q \\ \dot{p} \\ \dot{q} \\ w \\ \dot{w} \end{pmatrix}, \qquad B = n^2 \begin{pmatrix} 0 \\ 0 \\ P \\ \bar{P} \\ 0 \\ W \end{pmatrix} \tag{9.2.24}$$

one may rewrite these equations as a system of the first-order differential equations

$$\dot{V} = AV + B\,. \tag{9.2.25}$$

The matrix A is here of block form:

$$A = \begin{pmatrix} A_1 & 0 \\ 0 & A_2 \end{pmatrix} \tag{9.2.26}$$

with

$$A_1 = \begin{pmatrix} 0 & 0 & 1 & 0 \\ 0 & 0 & 0 & 1 \\ \dfrac{3n^2a^3}{2\rho^3} & \dfrac{3n^2a^3}{2\rho^3} & -in\left(1+\dfrac{a^3}{\rho^3}\right) & -in\left(1-\dfrac{a^3}{\rho^3}\right) \\ \dfrac{3n^2a^3}{2\rho^3} & \dfrac{3n^2a^3}{2\rho^3} & in\left(1-\dfrac{a^3}{\rho^3}\right) & in\left(1+\dfrac{a^3}{\rho^3}\right) \end{pmatrix}, \tag{9.2.27}$$

and

$$A_2 = \begin{pmatrix} 0 & 1 \\ -\dfrac{n^2a^3}{\rho^3} & 0 \end{pmatrix}. \tag{9.2.28}$$

Based on the results of the previous section one may perform a transformation

$$V = JX \tag{9.2.29}$$

to a vector X of new variables

$$X = \begin{pmatrix} \xi \\ h \\ u \\ \bar{u} \\ v \\ \bar{v} \end{pmatrix}. \tag{9.2.30}$$

The matrix J is of the same form as A and has the blocks

$$J_1 = \begin{pmatrix} 1 & -\frac{2}{3} & \lambda & \mu \\ -1 & -\frac{2}{3} & \bar{\mu} & \bar{\lambda} \\ 0 & in & in\gamma & in\delta \\ 0 & -in & -in\bar{\delta} & -in\bar{\gamma} \end{pmatrix}. \tag{9.2.31}$$

and

$$J_2 = \begin{pmatrix} \varsigma & \varsigma \\ \mathrm{i}n\nu & -\mathrm{i}n\bar{\nu} \end{pmatrix} . \tag{9.2.32}$$

Independently of the previous section, it may be useful to give the matrix derivation of the reduction of (9.2.25) by means of (9.2.29) to Jordan form. By easy manipulation with (9.1.16)–(9.1.22) one finds the determinants

$$\det(J_1) = \tfrac{2}{3}n^2\eta\,, \qquad \det(J_2) = -2\,\mathrm{i}\,n\eta \tag{9.2.33}$$

and the blocks of the inverse matrix

$$J_1^{-1} = \frac{3}{2n\eta} \begin{pmatrix} n\theta & -n\bar{\theta} & \mathrm{i}\phi & \mathrm{i}\bar{\phi} \\ n\chi & n\chi & \mathrm{i}\psi & -\mathrm{i}\bar{\psi} \\ -n\bar{\nu} & -n\bar{\nu} & \mathrm{i}(-\frac{1}{3}\bar{\varsigma} + \frac{2}{3}\bar{\nu}) & \mathrm{i}(-\frac{1}{3}\bar{\varsigma} - \frac{2}{3}\bar{\nu}) \\ -n\nu & -n\nu & \mathrm{i}(\ \frac{1}{3}\varsigma + \frac{2}{3}\nu) & \mathrm{i}(\ \frac{1}{3}\varsigma - \frac{2}{3}\nu) \end{pmatrix} \tag{9.2.34}$$

and

$$J_2^{-1} = \frac{1}{2n\eta} \begin{pmatrix} n\bar{\nu} & -\mathrm{i}\bar{\varsigma} \\ n\nu & \mathrm{i}\varsigma \end{pmatrix} . \tag{9.2.35}$$

In addition to (9.1.16)–(9.1.22) one has here

$$\theta = \frac{1}{3}\eta - \frac{1}{2}e\frac{\rho}{a}(\sigma - \sigma^{-1})\,, \tag{9.2.36}$$

$$\chi = -2\eta - \frac{3}{2}\frac{e}{\eta}(2e + \sigma + \sigma^{-1})\,, \tag{9.2.37}$$

$$\phi = -\frac{2}{3}\eta\frac{\rho^2}{a^2} + \frac{1}{3}e\frac{\rho}{a}(\sigma - \sigma^{-1}) \tag{9.2.38}$$

and

$$\psi = \frac{1+e^2}{\eta} + \left(\frac{1}{\eta} + \frac{1}{2}\frac{\rho}{a}\right)e\sigma + \left(\frac{1}{\eta} - \frac{1}{2}\frac{\rho}{a}\right)e\sigma^{-1} . \tag{9.2.39}$$

These coefficients may again be retained in closed form or expanded into (e, M) or (q, w) series. Differentiation of (9.1.16)–(9.1.22) results in

$$\begin{aligned}
\dot{\lambda} &= \mathrm{i}\,n(\gamma - \lambda)\,, \\
\dot{\mu} &= \mathrm{i}\,n(\delta + \mu)\,, \\
\mathrm{i}\dot{\gamma} &= n\left[\left(2 + \frac{a^3}{\rho^3}\right)\gamma + \left(\frac{a^3}{\rho^3} - 1\right)\bar{\delta} + \frac{3}{2}\frac{a^3}{\rho^3}(\lambda + \bar{\mu})\right] , \\
\mathrm{i}\dot{\delta} &= n\left[\frac{a^3}{\rho^3}\delta + \left(\frac{a^3}{\rho^3} - 1\right)\bar{\gamma} + \frac{3}{2}\frac{a^3}{\rho^3}(\bar{\lambda} + \mu)\right] , \\
\dot{\varsigma} &= \mathrm{i}\,n(\nu - \varsigma)
\end{aligned}$$

and

$$\mathrm{i}\dot{\nu} = n\left(\nu - \frac{a^3}{\rho^3}\varsigma\right).$$

From this it follows that

$$A_1 J_1 - \dot{J}_1 = \mathrm{i}n \begin{pmatrix} 0 & 1 & \lambda & -\mu \\ 0 & -1 & \bar{\mu} & -\bar{\lambda} \\ 0 & 0 & \mathrm{i}n\gamma & -\mathrm{i}n\delta \\ 0 & 0 & -\mathrm{i}n\bar{\delta} & \mathrm{i}n\bar{\gamma} \end{pmatrix} \tag{9.2.40}$$

and

$$A_2 J_2 - \dot{J}_2 = \mathrm{i}n \begin{pmatrix} \varsigma & -\bar{\varsigma} \\ \mathrm{i}n\nu & \mathrm{i}n\bar{\nu} \end{pmatrix}. \tag{9.2.41}$$

Transformation (9.2.29) brings system (9.2.25) to the form

$$\dot{X} = \mathrm{i}n[\mathcal{P}X + R(X, t)] \tag{9.2.42}$$

with

$$\mathrm{i}n\mathcal{P} = J^{-1}\left(AJ - \dot{J}\right) \tag{9.2.43}$$

and

$$\mathrm{i}nR = J^{-1}B. \tag{9.2.44}$$

Relations (9.2.40) and (9.2.41) enable us to find

$$\mathcal{P} = \begin{pmatrix} 0 & 1 & 0 & 0 & 0 & 0 \\ 0 & 0 & 0 & 0 & 0 & 0 \\ 0 & 0 & 1 & 0 & 0 & 0 \\ 0 & 0 & 0 & -1 & 0 & 0 \\ 0 & 0 & 0 & 0 & 1 & 0 \\ 0 & 0 & 0 & 0 & 0 & -1 \end{pmatrix}. \tag{9.2.45}$$

Hence, the matrix $\mathcal{P}$ is constant and is of Jordan form. As for the components of the vector of the right-hand sides

$$R = \begin{pmatrix} R_1 \\ R_2 \\ R_3 \\ R_4 \\ R_5 \\ R_6 \end{pmatrix} \tag{9.2.46}$$

from (9.2.44), (9.2.24), (9.2.31) and (9.2.32) we get

$$R_1 = \frac{3}{2\eta}\left(P\phi + \bar{P}\bar{\phi}\right), \tag{9.2.47}$$

$$R_2 = \frac{3}{2\eta}\left(P\psi - \bar{P}\bar{\psi}\right), \tag{9.2.48}$$

$$R_3 = \frac{1}{2\eta}\left[(-\varsigma + 2\bar{\nu})P - (\varsigma + 2\bar{\nu})\bar{P}\right], \tag{9.2.49}$$

$$R_5 = -\frac{1}{2\eta}\varsigma W \tag{9.2.50}$$

and

$$R_4 = -\bar{R}_3, \qquad R_6 = -\bar{R}_5. \tag{9.2.51}$$

Equations (9.2.42) are typical for the perturbed two-body problem. It should be recalled once more that the right-hand member R is holomorphic with respect to X with the zero- and first-degree terms proportional to the small perturbing parameter. In many problems of celestial mechanics R depends on t by means of periodic or quasiperiodic functions and it may be necessary to find the solution of (9.2.42) without secular terms.

9.3 Solution Techniques

System (9.2.42) may be solved by the technique of the general planetary theory (Brumberg, 1970; Brumberg and Chapront, 1973). In fact one looks not for a straightforward solution but for a nearly identical transformation of the old variables X to new variables Y

$$X = Y + \Gamma(Y, t) \tag{9.3.1}$$

so that the right-hand member F of the new equations

$$\dot{Y} = in\left[\mathcal{P}Y + F(Y, t)\right] \tag{9.3.2}$$

would be as simple as possible. Hence Γ and F should not contain secular terms.

Substituting (9.3.1) into (9.2.42) and comparing with (9.3.2) one finds that

$$F = (E + \Gamma_Y)^{-1}\left(R + \mathcal{P}\Gamma - \Gamma_Y\mathcal{P}Y - \frac{1}{in}\Gamma_t\right),$$

E being the unit matrix, Γ_Y the Jacobian of Γ with respect to Y and Γ_t the partial time derivative of Γ. The vector functions Γ and F are holomorphic in Y and are determined step by step. Introducing the vector U,

$$U = (E + \Gamma_Y)^{-1}\left[R + \Gamma_Y\left(\frac{1}{in}\Gamma_t + \Gamma_Y\mathcal{P}Y - \mathcal{P}\Gamma\right)\right], \tag{9.3.3}$$

one may rewrite F in the form

$$F = U + \mathcal{P}\Gamma - \Gamma_Y \mathcal{P} Y - \frac{1}{in}\Gamma_t \,. \tag{9.3.4}$$

If Γ is known up to the terms of some total degree then the function U will be determined up to the terms of one degree more. The corresponding more accurate value of Γ may be found from the equation

$$\frac{1}{in}\Gamma_t + \Gamma_Y \mathcal{P} Y - \mathcal{P}\Gamma = U^+ \,, \tag{9.3.5}$$

where U^+ contains all those terms of U which permit one to integrate (9.3.5) without secular terms. The remaining critical terms of U are denoted by U^*; hence

$$U = U^* + U^+ \,. \tag{9.3.6}$$

Expressions (9.3.3) and (9.3.4) now take the form

$$U = R - \Gamma_Y U^* \tag{9.3.7}$$

and

$$F = U^* \,. \tag{9.3.8}$$

Relations (9.3.5)–(9.3.7) serve for the step-by-step determination of U and Γ. The most cumbersome operation of this algorithm is the substitution of (9.3.1) into the right-hand member $R(X, t)$. This should be done by using some appropriate CAS.

The components of Y and Γ satisfy relations of the type

$$\bar{Y}_1 = -Y_1 \,, \qquad \bar{Y}_2 = Y_2 \,, \qquad Y_4 = \bar{Y}_3 \,, \qquad Y_6 = \bar{Y}_5 \,. \tag{9.3.9}$$

For the components of U one has

$$\bar{U}_1 = U_1 \,, \qquad \bar{U}_2 = -U_2 \,, \qquad U_4 = -\bar{U}_3 \,, \qquad U_6 = -\bar{U}_5 \,. \tag{9.3.10}$$

Therefore, for the components $m = 1, 2, 3$ and 5 equation (9.3.5) leads to

$$\frac{1}{in}\frac{\partial \Gamma_m}{\partial t} + \frac{\partial \Gamma_m}{\partial Y_1} Y_2 + \frac{\partial \Gamma_m}{\partial Y_3} Y_3 - \frac{\partial \Gamma_m}{\partial Y_4} Y_4 + \frac{\partial \Gamma_m}{\partial Y_5} Y_5 - \frac{\partial \Gamma_m}{\partial Y_6} Y_6 - \\ - (\mathcal{P}\Gamma)_m = U_m^+ \tag{9.3.11}$$

with

$$(\mathcal{P}\Gamma)_1 = \Gamma_2 \,, \quad (\mathcal{P}\Gamma)_2 = 0 \,, \quad (\mathcal{P}\Gamma)_3 = \Gamma_3 \,, \quad (\mathcal{P}\Gamma)_5 = \Gamma_5 \,. \tag{9.3.12}$$

U_m and Γ_m have the form of the power series

$$U_m = \sum U^{(m)}_{klpqrs}(t) Y_1^k Y_2^l Y_3^p Y_4^q Y_5^r Y_6^s \tag{9.3.13}$$

and

$$\Gamma_m = \sum \Gamma^{(m)}_{klpqrs}(t) Y_1^k Y_2^l Y_3^p Y_4^q Y_5^r Y_6^s \,, \tag{9.3.14}$$

where the summation is performed over all non-negative values of k, l, p, q, r and s.

It is sufficient to deal only with a particular solution of equations (9.3.2) when $Y_1 = Y_2 = 0$ identically. Omitting the zero values of the indices k and l in the coefficients of (9.3.13) and (9.3.14), which are unnecessary in this case, one gets from (9.3.11)

$$\frac{1}{in}\dot{\Gamma}^{(1)}_{pqrs} + (p-q+r-s)\Gamma^{(1)}_{pqrs} - \Gamma^{(2)}_{pqrs} = U^{(1)}_{pqrs} \,, \tag{9.3.15}$$

$$\frac{1}{in}\dot{\Gamma}^{(2)}_{pqrs} + (p-q+r-s)\Gamma^{(2)}_{pqrs} = U^{(2)}_{pqrs} \,, \tag{9.3.16}$$

$$\frac{1}{in}\dot{\Gamma}^{(3)}_{pqrs} + (p-q+r-s-1)\Gamma^{(3)}_{pqrs} = U^{(3)}_{pqrs} \tag{9.3.17}$$

and

$$\frac{1}{in}\dot{\Gamma}^{(5)}_{pqrs} + (p-q+r-s-1)\Gamma^{(5)}_{pqrs} = U^{(5)}_{pqrs} \,. \tag{9.3.18}$$

The condition for the existence of a solution without secular terms is

$$U_2^* = 0 \,, \tag{9.3.19}$$

i.e. the component U_2 should not contain critical terms. Under this condition equation (9.3.16) may be solved without secular terms. The component U_1 may have critical terms but in equation (9.3.15) they cancel out by an adequate choice of the arbitrary constant in Γ_2 introduced in integrating (9.3.16). Hence, the integration of (9.3.15) is performed without introducing secular terms. Since $F_1 = F_2 = 0$ in the right-hand members of (9.3.2) these equations actually admit a solution $Y_1 = Y_2 = 0$.

Using the mean longitude $\lambda = nt + \varepsilon$ as an argument, the solution of (9.3.15)–(9.3.18) may be expressed by means of the Hansen quadratures

$$\Gamma^{(2)}_{pqrs} = \mathrm{i} \int U^{(2)}_{pqrs} \exp\left[\mathrm{i}(p-q+r-s)(\lambda - \tilde{\lambda})\right] d\lambda - U^{*(1)}_{pqrs} \,, \tag{9.3.20}$$

$$\Gamma^{(1)}_{pqrs} = \mathrm{i} \int \left(U^{(1)}_{pqrs} + \Gamma^{(2)}_{pqrs}\right) \exp\left[\mathrm{i}(p-q+r-s)(\lambda - \tilde{\lambda})\right] d\lambda \,, \tag{9.3.21}$$

$$\Gamma^{(3)}_{pqrs} = \mathrm{i} \int U^{(3)}_{pqrs} \exp\left[\mathrm{i}(p-q+r-s-1)(\lambda - \tilde{\lambda})\right] d\lambda \tag{9.3.22}$$

and

$$\Gamma^{(5)}_{pqrs} = \mathrm{i} \int U^{(5)}_{pqrs} \exp\left[\mathrm{i}(p-q+r-s-1)(\lambda - \tilde{\lambda})\right] d\lambda \,. \tag{9.3.23}$$

The quantity $\tilde{\lambda}$ is considered to be constant during integration, restoring the meaning of λ after integration. Critical terms in U_m are defined by the relation

$$\begin{aligned} U^{*(m)}_{pqrs} = {} & \exp\left[-\mathrm{i}(p-q+r-s-\nabla_m)\lambda\right]\times \\ & \times M_\lambda\left[U^{(m)}_{pqrs}\exp\mathrm{i}(p-q+r-s-\nabla_m)\lambda\right] \end{aligned} \tag{9.3.24}$$

with

$$\nabla_m = \begin{cases} 0, & m=1,\ 2, \\ 1, & m=3,\ 5, \end{cases} \tag{9.3.25}$$

M_λ being a λ-averaging operator excluding all short-period terms from any function under its symbol.

Thus, the vector F of the right-hand sides of (9.3.2) has the components

$$\begin{aligned} F_1 &= 0, \\ F_2 &= 0, \\ F_3 &= U_3^*, \\ F_4 &= -\bar{F}_3, \\ F_5 &= U_5^*, \\ F_6 &= -\bar{F}_5 \end{aligned} \tag{9.3.26}$$

with U_m^* ($m=3$ and $m=5$) being determined by series (9.3.13) with coefficients (9.3.24) for $k=l=0$. Designating

$$Y_3 = a\,, \qquad Y_5 = b \tag{9.3.27}$$

one may explicitly rewrite system (9.3.2) as follows:

$$\begin{aligned} \dot{a} &= \mathrm{i}\,n(a+F_3)\,, \\ \dot{b} &= \mathrm{i}\,n(b+F_5)\,. \end{aligned} \tag{9.3.28}$$

Going from a and b to the slowly changing variables α and β

$$\begin{aligned} a &= \alpha\exp\mathrm{i}\lambda\,, \\ b &= \beta\exp\mathrm{i}\lambda \end{aligned} \tag{9.3.29}$$

one gets a secular system

$$\begin{aligned} \dot{\alpha} &= \mathrm{i}\,n\Phi\,, \\ \dot{\beta} &= \mathrm{i}\,n\Psi \end{aligned} \tag{9.3.30}$$

with the right-hand members

$$\Phi = F_3\exp(-\mathrm{i}\lambda)\,, \tag{9.3.31}$$

$$\Psi = F_5 \exp(-\mathrm{i}\lambda) \tag{9.3.32}$$

or

$$\begin{pmatrix} \Phi \\ \Psi \end{pmatrix} = \sum M_\lambda \left[U^{(m)}_{pqrs} \exp \mathrm{i}(p - q + r - s - 1)\lambda\right] \alpha^p \bar{\alpha}^q \beta^r \bar{\beta}^s \,, \tag{9.3.33}$$

where $m = 3$ for Φ and $m = 5$ for Ψ. In the general case equations (9.3.30) form a polynomial system with slowly changing or even constant coefficients. The solution of this system introduces four supplementary arbitrary constants to be determined in the usual way from initial data or observational results. It is of importance that the mean motion is fixed and does not change in constructing transformation (9.3.1) and solving final system (9.3.30).

This technique is applied in the next chapter to construct the GPT (general planetary theory).

10 The General Planetary Theory

10.1 The Basic Theory

The construction of a general planetary theory (GPT), i.e. a theory representing the coordinates of the major planets in a purely trigonometric form and valid, at least formally, for an indefinite interval of time, has always been considered one of the most important problems of celestial mechanics. Laplace was the first to propose solving the equations of planetary motion in a purely trigonometric form with respect to time. But technical difficulties associated with a trigonometric form forced him to develop another form of planetary theory involving secular and mixed terms. Having failed to find efficient methods for the practical construction of a GPT, Le Verrier developed his famous analytical theories of motion of the planets in just the classical form first indicated by Laplace. A mathematical form of a GPT was rigorously proved for the first time by Newcomb (1876). Newcomb considered his technique to be only an existence theorem although he actually used Newton-type quadratic iterations underlying the present KAM theory (Section 8.4). Later on the trigonometric representation of GPT was advanced by Dziobek (1888), Poincaré (1905) and Charlier (1902, 1907). The most serious practical attempt to construct a GPT was made by Gyldén (1893, 1908). Gyldén created his own world of the art of celestial mechanics, where the planets move on periplegmatic orbits with constant or variable diastemae (almost periodic motions in modern, more prosaic terminology). Very regrettably, his work remained uncompleted. In spite of significant contributions to the development of classic-type planetary theories, Hill regarded them as only a temporary makeshift until efficient methods for constructing a GPT could be further elaborated (Hill, 1905, 1907).

All these investigations demonstrate that the GPT trigonometric series for the heliocentric rectangular coordinates x_i, y_i, z_i ($i = 1, 2, \ldots, N$) of the N planets and the mutual distances Δ_{ij} between them are of the form

$$x_i + \mathrm{i}\, y_i = \sum A^{(i)}_{\alpha\alpha'\alpha''} \exp \mathrm{i}(\alpha w + \alpha' w' + \alpha'' w''), \tag{10.1.1}$$

$$z_i = \sum B^{(i)}_{\alpha\alpha'\alpha''} \sin(\alpha w + \alpha' w' + \alpha'' w'') \tag{10.1.2}$$

and

$$\Delta_{ij} = \sum C^{(ij)}_{\alpha\alpha'\alpha''} \cos(\alpha w + \alpha' w' + \alpha'' w''). \tag{10.1.3}$$

The summation is performed here over all integer values of N-multi-indices α, α' and α''. N-vectors w, w' and w'' are linear functions of time t with N-vector frequencies n, n' and n'', respectively. The components of n represent the mean planetary motions. The components of n' and n'' are caused by the secular motions of the planetary perihelia and nodes, respectively. One of the components of n'' is zero. The sum of all the indices α, α' and α'' is equal to 1 in (10.1.1) and zero in (10.1.2) and (10.1.3). The sum of all the indices α'' is odd in (10.1.2) and even in (10.1.1) and (10.1.3). The coefficients of these series are real functions of the semi-major axes a_j, eccentricities e_j and inclinations i_j. In virtue of the D'Alambert characteristics they are of order $\sum |\alpha'_j|$ and $\sum |\alpha''_j|$ with respect to the eccentricities and inclinations, respectively. In addition,

$$B_{\alpha\alpha'\alpha''} = -B_{-\alpha,-\alpha',-\alpha''}\,, \qquad C_{\alpha\alpha'\alpha''} = C_{-\alpha,-\alpha',-\alpha''}\,. \tag{10.1.4}$$

Form (10.1.1)–(10.1.3) is mainly of theoretical interest. Series (10.1.2)–(10.1.3) cannot be efficiently used in practice owing to the large difference in the magnitudes of n, on the one hand, and n' and n'', on the other (in satellite problems, for example, in the lunar theory, such a difference between the basic frequencies is much smaller, enabling one to use purely trigonometric series). Therefore, to make a GPT feasible it is necessary to present it in a form that enables one to separate the short-period (w) and long-period (w' and w'') trigonometric arguments. The GPT elaborated by Brumberg (1970), Brumberg and Chapront (1973) and Brumberg et al., (1978) is based on the ideas of Hill (the introduction of a quasi-periodic intermediary generalizing Hill's variational curve in the theory of motion of the Moon), von Zeipel (the separation of the fast and slowly changing variables) and Birkhoff (the development of a normalization transformation of the dynamical system theory). This theory was presented in heliocentric rectangular coordinates. The alternative form of a GPT was developed by Duriez (1977, 1978) using osculating elements. The construction of a GPT in rectangular coordinates is likely to be a little bit simpler, since $6N$ osculating elements are replaced by $4N$ Laplace-type variables (N mean motions n_i and N initial values of the vector w of the mean longitudes remain fixed as the parameters of the intermediary). In this respect the paper by Cherry (1924) could also be mentioned, which concerns some of the disadvantages of using osculating elements. Anyway, the difference in using coordinates or elements is revealed only when treating short-period perturbations. In dealing with long-period perturbations one always use $4N$ Laplace-type elements to describe the secular notions of perihelia and nodes. In what follows, we shall express the GPT in terms of heliocentric rectangular coordinates.

The differential equations of motion in terms of the variables p_i, q_i and w_i introduced by (7.2.2) and (7.2.3) have form (7.2.5) and (7.2.6) with right-hand sides (7.2.7) and (7.2.8). These equations are treated in the GPT by the technique of Chapter 9 with three modifications:

1. the extension of the order of the system from 6 to $6N$;
2. the determination of the zero- and first-degree terms with respect to planetary eccentricities and inclinations as separate steps of the general algorithm;
3. the replacement of a Keplerian intermediary by a quasi-periodic intermediary corresponding to the zero eccentricities and inclinations of the planets.

As can be seen from the technique of Chapter 9 the basic GPT series have the form

$$p_i = \sum_{m=0}^{\infty} \sum_{p+q+r+s=m} p_{pqrs}^{(i)}(t) \prod_{j=1}^{N} a_j^{p_j} \bar{a}_j^{q_j} b_j^{r_j} \bar{b}_j^{s_j} \tag{10.1.5}$$

and

$$w_i = \sum_{m=1}^{\infty} \sum_{p+q+r+s=m} w_{pqrs}^{(i)}(t) \prod_{j=1}^{N} a_j^{p_j} \bar{a}_j^{q_j} b_j^{r_j} \bar{b}_j^{s_j} . \tag{10.1.6}$$

Here a_j and b_j are the complex variables proportional to the eccentricities and inclinations, respectively. The bar again marks a conjugate quantity. The summation is performed over all non-negative values of N-indices $p = (p_j)$, $q = (q_j)$, $r = (r_j)$ and $s = (s_j)$. The coefficients in (10.1.5) and (10.1.6) are 2π-periodic functions of the differences of the mean longitudes $\lambda_i - \lambda_j$, $j = 1, 2, \ldots, N$, $j \neq i$. Hence, these series have $4N$ power variables and $N(N-1)/2$ angular arguments. The terms with $m = 0$ in (10.1.5) represent an intermediate solution that depends on $2N$ constants n_i and ε_i occurring in (7.2.4). The sum of all the r_j and s_j is even in (10.1.5) and odd in (10.1.6). The terms with $m = 1$ describe the linear theory with respect to eccentricities in (10.1.5) and inclinations in (10.1.6).

Series (10.1.5) and (10.1.6) are in fact not a solution of the equations of motion of N planets but a transformation to the secular system describing the evolution of the planetary orbits. Applying (7.3.3) now to each planet,

$$\begin{aligned} a_i &= \alpha_i \exp \mathrm{i}\lambda_i , \\ b_i &= \beta_i \exp \mathrm{i}\lambda_i , \end{aligned} \tag{10.1.7}$$

one has an autonomous system determining α_i and β_i. In the vector notation $\alpha = (\alpha_j)$, $\beta = (\beta_j)$ this system has the form

$$\begin{aligned} \dot{\alpha} &= \mathrm{i}\mathcal{N}\big[A\alpha + \Phi(\alpha, \bar{\alpha}, \beta, \bar{\beta})\big], \\ \dot{\beta} &= \mathrm{i}\mathcal{N}\big[B\beta + \Psi(\alpha, \bar{\alpha}, \beta, \bar{\beta})\big] \end{aligned} \tag{10.1.8}$$

with an $N \times N$ diagonal matrix $\mathcal{N}$ of the mean motions, $N \times N$ constant real matrices A and B and N-vector real functions Φ and Ψ expanded in power series in α_i, $\bar{\alpha}_i$, β_i and $\bar{\beta}_i$ starting with the third-degree terms. To complete this system one should add the equations for the conjugate variables. Then one has a secular system of order $4N$, and its solution gives the remaining $4N$ constants of integration of the equations of motion of N planets. Neglecting functions Φ and Ψ and restricting A and B to only the first-order terms with respect to μ one obtains the well-known secular system of Lagrange and Laplace.

The solution of (10.1.8) in a purely trigonometric form may be obtained by the technique of Section 4.2 . Substituting this literal trigonometric solution into (10.1.5) and (10.1.6) one might arrive at series (10.1.1) and (10.1.2). But such a substitution is quite useless for practical applications. Instead, it is reasonable to use system (10.1.8) for determining the numerical values of α_i and β_i for any moment of time. Then one can compute the rectangular coordinates of the planets by evaluating series (10.1.5) and (10.1.6) with (10.1.7). These series together with system (10.1.8) provide the separation of fast and slowly changing variables in GPT.

The computation of the coefficients of (10.1.5) and (10.1.6) is facilitated by representing them as series in powers of μ:

$$
\begin{aligned}
p^{(i)}_{pqrs}(t) &= \sum_{k=0}^{\infty} \mu^k \, {}_k p^{(i)}_{pqrs}(t), \\
w^{(i)}_{pqrs}(t) &= \sum_{k=0}^{\infty} \mu^k \, {}_k w^{(i)}_{pqrs}(t).
\end{aligned}
\tag{10.1.9}
$$

Nevertheless, in some respects the iteration techniques of a GPT might be preferable (Brumberg, 1974). In particular, in Section 7.2 it is shown that for an intermediate orbit the corresponding terms with $m = 0$ in (10.1.5) may be easily determined by iteration techniques.

For $k = 0$ (the two-body problem) the coefficients in (10.1.9) do not depend on t and are the same for each planet. Their values (up to $m = 5$ inclusive) are listed in (7.3.8)–(7.3.17). The coefficients of secular system (10.1.8) are expanded in series like (10.1.9) but starting with $k = 1$. In this respect it should be noted that k-order approximations have different meaning in a GPT and classic-type planetary theories. For example, a first-order GPT means that one takes into account in the solution of (10.1.8) terms such as μt, $\mu^2 t^2$, $\mu^3 t^3$, ... but neglects terms of the order $\mu^2 t$, $\mu^3 t^2$, ... The first-order classic-type theory neglects terms of the order $\mu^2 t^2$, ... More detailed comparison of these two types of planetary theory is given by Duriez (1978).

For $k > 0$ the coefficients of (10.1.9) are written in the form (omitting for clarity the power sub-indices p, q, r and s)

$$
\begin{aligned}
\underset{1}{p}^{(i)} &= \sum_{j=1}^{N}{}^{(i)} \underset{1}{p}_j^{(i)}, \\
\underset{2}{p}^{(i)} &= \sum_{j=1}^{N}{}^{(i)} \underset{2}{p}_j^{(i)} + \frac{1}{2} \sum_{j=1}^{N}{}^{(i)} \sum_{k=1}^{N}{}^{(i,j)} \underset{2}{p}_{jk}^{(i)}, \\
&\dots
\end{aligned}
\tag{10.1.10}
$$

and similarly for w coordinates. These terms describe the action of planet j on planet i, or the combined action of planets j and k on planet i, etc. All these time-dependent coefficients can now be constructed in three different forms.

The first form, version (a) of the GPT, is given by exponential series in the arguments $\mathrm{i}(\lambda_i - \lambda_j)$ with real coefficients (Brumberg and Chapront, 1973; Brumberg et al., 1978). Putting

$$
\zeta_{ij} = \exp \mathrm{i}(\lambda_i - \lambda_j) \tag{10.1.11}
$$

one has to deal with one-argument exponential series for $n = 1, 2, \dots$

$$
\underset{n}{p}_j^{(i)} = \sum_{\sigma=-\infty}^{\infty} \underset{n}{\tilde{p}}_\sigma^{(ij)} \zeta_{ij}^\sigma, \tag{10.1.12}
$$

two-argument exponential series for $n = 2, 3, \dots$

$$
\underset{n}{p}_{jk}^{(i)} = \sum_{\sigma=-\infty}^{\infty} \sum_{\rho=-\infty}^{\infty} \underset{n}{\tilde{p}}_{\sigma\rho}^{(ijk)} \zeta_{ij}^\sigma \zeta_{ik}^\rho, \tag{10.1.13}
$$

and so on. The starting point for this version is expansion (2.3.4) for function (2.3.1). Such ζ-series are quite simple to deal with but they are too lengthy (at least, for pairs of planets with a large ratio of the semi-major axes). The coefficients of (10.1.12) and (10.1.13) depend on the semi-major axes in a rather complicated manner so that version (a) implies the use of numerical values of the semi-major axes and planetary masses.

The second form, version (b) of the GPT, reduces to closed-form expansions involving Jacobi elliptic functions and integrals with moduli k_{ij}, amplitudes φ_{ij} and arguments u_{ij} given by (6.3.22). It also involves some quadratures of these functions. This form, which is valid only for the first-order theory ($k = 1$), was proposed by Brumberg (1994). The starting point of version (b) is representation (6.3.15) for function (2.3.1).

Finally, the third form, version (c) of the GPT, implies exponential series in the arguments $\mathrm{i}\pi u_{ij}/K(k_{ij})$ (Brumberg, 1992). Introducing τ_{ij} as in (6.3.26) one has to deal with one-argument series for $n = 1, 2, \dots,$

$$
\underset{n}{p}_j^{(i)} = \sum_{\sigma=-\infty}^{\infty} \underset{n}{p}_\sigma^{(ij)} \tau_{ij}^\sigma, \tag{10.1.14}
$$

two-argument series for $n = 2, 3, \ldots,$

$$\underset{n}{p}_{jk}^{(i)} = \sum_{\sigma=-\infty}^{\infty} \sum_{\rho=-\infty}^{\infty} \underset{n}{p}_{\sigma\rho}^{(ijk)} \tau_{ij}^{k} \tau_{ik}^{\rho}, \tag{10.1.15}$$

and so on. The coefficients of these series may be expanded in powers of the Jacobi nomes given by (6.3.27). As indicated in (6.3.29) the maximum value of q_{ij} for the Solar System is 0.215, which is comparatively small. Therefore, these τ-series are much shorter than the ζ-series of version (a). In contrast to version (b) they may be constructed for any order $k = 1, 2, \ldots$ in series (10.1.9). Moreover, it is feasible to construct a GPT in version (c) while retaining all the parameters in a purely analytical form.

In what follows we give some details of the construction of a GPT in these three versions.

10.2 Right-Hand Members

The right-hand members (7.2.7) and (7.2.8) can be expressed in terms of p_i, q_i and w_i ($i = 1, 2, \ldots, N$) either in the closed form (7.2.33) and (7.2.34) adapted for the iteration process or in the form of power series. These series resulting from expansion (6.2.58) of force function (6.2.53) are as follows:

$$P_i = \sum_{k=0}^{\infty}\sum_{l=0}^{\infty}\sum_{m=0}^{\infty} \varphi_{klm} p_i^k q_i^l w_i^{2m} + \\ + \mu \sum_{j=1}^{N}{}^{(i)} \kappa_{ij} \sum_{k=0}^{\infty}\sum_{l=0}^{\infty}\sum_{r=0}^{\infty}\sum_{s=0}^{\infty}\sum_{m=0}^{\infty}\sum_{t=-m}^{m} \psi_{klrsmt}^{(ij)} \times \\ \times p_i^k q_i^l p_j^r q_j^s w_i^{m-t} w_j^{m+t} \tag{10.2.1}$$

and

$$W_i = \sum_{k=0}^{\infty}\sum_{l=0}^{\infty}\sum_{m=0}^{\infty} \rho_{klm} p_i^k q_i^l w_i^{2m+1} + \\ + \mu \sum_{j=1}^{N}{}^{(i)} \kappa_{ij} \sum_{k=0}^{\infty}\sum_{l=0}^{\infty}\sum_{r=0}^{\infty}\sum_{s=0}^{\infty}\sum_{m=0}^{\infty}\sum_{t=-m-1}^{m} \theta_{klrsmt}^{(ij)} \times \\ \times p_i^k q_i^l p_j^r q_j^s w_i^{m-t} w_j^{m+t+1} \tag{10.2.2}$$

with

$$\varphi_{klm} = -\left(\delta_{k0}\delta_{l0} + \tfrac{1}{2}\delta_{k1}\delta_{l0} + \tfrac{3}{2}\delta_{k0}\delta_{l1}\right)\delta_{m0} + 2(l+1)A_{k,l+1,m}, \tag{10.2.3}$$

$$\psi_{klrsmt}^{(ij)} = 2(l+1)B_{k,l+1,r,s,m,t}^{(ij)}, \tag{10.2.4}$$

$$\rho_{klm} = \delta_{k0}\delta_{l0}\delta_{m0} + 2(m+1)A_{k,l,m+1} \tag{10.2.5}$$

and

$$\theta^{(ij)}_{klrsmt} = (m+1-t)B^{(ij)}_{k,l,r,s,m+1,t} \,. \tag{10.2.6}$$

The coefficients A_{klm} and $B^{(ij)}_{klrsmt}$ are determined by (6.2.59) and (6.2.60), respectively. It is evident that

$$\varphi_{000} = \varphi_{100} = \varphi_{010} = 0$$

and

$$\rho_{000} = 0 \,.$$

Expressions (7.2.33) or (10.2.1) are used for determining the intermediate quasi-periodic solution

$$\begin{aligned} p_i &= p_i^{(0)} \,, \\ w_i &= 0 \,. \end{aligned} \tag{10.2.7}$$

Putting now

$$p_i = p_i^{(0)} + \delta p_i \tag{10.2.8}$$

it is necessary to determine the right-hand members in terms of δp_i and w_i. Explicitly separating the zero- and first-degree terms with respect to eccentricities and inclinations one has

$$P_i = P_i^{(0)} - \sum_{j=1}^{N} (K_{ij}\delta p_j + L_{ij}\delta q_j) + P_i^* \tag{10.2.9}$$

and

$$W_i = -\sum_{j=1}^{N} M_{ij} w_j + W_i^* \tag{10.2.10}$$

with

$$\begin{aligned} M_{ii} &= -2K_{ii} \,, \\ M_{ij} &= -2K_{ij}\zeta_{ij} \quad (i \neq j). \end{aligned} \tag{10.2.11}$$

Assuming particular solution (10.2.7) to be known one can find a set of functions that depend only on this intermediary (Brumberg, 1974):

$$\begin{aligned}
A_i &= \left(1 - p_i^{(0)}\right)\left(\frac{a_i}{r_i^{(0)}}\right)^5, \qquad B_i = \left(\frac{a_i}{r_i^{(0)}}\right)^3,\\
C_i &= \left(1 - p_i^{(0)}\right)\left(\frac{a_i}{r_i^{(0)}}\right)^3,\\
D_i &= \left(1 - q_i^{(0)}\right)\left(\frac{a_i}{r_i^{(0)}}\right)^2,\\
E_i &= \left(\frac{a_i}{r_i^{(0)}}\right)^2, \qquad F_i = \left(1 - p_i^{(0)}\right)^2\left(\frac{a_i}{r_i^{(0)}}\right)^5
\end{aligned} \tag{10.2.12}$$

and

$$\begin{aligned}
A_{ij} &= \left[1 - p_i^{(0)} - \frac{a_j}{a_i}\left(1 - p_j^{(0)}\right)\zeta_{ij}^{-1}\right]\left(\frac{a_i}{\Delta_{ij}^{(0)}}\right)^5,\\
B_{ij} &= \left(\frac{a_i}{\Delta_{ij}^{(0)}}\right)^3,\\
C_{ij} &= \left[1 - p_i^{(0)} - \frac{a_j}{a_i}\left(1 - p_j^{(0)}\right)\zeta_{ij}^{-1}\right]\left(\frac{a_i}{\Delta_{ij}^{(0)}}\right)^3,\\
D_{ij} &= \left[1 - q_i^{(0)} - \frac{a_j}{a_i}\left(1 - q_j^{(0)}\right)\zeta_{ij}\right]\left(\frac{a_i}{\Delta_{ij}^{(0)}}\right)^2,\\
E_{ij} &= \left(\frac{a_i}{\Delta_{ij}^{(0)}}\right)^2,\\
F_{ij} &= \left[1 - p_i^{(0)} - \frac{a_j}{a_i}\left(1 - p_j^{(0)}\right)\zeta_{ij}^{-1}\right]^2\left(\frac{a_i}{\Delta_{ij}^{(0)}}\right)^5.
\end{aligned} \tag{10.2.13}$$

Computation of these functions involves no difficulties. Indeed, having found the auxiliary series

$$U_i(k,l) = \left(1 - p_i^{(0)}\right)^k \left(1 - q_i^{(0)}\right)^l$$

and

$$U_{ij}(k,l) = \left[1 - p_i^{(0)} - \frac{a_j}{a_i}\left(1 - p_j^{(0)}\right)\zeta_{ij}^{-1}\right]^k \left[1 - q_i^{(0)} - \frac{a_j}{a_i}\left(1 - q_j^{(0)}\right)\zeta_{ij}\right]^l$$

for the values $(k,l) = (3,5)$, $(3,3)$, $(1,3)$, $(1,0)$, $(1,1)$ and $(1,5)$ one has, in computing (10.2.12) and (10.2.13), to raise these series to the power $-M/2$ with $M = 1$ or $M = 2$. The operation

$$V = U^{-\frac{M}{2}}$$

has been considered in Section 1.2 . The initial value $V = V_0$ corresponds to unperturbed planar circular orbits, i.e. $V_0 = 1$ for (10.2.12) and

$$V_0 = \gamma\left(0,\ -\frac{1}{2}kM,\ -\frac{1}{2}lM,\ 0,\ \frac{a_j}{a_i},\ \zeta_{ij}\right)$$

for (10.2.13). One can also use the checking relations

$$A_i = C_i E_i\,, \qquad B_i = C_i D_i\,, \qquad A_i = D_i F_i \tag{10.2.14}$$

and similar ones for the two-index functions. The coefficients of the linear terms in (10.2.9) and (10.2.10) are expressed by

$$\begin{aligned} K_{ii} &= \frac{1}{2}\left(1 - B_i - \mu \sum_{j=1}^{N}{}^{(i)} \kappa_{ij} B_{ij}\right), \\ K_{ij} &= \frac{1}{2}\mu\kappa_{ij}\left[\frac{a_j}{a_i}B_{ij} - \left(\frac{a_i}{a_j}\right)^2 B_j\right]\zeta_{ij}^{-1} \quad (i \neq j) \end{aligned} \tag{10.2.15}$$

and

$$\begin{aligned} L_{ii} &= \frac{3}{2}\left(1 - F_i - \mu \sum_{j=1}^{N}{}^{(i)} \kappa_{ij} F_{ij}\right), \\ L_{ij} &= \frac{3}{2}\mu\kappa_{ij}\left[\frac{a_j}{a_i}\zeta_{ij}F_{ij} - \left(\frac{a_i}{a_j}\right)^2 \zeta_{ij}^{-1} F_j\right] \quad (i \neq j) \end{aligned} \tag{10.2.16}$$

with (10.2.11). If one further writes

$$S_i = D_i\delta p_i + \bar{D}_i\delta q_i - E_i(\delta p_i\delta q_i + w_i^2), \tag{10.2.17}$$

$$\begin{aligned} S_{ij} &= D_{ij}\left(\delta p_i - \frac{a_j}{a_i}\zeta_{ij}^{-1}\delta p_j\right) + \bar{D}_{ij}\left(\delta q_i - \frac{a_j}{a_i}\zeta_{ij}\delta q_j\right) - \\ &\quad - E_{ij}\left[\left(\delta p_i - \frac{a_j}{a_i}\zeta_{ij}^{-1}\delta p_j\right)\left(\delta q_i - \frac{a_j}{a_i}\zeta_{ij}\delta q_j\right) + \left(w_i - \frac{a_j}{a_i}w_j\right)^2\right] \end{aligned} \tag{10.2.18}$$

and introduces the functions $T_i = T(S_i)$, $T_{ij} = T(S_{ij})$ with

$$T(S) = (1 - S)^{-\frac{3}{2}} - 1 - \frac{3}{2}S = \sum_{k=2}^{\infty}\frac{(\frac{3}{2})_k}{(1)_k}S^k \tag{10.2.19}$$

then in virtue of the relations

$$\begin{aligned} \left(\frac{r_i}{a_i}\right)^2 &= \frac{1 - S_i}{E_i}, \\ \left(\frac{a_i}{r_i}\right)^3 &= B_i\left(1 + \frac{3}{2}S_i + T_i\right) \end{aligned} \tag{10.2.20}$$

and

$$\left(\frac{\Delta_{ij}}{a_i}\right)^2 = \frac{1-S_{ij}}{E_{ij}},$$
$$\left(\frac{a_i}{\Delta_{ij}}\right)^3 = B_{ij}\left(1+\frac{3}{2}S_{ij}+T_{ij}\right) \tag{10.2.21}$$

one obtains

$$\begin{aligned}P_i^* &= -\frac{3}{2}A_i\left(\delta p_i\delta q_i + w_i^2\right) - \frac{3}{2}B_iS_i\delta p_i + \left(C_i - B_i\delta p_i\right)T_i + \\ &+ \mu\sum_{j=1}^{N}{}^{(i)}\kappa_{ij}\left\{-\frac{3}{2}A_{ij}\left[\left(\delta p_i - \frac{a_j}{a_i}\zeta_{ij}^{-1}\delta p_j\right)\left(\delta q_i - \frac{a_j}{a_i}\zeta_{ij}\delta q_j\right)+\right.\right.\\ &\left. + \left(w_i - \frac{a_j}{a_i}w_j\right)^2\right] - \frac{3}{2}B_{ij}S_{ij}\left(\delta p_i - \frac{a_j}{a_i}\zeta_{ij}^{-1}\delta p_j\right) + \\ &+ \left[C_{ij} - B_{ij}\left(\delta p_i - \frac{a_j}{a_i}\zeta_{ij}^{-1}\delta p_j\right)\right]T_{ij} + \left(\frac{a_i}{a_j}\right)^2\left[-\frac{3}{2}A_j\left(\delta p_j\delta q_j + w_j^2\right) - \right.\\ &\left.\left. - \frac{3}{2}B_jS_j\delta p_j + \left(C_j - B_j\delta p_j\right)T_j\right]\zeta_{ij}^{-1}\right\}\end{aligned} \tag{10.2.22}$$

and

$$\begin{aligned}W_i^* &= -B_i\left(\tfrac{3}{2}S_i + T_i\right)w_i - \mu\sum_{j=1}^{N}{}^{(i)}\kappa_{ij}\left[B_{ij}\left(\tfrac{3}{2}S_{ij}+T_{ij}\right)\left(w_i - \frac{a_j}{a_i}w_j\right)+\right.\\ &\left. + \left(\frac{a_i}{a_j}\right)^2 B_j\left(\tfrac{3}{2}S_j + T_j\right)w_j\right].\end{aligned} \tag{10.2.23}$$

These expansions are well adapted to the iterative process (Brumberg, 1974). P_i^* and W_i^* may also be presented in the form of power series in δp_i, δq_i, w_i ($i = 1, 2, \ldots, N$) as follows:

$$\begin{aligned}P_i^* &= \sum_{\substack{k=0\\k+l+2m>1}}^{\infty}\sum_{l=0}^{\infty}\sum_{m=0}^{\infty} P_{klm}^{(i)}(\delta p_i)^k(\delta q_i)^l w_i^{2m} + \\ &+ \mu\sum_{j=1}^{N}{}^{(i)}\sum_{\substack{k=0\\k+l+r+s+2m>1}}^{\infty}\sum_{l=0}^{\infty}\sum_{r=0}^{\infty}\sum_{s=0}^{\infty}\sum_{m=0}^{\infty}\sum_{t=-m}^{m} P_{klrsmt}^{(ij)}\times \\ &\times(\delta p_i)^k(\delta q_i)^l(\delta p_j)^r(\delta q_j)^s w_i^{m-t}w_j^{m+t}\end{aligned} \tag{10.2.24}$$

and

$$\begin{aligned}W_i^* &= \sum_{\substack{k=0\\k+l+2m>0}}^{\infty}\sum_{l=0}^{\infty}\sum_{m=0}^{\infty} W_{klm}^{(i)}(\delta p_i)^k(\delta q_i)^l w_i^{2m+1} + \\ &+ \mu\sum_{j=1}^{N}{}^{(i)}\sum_{\substack{k=0\\k+l+r+s+2m>0}}^{\infty}\sum_{l=0}^{\infty}\sum_{r=0}^{\infty}\sum_{s=0}^{\infty}\sum_{m=0}^{\infty}\sum_{t=-m-1}^{m} W_{klrsmt}^{(ij)}\times \\ &\times(\delta p_i)^k(\delta q_i)^l(\delta p_j)^r(\delta q_j)^s w_i^{m-t}w_j^{m+t+1}\end{aligned} \tag{10.2.25}$$

with the coefficients

$$P^{(i)}_{klm} = (-1)^m \frac{\left(\frac{3}{2}\right)_m \left(\frac{1}{2}+m\right)_k \left(\frac{3}{2}+m\right)_l}{(1)_m(1)_k(1)_l} \times \\ \times \left(1-p_i^{(0)}\right)^{-\frac{1}{2}-m-k} \left(1-q_i^{(0)}\right)^{-\frac{3}{2}-m-l} - \\ - \left[\left(1+\tfrac{1}{2}p_i^{(0)}+\tfrac{3}{2}q_i^{(0)}\right)\delta_{k0}\delta_{l0}+\tfrac{1}{2}\delta_{k1}\delta_{l0}+\tfrac{3}{2}\delta_{k0}\delta_{l1}\right]\delta_{m0}\,, \qquad (10.2.26)$$

$$P^{(ij)}_{klrsmt} = \\ = \frac{(-1)^t \left(\frac{3}{2}\right)_m \left(\frac{1}{2}+m\right)_{k+r} \left(\frac{3}{2}+m\right)_{l+s} (1+r)_k(1+s)_l(1+m-t)_{m+t}}{(1)_m(1)_{k+r}(1)_{l+s}(1)_k(1)_l(1)_{m+t}} \times \\ \times \kappa_{ij}\Bigg[\Gamma\Big(r+s+m+t, -\tfrac{1}{2}-m-k-r, -\tfrac{3}{2}-m-l-s, -r+s, \\ \frac{a_j}{a_i}, p_i^{(0)}, p_j^{(0)}, \zeta_{ij}\Big) + \\ + \delta_{k0}\delta_{l0}\delta_{tm}\left(\frac{a_i}{a_j}\right)^2 \zeta_{ij}^{-1}\left(1-p_j^{(0)}\right)^{-\frac{1}{2}-m-r}\left(1-q_j^{(0)}\right)^{-\frac{3}{2}-m-s}\Bigg], \qquad (10.2.27)$$

$$W^{(i)}_{klm} = (-1)^{m+1} \frac{\left(\frac{3}{2}\right)_m \left(\frac{3}{2}+m\right)_k \left(\frac{3}{2}+m\right)_l}{(1)_m(1)_k(1)_l} \times \\ \times \left(1-p_i^{(0)}\right)^{-\frac{3}{2}-m-k} \left(1-q_i^{(0)}\right)^{-\frac{3}{2}-m-l} + \delta_{k0}\delta_{l0}\delta_{m0} \qquad (10.2.28)$$

and

$$W^{(ij)}_{klrsmt} = \\ = \frac{(-1)^t \left(\frac{3}{2}\right)_m \left(\frac{3}{2}+m\right)_{k+r} \left(\frac{3}{2}+m\right)_{l+s} (1+r)_k(1+s)_l(1+m-t)_{m+t+1}}{(1)_m(1)_{k+r}(1)_{l+s}(1)_k(1)_l(1)_{m+t+1}} \times \\ \times \kappa_{ij}\Bigg[\Gamma\Big(r+s+m+t+1, -\tfrac{3}{2}-m-k-r, -\tfrac{3}{2}-m-l-s, -r+s, \\ \frac{a_j}{a_i}, p_i^{(0)}, p_j^{(0)}, \zeta_{ij}\Big) - \\ - \delta_{k0}\delta_{l0}\delta_{tm}\left(\frac{a_i}{a_j}\right)^2 \left(1-p_j^{(0)}\right)^{-\frac{3}{2}-m-r}\left(1-q_j^{(0)}\right)^{-\frac{3}{2}-m-s}\Bigg]. \qquad (10.2.29)$$

Here we have introduced a function

$$\Gamma(n,x,y,\nu,\alpha,u,v,\zeta) = \\ = \alpha^n\left[1-u-\alpha(1-v)\zeta^{-1}\right]^x\left[1-\bar{u}-\alpha(1-\bar{v})\zeta\right]^y(-\zeta)^\nu\,, \qquad (10.2.30)$$

which is a straightforward generalization of (2.3.1). Again if $\alpha > 1$ one should use the relation

$$\Gamma(n,x,y,\nu,\alpha,u,v,\zeta) = \Gamma(-n-x-y,y,x,\nu-x+y,\alpha^{-1},\bar{v},\bar{u},\zeta)\,. \qquad (10.2.31)$$

Evidently, one also has

$$\Gamma(n, x, y, \nu, \alpha, \bar{u}, \bar{v}, \zeta^{-1}) = \Gamma(n, y, x, -\nu, \alpha, u, v, \zeta). \tag{10.2.32}$$

Expressions (10.2.26)–(10.2.29) result immediately from defining relations (7.2.24) and (7.2.25), taking into account decomposition (10.2.8) and the expansions

$$\left(\frac{a_i}{r_i}\right)^n = \sum_{k=0}^{\infty}\sum_{l=0}^{\infty}\sum_{m=0}^{\infty} \alpha_{klm}^{(n)} \left(1 - p_i^{(0)}\right)^{-\frac{n}{2}-m-k} \left(1 - q_i^{(0)}\right)^{-\frac{n}{2}-m-l} \times \\ \times (\delta p_i)^k (\delta q_i)^l w_i^{2m} \tag{10.2.33}$$

with (6.2.46) and

$$\left(\frac{a_i}{\Delta_{ij}}\right)^n = \sum_{k=0}^{\infty}\sum_{l=0}^{\infty}\sum_{r=0}^{\infty}\sum_{s=0}^{\infty}\sum_{m=0}^{\infty}\sum_{t=-m}^{m} (\delta p_i)^k (\delta q_i)^l (\delta p_j)^r (\delta q_j)^s w_i^{m-t} w_j^{m+t} \times \\ \times \frac{(-1)^t \left(\frac{n}{2}\right)_m \left(\frac{n}{2}+m\right)_{k+r} \left(\frac{n}{2}+m\right)_{l+s} (1+r)_k (1+s)_l (1+m-t)_{m+t}}{(1)_m (1)_{k+r} (1)_{l+s} (1)_k (1)_l (1)_{m+t}} \times \\ \times \Gamma\Big(r+s+m+t, -\frac{n}{2}-m-k-r, -\frac{n}{2}-m-l-s, -r+s, \\ \frac{a_j}{a_i}, p_i^{(0)}, p_j^{(0)}, \zeta_{ij}\Big). \tag{10.2.34}$$

The coefficients K_{ij}, L_{ij}, M_{ij}, $P_{klm}^{(i)}$, $P_{klrsmt}^{(ij)}$, $W_{klm}^{(i)}$ and $W_{klrsmt}^{(ij)}$ of right-hand members (10.2.9) and (10.2.10) may also be represented as series in powers of the intermediate solution $p_i^{(0)}$ and $q_i^{(0)}$. The linear-term coefficients are, evidently,

$$K_{ii} = -P_{100}^{(i)} - \mu \sum_{j=1}^{N}{}^{(i)} P_{100000}^{(ij)}, \\ K_{ij} = -\mu P_{001000}^{(ij)}, \quad (i \neq j) \tag{10.2.35}$$

$$L_{ii} = -P_{010}^{(i)} + \mu \sum_{j=1}^{N}{}^{(i)} P_{010000}^{(ij)}, \\ L_{ij} = -\mu P_{000100}^{(ij)}, \quad (i \neq j) \tag{10.2.36}$$

$$M_{ii} = -W_{000}^{(i)} - \mu \sum_{j=1}^{N}{}^{(i)} W_{00000-1}^{(ij)}, \\ M_{ij} = -\mu W_{000000}^{(ij)}. \quad (i \neq j) \tag{10.2.37}$$

Function (10.2.30) for small u and v may be easily expanded in the power series

$$\Gamma(n,x,y,\nu,\alpha,u,v,\zeta) = \sum_{k=0}^{\infty}\sum_{l=0}^{\infty}\sum_{r=0}^{\infty}\sum_{s=0}^{\infty} \frac{(-x)_{k+r}(-y)_{l+s}(1+r)_k(1+s)_l}{(1)_{k+r}(1)_{l+s}(1)_k(1)_l} \times$$
$$\times\, \gamma(n+r+s, x-k-r, y-l-s, \nu-r+s, \alpha, \zeta) u^k \bar{u}^l v^r \bar{v}^s \,. \qquad (10.2.38)$$

Hence, we get

$$P^{(i)}_{\alpha\beta m} = \sum_{k=0}^{\infty}\sum_{l=0}^{\infty} \frac{(1+k)_\alpha(1+l)_\beta}{(1)_\alpha(1)_\beta} \varphi_{\alpha+k,\beta+l,m}\, p_i^k q_i^l \,, \qquad (10.2.39)$$

$$P^{(ij)}_{\alpha\beta\sigma\nu mt} = \kappa_{ij} \sum_{k=0}^{\infty}\sum_{l=0}^{\infty}\sum_{r=0}^{\infty}\sum_{s=0}^{\infty} \frac{(1+k)_\alpha(1+l)_\beta(1+r)_\sigma(1+s)_\nu}{(1)_\alpha(1)_\beta(1)_\sigma(1)_\nu} \times$$
$$\times\, \psi^{(ij)}_{\alpha+k,\beta+l,\sigma+r,\nu+s,m,t}\, p_i^k q_i^l p_j^r q_j^s \,, \qquad (10.2.40)$$

$$W^{(i)}_{\alpha\beta m} = \sum_{k=0}^{\infty}\sum_{l=0}^{\infty} \frac{(1+k)_\alpha(1+l)_\beta}{(1)_\alpha(1)_\beta} \rho_{\alpha+k,\beta+l,m}\, p_i^k q_i^l \,, \qquad (10.2.41)$$

and

$$W^{(ij)}_{\alpha\beta\sigma\nu mt} = \kappa_{ij} \sum_{k=0}^{\infty}\sum_{l=0}^{\infty}\sum_{r=0}^{\infty}\sum_{s=0}^{\infty} \frac{(1+k)_\alpha(1+l)_\beta(1+r)_\sigma(1+s)_\nu}{(1)_\alpha(1)_\beta(1)_\sigma(1)_\nu} \times$$
$$\times\, \theta^{(ij)}_{\alpha+k,\beta+l,\sigma+r,\nu+s,m,t}\, p_i^k q_i^l p_j^r q_j^s \,. \qquad (10.2.42)$$

Expressions (10.2.39)–(10.2.42) should be evaluated by substituting the coordinates of intermediary (10.2.7).

10.3 The Intermediary

In version (a) of the GPT the particular quasi-periodic intermediate solution is given by ζ-series (10.1.10)–(10.1.13) or in the general form

$$p_i^{(0)} = \sum p_k^{(i)} \exp \mathrm{i}(k\lambda) \qquad (10.3.1)$$

with

$$(k\lambda) = k_1\lambda_1 + \ldots + k_N\lambda_N \qquad (10.3.2)$$

and the condition

$$k_1 + k_2 + \ldots + k_N = 0 \qquad (10.3.3)$$

for the summation multi-index k. Such series have been considered in Section 7.2 in combination with iteration techniques. If at some step of the process of approximation the right-hand members P_i are represented by series like (10.3.1) with coefficients $P_k^{(i)}$ then the corresponding coefficients of the intermediary result directly from equations (7.2.5):

$$p_0^{(i)} = -\tfrac{1}{3} P_0^{(i)}\,, \tag{10.3.4}$$

and

$$p_k^{(i)} = n_i^2 \frac{\left[(kn)^2 - 2n_i(kn) + \frac{3}{2}n_i^2\right] P_k^{(i)} - \frac{3}{2} n_i^2 P_{-k}^{(i)}}{(kn)^2\left[n_i^2 - (kn)^2\right]}\,. \tag{10.3.5}$$

Such an intermediary has been analyzed in detail by Brumberg et al. (1975).

The starting relations for version (b) of the GPT (Brumberg, 1994) were given in Section 6.3 . First of all, to construct the first-order intermediary one has to deal with quadratures as follows:

$$\int \frac{d\varphi}{\delta^3} = \frac{E(\varphi,k)}{k'^2} - \frac{k^2}{k'^2}\frac{\sin\varphi\cos\varphi}{\delta}\,, \tag{10.3.6}$$

$$\int \frac{\cos 2\varphi}{\delta^3}\, d\varphi = \frac{2}{k^2}F(\varphi,k) - \frac{2-k^2}{k^2k'^2}E(\varphi,k) + \frac{2-k^2}{k'^2}\frac{\sin\varphi\cos\varphi}{\delta}\,, \tag{10.3.7}$$

$$\int \frac{\sin 2\varphi}{\delta^3}\, d\varphi = \frac{2}{k^2}\left(\frac{1}{\delta}\right)^+, \quad \left(\frac{1}{\delta}\right)^+ = \frac{1}{\delta} - \frac{2}{\pi}K(k)\,, \tag{10.3.8}$$

$$\int \frac{\cos 4\varphi}{\delta^3}\, d\varphi = -\frac{8(2-k^2)}{k^4}F(\varphi,k) + \frac{16-16k^2+k^4}{k^4k'^2}E(\varphi,k) - \frac{8-8k^2+k^4}{k^2k'^2}\frac{\sin\varphi\cos\varphi}{\delta}\,, \tag{10.3.9}$$

$$\int \frac{\sin 4\varphi}{\delta^3}\, d\varphi = -\frac{4(2-k^2)}{k^4}\left(\frac{1}{\delta}\right)^+ - \frac{8}{k^4}\delta^+, \quad \delta^+ = \delta - \frac{2}{\pi}E(k)\,, \tag{10.3.10}$$

$$\int \frac{d\varphi}{\delta^5} = -\frac{1}{3k'^2}F(\varphi,k) + \frac{2(2-k^2)}{3k'^4}E(\varphi,k) - \frac{k^2}{3k'^2}\left[\frac{1}{\delta^2} + \frac{2(2-k^2)}{k'^2}\right]\frac{\sin\varphi\cos\varphi}{\delta}\,, \tag{10.3.11}$$

$$\int \frac{\cos 2\varphi}{\delta^5}\, d\varphi = \frac{2-k^2}{3k^2k'^2}F(\varphi,k) - \frac{2(1-k^2+k^4)}{3k^2k'^4}E(\varphi,k) + \frac{1}{3k^2}\left[\frac{2-k^2}{\delta^2} + \frac{2(1-k^2+k^4)}{k'^2}\right]\frac{\sin\varphi\cos\varphi}{\delta}\,, \tag{10.3.12}$$

$$\int \frac{\sin 2\varphi}{\delta^5}\, d\varphi = \frac{2}{3k^2}\left(\frac{1}{\delta^3}\right)^+, \quad \left(\frac{1}{\delta^3}\right)^+ = \frac{1}{\delta^3} - \frac{2E(k)}{\pi k'^2}\,, \tag{10.3.13}$$

$$\int \frac{\cos 4\varphi}{\delta^5}\, d\varphi = \frac{16-16k^2-k^4}{3k^4k'^2}F(\varphi,k) - \frac{2(8-12k^2+2k^4+k^6)}{3k^4k'^4}E(\varphi,k) + \frac{1}{3k^2k'^2}\left[-\frac{8-8k^2+k^4}{\delta^2} + \frac{2(8-12k^2+2k^4+k^6)}{k'^2}\right]\frac{\sin\varphi\cos\varphi}{\delta} \tag{10.3.14}$$

and

$$\int \frac{\sin 4\varphi}{\delta^5}\,d\varphi = -\frac{4(2-k^2)}{3k^4}\left(\frac{1}{\delta^3}\right)^+ + \frac{8}{k^4}\left(\frac{1}{\delta}\right)^+ . \tag{10.3.15}$$

This list extends the similar list of integrals given by Williams et al. (1987). In (10.3.8), (10.3.10), (10.3.13) and (10.3.15) the arbitrary constants of integration are added in such a way that the mean values (with respect to φ) of the corresponding expressions are zero. In general, designating the mean value by angle brackets one has

$$\langle \delta^{-n} \rangle = \tfrac{1}{2}(a+a')^n c_n^{(0)}(a,a')$$

with values (6.3.6), (6.3.8) and (6.3.11). Then, from (10.3.7)–(10.3.10) and (10.3.12)–(10.3.15) we have

$$\left\langle \frac{\exp i2\varphi}{\delta^3} \right\rangle = \frac{4K(k)}{\pi k^2} - \frac{2(2-k^2)}{\pi k^2 k'^2}E(k)\,, \tag{10.3.16}$$

$$\left\langle \frac{\exp i4\varphi}{\delta^3} \right\rangle = -\frac{16(2-k^2)}{\pi k^4}K(k) + \frac{2(16-16k^2+k^4)}{\pi k^4 k'^2}E(k)\,, \tag{10.3.17}$$

$$\left\langle \frac{\exp i2\varphi}{\delta^5} \right\rangle = \frac{2(2-k^2)}{3\pi k^2 k'^2}K(k) - \frac{4(1-k^2+k^4)}{3\pi k^2 k'^4}E(k) \tag{10.3.18}$$

and

$$\left\langle \frac{\exp i4\varphi}{\delta^5} \right\rangle = \frac{2(16-16k^2-k^4)}{3\pi k^4 k'^2}K(k) - \frac{4(8-12k^2+2k^4+k^6)}{3\pi k^4 k'^4}E(k)\,. \tag{10.3.19}$$

Taking into account (6.3.18) and the differential relation

$$d\varphi = \mathrm{dn}\,u\,du \tag{10.3.20}$$

one can rewrite expressions (10.3.6)–(10.3.15) in terms of elliptic functions. To separate the secular and periodic parts of the elliptic integrals one can use the relations

$$E(\varphi, k) = \frac{E}{K}u + Z(u)\,, \tag{10.3.21}$$

$$\mathrm{am}\,u = \frac{\pi}{2K}u + \mathrm{am}^+ u \tag{10.3.22}$$

with the purely periodic Jacobi zeta function $Z(u)$

$$Z(u) = \frac{2\pi}{K}\sum_{m=1}^{\infty}\frac{q^m}{1-q^{2m}}\sin m\frac{\pi u}{K} \tag{10.3.23}$$

and the purely periodic part $\mathrm{am}^+ u$ of the amplitude function determined from (2.5.27). From (10.3.21) and (10.3.22) it follows that

$$F^+(\varphi,\,k) \equiv F(\varphi,\,k) - \frac{2K}{\pi}\varphi = -\frac{2K}{\pi}\,\mathrm{am}^+\,u \tag{10.3.24}$$

and

$$E^+(\varphi,\,k) \equiv E(\varphi,\,k) - \frac{2E}{\pi}\varphi = Z(u) - \frac{2E}{\pi}\,\mathrm{am}^+\,u \tag{10.3.25}$$

with purely periodic right-hand members. Numerical techniques to evaluate these and many other quantities involving elliptic functions and integrals have been proposed by Fukushima (1991). In what follows, dealing with π-periodic functions $f = f(\varphi)$ we shall separate the constant part of such functions by writing

$$f(\varphi) = f^* + f^+(\varphi) \tag{10.3.26}$$

with

$$f^* = \langle f\rangle = \frac{1}{\pi}\int_0^\pi f(\varphi)\,d\varphi = \frac{1}{\pi}\int_0^{2K} f(\mathrm{am}\,u)\,dnu\,du\,. \tag{10.3.27}$$

Right-hand members (10.2.1) for the first-order intermediary are reduced to the simple expression

$$P_i = \mu\sum_{j=1}^{N}{}^{(i)}\kappa_{ij}\left[\left(\frac{a_i}{a_j}\right)^2\zeta_{ij}^{-1} + \frac{a_i^2}{\Delta_{ij}^3}(a_i - a_j\zeta_{ij}^{-1})\right], \tag{10.3.28}$$

where Δ_{ij} should be taken for the coplanar circular orbits. The first-order intermediary $\underset{1}{p}^{(i)}$ of form (10.1.10) may be obtained by quadratures (7.2.19). Omitting the technical details given by Brumberg (1994) one has as the final, closed-form intermediate solution

$$\langle \underset{1}{p}_j^{(i)}\rangle = -\frac{\kappa_{ij}}{3\pi}\left[\frac{a_i}{a_i+a_j}K(k_{ij}) + \frac{a_i}{a_i-a_j}E(k_{ij})\right] \tag{10.3.29}$$

and

$$\begin{aligned}
\underset{1}{p}_j^{(i)+} &= \kappa_{ij}m_{ij}\Bigg\{\frac{m_{ij}}{1-m_{ij}^2}\left(\frac{a_i}{a_j}\right)^2\Big[(1+2m_{ij})\cos 2\varphi_{ij} + \\
&\quad + \mathrm{i}(1+2m_{ij}+3m_{ij}^2)\sin 2\varphi_{ij}\Big] - \\
&\quad - \frac{2a_i}{a_i+a_j}\left[\left(\frac{1}{\delta_{ij}}\right)^+ - \mathrm{i}(1-3m_{ij})F^+(\varphi_{ij}\,,\,k_{ij})\right] - \\
&\quad - \mathrm{i}\frac{a_i}{a_i-a_j}\left[k_{ij}^2\frac{\sin 2\varphi_{ij}}{\delta_{ij}} - 2E^+(\varphi_{ij}\,,\,k_{ij})\right] - \\
&\quad - \frac{1}{4}\mathrm{i}\frac{a_i^2k_{ij}^2}{a_j(a_i+a_j)}\left[3\Phi_3^+(\varphi_{ij}\,,k_{ij}\,,0\,,2m_{ij}) + \Phi_3^+(\varphi_{ij}\,,k_{ij}\,,0\,,-2m_{ij})\right] - \\
&\quad - \frac{1}{8}\mathrm{i}\frac{a_ik_{ij}^2}{a_i+a_j}\Big[9\Phi_3^+(\varphi_{ij}\,,k_{ij}\,,2\,,2m_{ij}) - 3\Phi_3^+(\varphi_{ij}\,,k_{ij}\,,-2\,,2m_{ij}) + \\
&\quad + 3\Phi_3^+(\varphi_{ij}\,,k_{ij}\,,-2\,,-2m_{ij}) - \Phi_3^+(\varphi_{ij}\,,k_{ij}\,,2\,,-2m_{ij})\Big]\Bigg\}
\end{aligned} \tag{10.3.30}$$

with

$$m_{ij} = \frac{n_i}{n_i - n_j} . \tag{10.3.31}$$

Here one meets integrals of the type

$$\Phi_n(\varphi, k, s, \alpha) = \int \frac{1}{\delta^n} \exp \mathrm{i}\left[s\varphi + \alpha(\tilde{\varphi} - \varphi)\right] d\varphi \tag{10.3.32}$$

with integers n, s and irrational α. The quantity $\tilde{\varphi}$ is a Hansen-type variable to be considered constant under integration and taken as φ outside the integral sign. The function Φ_n satisfies the differential equation

$$\frac{d\Phi_n}{d\varphi} - \mathrm{i}\,\alpha\Phi_n = \frac{\exp \mathrm{i}s\varphi}{\delta^n} . \tag{10.3.33}$$

With respect to φ this function is again split as in (10.3.26) and expression (10.3.30) explicitly contains only a purely trigonometric part Φ_n^+ of this function. The constant part of this function may be easily evaluated by expanding δ^{-n} with the aid of (6.3.1), (6.3.4b) and (6.3.16):

$$\frac{1}{\delta^n} = \frac{1}{2}(a + a')^n \sum_{j=-\infty}^{\infty} (-1)^j c_n^{(j)}(a, a') \exp \mathrm{i}\,2j\varphi \tag{10.3.34}$$

and, hence,

$$\Phi_n(\varphi, k, s, \alpha) = \frac{1}{2}\mathrm{i}(a + a')^n \sum_{j=-\infty}^{\infty} (-1)^j c_n^{(j)}(a, a') \frac{\exp \mathrm{i}(2j + s)\varphi}{\alpha - s - 2j} . \tag{10.3.35}$$

This formula is not efficient for actually calculating Φ_n for large k. But it is useful in two respects. It shows that owing to the adequate choice of the constant of integration in (10.3.32) α does not affect the angular frequencies in (10.3.35). Then, from (10.3.35) it follows that

$$\Phi_n^*(\varphi, k, s, \alpha) = \begin{cases} 0, & s \text{ is odd}, \\ \frac{\mathrm{i}}{2\alpha}(-1)^{\frac{s}{2}}(a + a')^n c_n^{(s/2)}(a, a'), & s \text{ is even}. \end{cases} \tag{10.3.36}$$

Using (6.3.4b) one may also write

$$\Phi_n^*(\varphi, k, 2j, \alpha) = \frac{\mathrm{i}}{\alpha}(-1)^{|j|} \frac{\left(\frac{n}{2}\right)_{|j|}}{(1)_{|j|}} \left(\frac{k}{2}\right)^{2|j|} F\left(\frac{1}{2} + |j|, \frac{n}{2} + |j|, 1 + 2|j|, k^2\right) . \tag{10.3.37}$$

In particular,

$$\Phi_3^*(\varphi, k, 0, \alpha) = \mathrm{i}\,\frac{2E(k)}{\pi\alpha k'^2} , \tag{10.3.38}$$

and

$$\Phi_3^*(\varphi,k,2,\alpha)=\Phi_3^*(\varphi,k,-2,\alpha)=\frac{2\mathrm{i}}{\pi\alpha k^2}\left[2K(k)-\frac{2-k^2}{k'^2}E(k)\right]. \quad (10.3.39)$$

From definition (10.3.32) it follows that

$$\bar{\Phi}_n(\varphi,k,s,\alpha)=\Phi_n(\varphi,k,-s,-\alpha) \quad (10.3.40)$$

so that the number of functions Φ_n^+ with different parameters in (10.3.30) may be halved by using the conjugate functions.

In version (c) of the GPT the intermediate solution is constructed in the same manner as in version (a), replacing ζ-series by τ-series. One-argument trigonometric ζ-series for $P_j^{(i)}$ may be found, for example, by iteration formula resulting from (7.2.19):

$$\begin{aligned} p_j^{(i)} = &-\frac{1}{3}P_j^{(i)*}-\frac{1}{2}\mathrm{i}\,m_{ij}\int\Big[3\left(3P_j^{(i)+}-\bar{P}_j^{(i)+}\right)\exp\mathrm{i}\,2m_{ij}(\tilde{\varphi}_{ij}-\varphi_{ij})+ \\ &+\left(3\bar{P}_j^{(i)+}-P_j^{(i)+}\right)\exp\mathrm{i}\,2m_{ij}(\varphi_{ij}-\tilde{\varphi}_{ij})\Big]\,d\varphi_{ij}+ \\ &+\mathrm{i}\,4m_{ij}\int P_j^{(i)+}\,d\varphi_{ij}-6m_{ij}^2\int\!\!\int\left(P_j^{(i)+}-\bar{P}_j^{(i)+}\right)d\varphi_{ij}\,d\varphi_{ij} \end{aligned} \quad (10.3.41)$$

with value (10.3.31), differential relationships (6.3.30) and one-argument τ-series $P_j^{(i)}$ of the right-hand members. The τ-series for the first-order intermediary may also be obtained just from representation (10.3.30) treated as being of closed form with respect to the functions $\mathrm{i}\sin 2\varphi$, $\cos 2\varphi$, $\mathrm{i}\sin 2\varphi/\operatorname{dn}u$, $1/\operatorname{dn}u$, $\mathrm{i}\operatorname{am}^+u$, $\mathrm{i}\,Z(u)$ and $\mathrm{i}\Phi_n(\varphi,k,s,\alpha)$. The τ-expansions for the first six functions follow directly from standard series (2.5.70), (2.5.69), (2.5.77), (2.5.74), (2.5.27) and (10.3.23), respectively. As for integral (10.3.32), it may be rewritten in terms of the argument u in the form

$$\Phi_n(\varphi,k,s,\alpha)=\exp\mathrm{i}(\alpha\operatorname{am}u)\int\frac{1}{(\operatorname{dn}u)^{n-1}}\exp\left[\mathrm{i}(s-\alpha)\operatorname{am}u\right]du. \quad (10.3.42)$$

The function $1/(\operatorname{dn}u)^n$ may be easily expanded in the series

$$\left(\frac{1}{\operatorname{dn}u}\right)^n=\sum_{m=-\infty}^{\infty}D_m^{(n)}\exp\mathrm{i}\,m\frac{\pi u}{K},\quad D_{-m}^{(n)}=D_m^{(n)}. \quad (10.3.43)$$

Indeed, starting with series (2.5.74) and (2.5.78) for $1/\operatorname{dn}u$ and $1/(\operatorname{dn}u)^2$ one may find the expansions for $1/(\operatorname{dn}u)^n$ successively for $n=3,4,\ldots$ with the aid of the relation

$$\frac{(n-1)k'^2}{(\operatorname{dn}u)^n}=\frac{(n-2)(2-k^2)}{(\operatorname{dn}u)^{n-2}}-\frac{n-3}{(\operatorname{dn}u)^{n-4}}-\frac{1}{n-2}\frac{d^2}{du^2}\frac{1}{(\operatorname{dn}u)^{n-2}}. \quad (10.3.44)$$

On the other hand, proceeding from expansion (2.5.27) for am u one may generalize expansion (2.5.26) for any real number p (Brumberg, 1992; Klioner, 1992):

$$\exp \mathrm{i}(p \operatorname{am} u) = \sum_{k=-\infty}^{\infty} A_k^{(p)} \exp \mathrm{i}(p+2k)\frac{\pi u}{2K}, \quad A_{-k}^{(p)} = A_k^{(-p)}, \qquad (10.3.45)$$

p being a real number. The problem of determining $A_k^{(p)}$ is a particular case of the problem considered in Section 7.1 of producing expansion (7.1.39) for $\exp \mathrm{i} px$ with x defined by expansion (7.1.35). In the present case

$$x = \operatorname{am} u, \quad s_1 = \frac{\pi u}{2K}, \quad \frac{n'}{n} d_m = \frac{2}{m}\frac{q^m}{1+q^{2m}}, \quad w = \frac{\pi u}{K} \qquad (10.3.46)$$

and formula (7.1.40) remains valid for the coefficients $A_k^{(p)}$ as well. Substituting (10.3.43) and (10.3.45) into (10.3.42) and integrating the product of two Fourier series we obtain

$$\Phi_n(\varphi, k, s, \alpha) = \mathrm{i} \sum_{m=-\infty}^{\infty} \Lambda_m(k, s, \alpha, n) \exp \mathrm{i}(s+2m)\frac{\pi u}{2K} \qquad (10.3.47)$$

with the coefficients

$$\Lambda_m(k, s, \alpha, n) = \frac{2K}{\pi} \sum_{l=-\infty}^{\infty} \frac{A_l^{(\alpha)}}{\alpha - s - 2m - 2l} \sum_{k=-\infty}^{\infty} A_k^{(s-\alpha)} D_{m-l-k}^{(n-1)}. \qquad (10.3.48)$$

The coefficients $D_k^{(n)}$ and $A_k^{(n)}$ are of the order $q^{|k|}$. This enables one to fix the finite ranges of summation in (10.3.48). One thus has

$$\Lambda_{-m}(k, s, \alpha, n) = -\Lambda_m(k, -s, -\alpha, n)$$

with the order of smallness $q^{|m|}$. The integer s occurring in (10.3.47) may have only even values in the GPT so that expansion (10.3.47) has the form of a τ-series. This completes the construction of the first-order intermediary in the form of τ-series.

Constructing the GPT intermediary beyond the first order and dealing with (10.3.41) for more complicated right-hand members such as $P_{jk}^{(i)}$, for example, one should integrate two-argument trigonometric series

$$I = \int f(u_{ij}, u_{ik})\, du_{ij} \qquad (10.3.49)$$

for the triplet of planets i, j and k. Of course, it is always possible to make the transformation from u_{ij} and u_{ik} to M_{ij} and M_{ik}, defined in accordance with (6.3.32), using the inversion of (6.3.33). Another way is again to apply the Hansen device as detailed in Section 7.1 . From (6.3.31) it follows that

$$2\,\mathrm{am}\,u_{ik} = 2\frac{n_i - n_k}{n_i - n_j}\,\mathrm{am}\,u_{ij} + \varepsilon_{ijk} \tag{10.3.50}$$

with

$$\varepsilon_{ijk} = \pi - (\varepsilon_i - \varepsilon_k) - \frac{n_i - n_k}{n_i - n_j}\left[\pi - (\varepsilon_i - \varepsilon_j)\right]. \tag{10.3.51}$$

As in (6.3.32) and (6.3.33) one has

$$w_{ik} + \sum_{m=1}^{\infty} d_m(q_{ik}) \sin m w_{ik} = x \tag{10.3.52}$$

with

$$x = w^*_{ijk} + \frac{n_i - n_k}{n_i - n_j} \sum_{m=1}^{\infty} d_m(q_{ij}) \sin m w_{ij}\,, \tag{10.3.53}$$

where w^*_{ijk} represents the linear function of w_{ij}

$$w^*_{ijk} = \frac{n_i - n_k}{n_i - n_j} w_{ij} + \varepsilon_{ijk}\,. \tag{10.3.54}$$

The conversion of (10.3.52) results in

$$w_{ik} = x + \sum_{m=1}^{\infty} c_m(q_{ik}) \sin m x\,. \tag{10.3.55}$$

Since any function of x may be expanded in trigonometric series in multiples of w_{ij} and w^*_{ijk}, the integration of (10.3.49) can be performed in a straightforward manner. Then w^*_{ijk} should be expressed from (10.3.52) and (10.3.53) in terms of w_{ij} and w_{ik} to restore the original form of the τ-series.

10.4 Linear Eccentricity and Inclination Terms

Having determined intermediary (10.2.7) one has to deal with the equations in $(\delta p_i,\, w_i)$ variables

$$\delta\ddot{p}_i + 2\,\mathrm{i}\,n_i \delta\dot{p}_i + n_i^2 \sum_{j=1}^{N}\left[\left(-\tfrac{3}{2}\delta_{ij} + K_{ij}\right)\delta p_j + \left(-\tfrac{3}{2}\delta_{ij} + L_{ij}\right)\delta q_j\right] = n_i^2 P_i^* \tag{10.4.1}$$

and

$$\ddot{w}_i + n_i^2 \sum_{j=1}^{N} (\delta_{ij} + M_{ij}) w_j = n_i^2 W_i^*\,. \tag{10.4.2}$$

The different expressions for the coefficients K_{ij}, L_{ij}, M_{ij} and the right-hand members P_i^*, W_i^* have been given in Section 10.2. The transformation

$$\begin{aligned} \delta p_i &= \xi_i - \tfrac{2}{3}h_i - \tfrac{1}{2}u_i + \tfrac{3}{2}\bar{u}_i\,, & w_i &= v_i + \bar{v}_i\,, \\ \delta\dot{p}_i &= \mathrm{i}n_i(h_i - \tfrac{1}{2}u_i - \tfrac{3}{2}\bar{u}_i)\,, & \dot{w}_i &= \mathrm{i}n_i(v_i - \bar{v}_i) \end{aligned} \tag{10.4.3}$$

resulting from (9.2.9) brings (10.4.1) and (10.4.2) to the system

$$\dot{X} = \mathrm{i}\mathcal{N}\Big[(\mathcal{P} + \mathcal{Q})X + R(X, t)\Big] \tag{10.4.4}$$

with $X = \|X_\kappa\|$ and $R = \|R_\kappa\|$ ($\kappa = 1, 2, \ldots, 6$), each column block representing the N-vector of the corresponding variables

$$X_1 = \xi, \qquad X_2 = h, \qquad X_3 = u, \qquad X_5 = v \tag{10.4.5}$$

and

$$\begin{aligned} R_1 &= -(P^* + \bar{P}^*)\,, & R_2 &= \tfrac{3}{2}(P^* - \bar{P}^*)\,, \\ R_3 &= \tfrac{1}{2}(P^* - 3\bar{P}^*)\,, & R_5 &= -\tfrac{1}{2}W^*\,. \end{aligned} \tag{10.4.6}$$

One thus has

$$\bar{X}_1 = -X_1, \qquad \bar{X}_2 = X_2, \qquad X_4 = \bar{X}_3, \qquad X_6 = \bar{X}_5 \tag{10.4.7}$$

and

$$\bar{R}_1 = R_1, \qquad \bar{R}_2 = -R_2, \qquad R_4 = -\bar{R}_3, \qquad R_6 = -\bar{R}_5\,. \tag{10.4.8}$$

The matrices $\mathcal{P} = \|\mathcal{P}_{\kappa\nu}\|$ and $\mathcal{Q} = \|\mathcal{Q}_{\kappa\nu}\|$ ($\kappa, \nu = 1, 2, \ldots, 6$) are composed of square blocks, each block being an $N \times N$ matrix. P is a constant Jordan matrix and its non-zero blocks are

$$\mathcal{P}_{12} = \mathcal{P}_{33} = \mathcal{P}_{55} = E, \qquad \mathcal{P}_{44} = \mathcal{P}_{66} = -E\,, \tag{10.4.9}$$

where E is an $N \times N$ unit matrix. Q is a quasi-periodic matrix consisting of 36 square blocks. 16 blocks with $(\kappa, 5)$, $(\kappa, 6)$, $(5, \kappa)$ and $(6, \kappa)$ for $\kappa = 1, 2, 3, 4$ are zero. The 12 significant blocks are as follows:

$$\begin{aligned} \mathcal{Q}_{11} &= K - L - \bar{K} + \bar{L}\,, & \mathcal{Q}_{12} &= -\tfrac{2}{3}(K + L + \bar{K} + \bar{L}), \\ \mathcal{Q}_{21} &= \tfrac{3}{2}(-K + L - \bar{K} + \bar{L}), & \mathcal{Q}_{22} &= K + L - \bar{K} - \bar{L}\,, \\ \mathcal{Q}_{13} &= \tfrac{1}{2}(-K + 3L + 3\bar{K} - \bar{L}), & \mathcal{Q}_{23} &= \tfrac{3}{4}(K - 3L + 3\bar{K} - \bar{L}), \\ \mathcal{Q}_{31} &= \tfrac{1}{2}(-K + L - 3\bar{K} + 3\bar{L}), & \mathcal{Q}_{32} &= \tfrac{1}{3}(K + L - 3\bar{K} - 3\bar{L}), \\ \mathcal{Q}_{33} &= \tfrac{1}{4}(K - 3L + 9\bar{K} - 3\bar{L}), & \mathcal{Q}_{34} &= \tfrac{1}{4}(-3K + L - 3\bar{K} + 9\bar{L}), \\ \mathcal{Q}_{55} &= \tfrac{1}{2}M\,, & \mathcal{Q}_{56} &= \tfrac{1}{2}M\,. \end{aligned} \tag{10.4.10}$$

The remaining 8 blocks satisfy the following relations of conjugation:

$$\begin{aligned} Q_{14} &= \bar{Q}_{13}, \quad Q_{24} = -\bar{Q}_{23}, \quad Q_{41} = \bar{Q}_{31}, \quad Q_{42} = -\bar{Q}_{32}, \\ Q_{43} &= -\bar{Q}_{34}, \quad Q_{44} = -\bar{Q}_{33}, \quad Q_{65} = -\bar{Q}_{56}, \quad Q_{66} = -\bar{Q}_{55}. \end{aligned} \tag{10.4.11}$$

Equations (10.4.4) are treated in just the same manner as equations (9.2.42). The only difference is that the expansions of the right-hand members $R(X, t)$ start with the second-degree terms in (10.4.4) and with the zero-degree terms in (9.2.42). The transformation

$$X = (E + S)Y + \Gamma(Y, t) \tag{10.4.12}$$

brings (10.4.4) to the system

$$\dot{Y} = \mathrm{i}\mathcal{N}\Big[HY + F(Y, t)\Big]. \tag{10.4.13}$$

The block vector Y of the new variables may be chosen in such a way that

$$Y_1 = Y_2 = 0, \qquad Y_3 = a, \qquad Y_5 = b, \tag{10.4.14}$$

a and b being the N-vectors of new variables. The blocks of Y and Γ satisfy the same relations as (10.4.7) whereas the blocks of F satisfy the same relations as (10.4.8). In (10.4.12) E represents the $6N \times 6N$ unit matrix but in dealing with non-block vectors and matrices E denotes the $N \times N$ unit matrix. In the same way $\mathcal{N}$ in (10.4.4) and (10.4.13) stands for the block diagonal $6N \times 6N$ matrix of the mean motions whereas when we are dealing with ordinary vectors and matrices $\mathcal{N}$ denotes the $N \times N$ diagonal matrix of the mean motions.

The problem of determining Γ and F will be considered in the next section. The determination of the linear terms with respect to the eccentricities and inclinations involves the matrices S and H. Since Q is proportional to μ, S may be easily determined by iterations with respect to μ from the system

$$G = Q(E + S) - \mathcal{N}^{-1}S\mathcal{N}G^*, \tag{10.4.15}$$

$$G = G^* + G^+ \tag{10.4.16}$$

and

$$\dot{S} + \mathrm{i}(S\mathcal{N}\mathcal{P} - \mathcal{N}\mathcal{P}S) = \mathrm{i}\mathcal{N}G^+, \tag{10.4.17}$$

starting with the value $S = 0$. The splitting of (10.4.16) into a "critical" part G^* and "non-critical" part G^+ is uniquely performed by imposing the condition of solving (10.4.17) without secular terms. The matrix S is of the same form as Q. Its 12 significant blocks satisfy the equations

$$\dot{S}_{21} = \mathrm{i}\mathcal{N}G_{21}, \tag{10.4.18}$$

$$\dot{S}_{11} = \mathrm{i}\mathcal{N}(S_{21} + G_{11}), \tag{10.4.19}$$

$$\dot{S}_{22} = \mathrm{i}(-S_{21}\mathcal{N} + \mathcal{N}G_{22}), \tag{10.4.20}$$

$$\dot{S}_{12} = -\mathrm{i}S_{11}\mathcal{N} + \mathrm{i}\mathcal{N}(S_{22} + G_{12}^{+}), \tag{10.4.21}$$

$$\dot{S}_{23} + \mathrm{i}S_{23}\mathcal{N} = \mathrm{i}\mathcal{N}G_{23}\,, \tag{10.4.22}$$

$$\dot{S}_{13} + \mathrm{i}S_{13}\mathcal{N} = \mathrm{i}\mathcal{N}(S_{23} + G_{13}), \tag{10.4.23}$$

$$\dot{S}_{31} - \mathrm{i}\mathcal{N}S_{31} = \mathrm{i}\mathcal{N}G_{31}\,, \tag{10.4.24}$$

$$\dot{S}_{32} - \mathrm{i}\mathcal{N}S_{32} = \mathrm{i}(-S_{31}\mathcal{N} + \mathcal{N}G_{32}), \tag{10.4.25}$$

$$\dot{S}_{33} + \mathrm{i}(S_{33}\mathcal{N} - \mathcal{N}S_{33}) = \mathrm{i}\mathcal{N}G_{33}^{+}\,, \tag{10.4.26}$$

$$\dot{S}_{34} - \mathrm{i}(S_{34}\mathcal{N} + \mathcal{N}S_{34}) = \mathrm{i}\mathcal{N}G_{34}\,, \tag{10.4.27}$$

$$\dot{S}_{55} + \mathrm{i}(S_{55}\mathcal{N} - \mathcal{N}S_{55}) = \mathrm{i}\mathcal{N}G_{55}^{+}\,, \tag{10.4.28}$$

$$\dot{S}_{56} - \mathrm{i}(S_{56}\mathcal{N} + \mathcal{N}S_{56}) = \mathrm{i}\mathcal{N}G_{56}\,. \tag{10.4.29}$$

The 8 remaining non-zero blocks are determined from the relations of conjugation that differ from (10.4.11) only in a reversal of sign. The matrix G has the same form as Q and S. The only non-zero blocks of G^* are

$$G_{12}^* = \langle G_{12}\rangle, \qquad G_{33}^* = DAD^{-1}\,, \qquad G_{55}^* = DBD^{-1} \tag{10.4.30}$$

and their conjugates

$$G_{44}^* = -D^{-1}AD\,, \qquad G_{66}^* = -D^{-1}BD\,.$$

In fact, from the structure of Q it is easily seen that

$$\langle G_{21}\rangle = \langle G_{11}\rangle = \langle G_{22}\rangle = 0\,. \tag{10.4.31}$$

The diagonal matrix D consists of the elements $\exp \mathrm{i}\lambda_i$ $(i = 1, 2, \ldots, N)$. The constant matrices A and B describing the linear part of the secular system (10.1.8) are determined by the relations

$$A = \langle D^{-1}G_{33}D\rangle, \qquad B = \langle D^{-1}G_{55}D\rangle\,. \tag{10.4.32}$$

After determining S and G one finds the matrix H of system (10.4.13):

$$H = \mathcal{P} + G^*\,. \tag{10.4.33}$$

In matrix designation formulae (10.4.3) for the N-vectors $\delta p = (\delta p_i)$ and $w = (w_i)$ take the form

$$\delta p = (-\tfrac{1}{2}E + c)\,a + (\tfrac{3}{2}E + d)\,\bar{a} + \Gamma_1 - \tfrac{2}{3}\Gamma_2 - \tfrac{1}{2}\Gamma_3 + \tfrac{3}{2}\bar{\Gamma}_3 \tag{10.4.34}$$

and

$$w = (E + f)b + (E + \bar{f})\bar{b} + \Gamma_5 + \bar{\Gamma}_5\,, \tag{10.4.35}$$

where the matrices c, d and f are composed of the blocks of the matrix S as follows:

$$c = S_{13} - \tfrac{2}{3}S_{23} - \tfrac{1}{2}S_{33} + \tfrac{3}{2}\bar{S}_{34}\,, \tag{10.4.36}$$

$$d = -\bar{S}_{13} - \tfrac{2}{3}\bar{S}_{23} + \tfrac{3}{2}\bar{S}_{33} - \tfrac{1}{2}S_{34} \tag{10.4.37}$$

and

$$f = S_{55} + \bar{S}_{56}\,. \tag{10.4.38}$$

Formulae (10.4.36)–(10.4.38) complete the determination of the linear perturbations with respect to the eccentricities and inclinations.

In version (a) each element $Q_{\kappa\nu}[i,j]$ $(\kappa,\nu = 1,2,\ldots,6;\ i,j = 1,2,\ldots,N)$ of Q is represented in form (10.3.1)–(10.3.3), i.e.

$$Q_{\kappa\nu}[i,j] = \sum Q_k^{(\kappa\nu ij)} \exp \mathrm{i}(k\lambda)\,. \tag{10.4.39}$$

The matrices G and S may be found, evidently, in the same form. The matrix G^* contains only the terms $G_k^{(\kappa\nu ij)}$ for the following values of the multi-index $k = (k_s)$:

$$k_s = 0\,, \quad \text{for } (\kappa,\nu) = (1,1),\ (1,2),\ (2,1),\ (2,2) \tag{10.4.40}$$

and

$$k_s = \delta_{si} - \delta_{sj}\,, \quad \text{for } (\kappa,\nu) = (3,3),\ (5,5)\,. \tag{10.4.41}$$

All other terms are included in G^+. Equations (10.4.18)–(10.4.29) imply the relations

$$S_k^{(21ij)} = \frac{n_i}{(kn)} G_k^{(21ij)}\,, \tag{10.4.42}$$

$$S_k^{(11ij)} = \frac{n_i}{(kn)} \left(G_k^{(11ij)} + S_k^{(21ij)} \right), \tag{10.4.43}$$

$$S_k^{(22ij)} = \frac{n_i}{(kn)} \left(G_k^{(22ij)} - \frac{n_j}{n_i} S_k^{(21ij)} \right), \tag{10.4.44}$$

$$S_k^{(12ij)} = \frac{n_i}{(kn)} \left(G_k^{(12ij)} - \frac{n_j}{n_i} S_k^{(11ij)} + S_k^{(22ij)} \right), \tag{10.4.45}$$

$$S_k^{(23ij)} = \frac{n_i}{(kn) + n_j} G_k^{(23ij)}\,, \tag{10.4.46}$$

$$S_k^{(13ij)} = \frac{n_i}{(kn) + n_j} \left(G_k^{(13ij)} + S_k^{(23ij)} \right), \tag{10.4.47}$$

$$S_k^{(31ij)} = \frac{n_i}{(kn) - n_i} G_k^{(31ij)}\,, \tag{10.4.48}$$

$$S_k^{(32ij)} = \frac{n_i}{(kn) - n_i} \left(G_k^{(32ij)} - \frac{n_j}{n_i} S_k^{(31ij)} \right), \tag{10.4.49}$$

$$S_k^{(\kappa\nu ij)} = \frac{n_i}{(kn) + n_j - n_i} G_k^{(\kappa\nu ij)}\,, \quad (\kappa,\nu) = (3,3),\ (5,5) \tag{10.4.50}$$

and

$$S_k^{(\kappa\nu ij)} = \frac{n_i}{(kn) - n_j - n_i} G_k^{(\kappa\nu ij)}\,, \quad (\kappa, \nu) = (3.4),\ (5,6)\,. \tag{10.4.51}$$

The zero values of the divisors in (10.4.42)–(10.4.45) and (10.4.50) cannot actually appear because they correspond to critical values (10.4.40) and (10.4.41) related to G^*.

In version (b) in dealing with the first-order theory it is convenient to represent each element of the block matrices in the form

$$\underset{1}{Q}_{\kappa\nu}[i,i] = \sum_{j=1}^{N} {}^{(i)}Q(\kappa,\nu,i,j,0) \tag{10.4.52}$$

for a diagonal element and

$$\underset{1}{Q}_{\kappa\nu}[i,j] = Q(\kappa,\nu,i,j,1) \tag{10.4.53}$$

for a non-diagonal element. Hence, for the block (κ, ν) of the matrix Q the quantity $Q(\kappa, \nu, i, j, \Delta)$ means the non-diagonal element (i, j) if $\Delta = 1$ and the j-component of the diagonal element (i, i) if $\Delta = 0$. For the one-block matrices K, L and M the indices κ and ν are absent. Thus, to the first order with respect to μ one has from (10.2.15), (10.2.16) and (10.2.11)

$$K(i,j,0) = -\left[\frac{1}{2}\kappa_{ij}\frac{a_i^3}{(a_i+a_j)^3}\frac{1}{\delta_{ij}^3} + \frac{3}{4}\underset{1}{p}_j^{(i)} + \frac{3}{4}\underset{1}{\bar{p}}_j^{(i)}\right], \tag{10.4.54}$$

$$K(i,j,1) = \frac{1}{2}\kappa_{ij}\left[-\left(\frac{a_i}{a_j}\right)^2 + \frac{a_i^2 a_j}{(a_i+a_j)^3}\frac{1}{\delta_{ij}^3}\right]\zeta_{ij}^{-1}, \tag{10.4.55}$$

$$L(i,j,0) = -\left[\frac{3}{2}\kappa_{ij}\frac{a_i^3}{(a_i+a_j)^5}\frac{1}{\delta_{ij}^5}(a_i - a_j\zeta_{ij}^{-1})^2 + \frac{3}{4}\underset{1}{p}_j^{(i)} + \frac{15}{4}\underset{1}{\bar{p}}_j^{(i)}\right], \tag{10.4.56}$$

$$L(i,j,1) = \frac{3}{2}\kappa_{ij}\left[-\left(\frac{a_i}{a_j}\right)^2 + \frac{a_i^2 a_j}{(a_i+a_j)^5}\frac{1}{\delta_{ij}^5}(a_j - a_i\zeta_{ij})^2\right]\zeta_{ij}^{-1}, \tag{10.4.57}$$

$$M(i,j,0) = -2K(i,j,0) \tag{10.4.58}$$

and

$$M(i,j,1) = -2K(i,j,1)\zeta_{ij}\,. \tag{10.4.59}$$

Let us remember that $\underset{1}{p}_j^{(i)}$ represents the sum of (10.3.29) and (10.3.30) and

$$\zeta_{ij} = -\exp(-\mathrm{i}\,2\varphi_{ij})\,. \tag{10.4.60}$$

In the first-order theory G coincides with Q as determined by (10.4.10). Each element $S(\kappa, \nu, i, j, \Delta)$ of the matrix S may be determined by quadratures based on equations (10.4.18)–(10.4.29). From equations (10.4.18)–(10.4.21) we have

$$S(2,1,i,j,\Delta) = -\mathrm{i}\,2m_{ij}\int G(2,1,i,j,\Delta)d\varphi_{ij}\,, \tag{10.4.61}$$

$$S(1,1,i,j,\Delta) = -\mathrm{i}\,2m_{ij}\int [G(1,1,i,j,\Delta)+S(2,1,i,j,\Delta)]\,d\varphi_{ij}\,, \tag{10.4.62}$$

$$S(2,2,i,j,\Delta) = -\mathrm{i}\,2m_{ij}\int \Big[G(2,2,i,j,\Delta)-\Big(1-\frac{\Delta}{m_{ij}}\Big)S(2,1,i,j,\Delta)\Big]\,d\varphi_{ij}, \tag{10.4.63}$$

$$S(1,2,i,j,\Delta) = -\mathrm{i}\,2m_{ij}\int \Big[\ G^{+}(1,2,i,j,\Delta)+S(2,2,i,j,\Delta)- \\ -\Big(1-\frac{\Delta}{m_{ij}}\Big)S(1,1,i,j,\Delta)\Big]\,d\varphi_{ij} \tag{10.4.64}$$

with m_{ij} determined by (10.3.31). It is easy to verify that $G^*(\kappa,\nu,i,j,\Delta)=0$ for $(\kappa,\nu)=(2,1),\ (1,1),\ (2,2)$ and

$$G^*(1,2,i,j,\Delta) = \tfrac{2}{3}(-1)^{\Delta}\kappa_{ij}a_i^2a_jc_3^{(1)}(a_i,a_j)$$

with value (6.3.9). In (10.4.61) one meets the single quadratures of $K(i,j,\Delta)$ and $L(i,j,\Delta)$. Along with this, double quadratures occur in (10.4.62) and (10.4.63). In (10.4.63) there are also triple quadratures.

From equations (10.4.22)–(10.4.25) it follows that

$$S(2,3,i,j,\Delta) = -\mathrm{i}\,2m_{ij}\int G(2,3,i,j,\Delta)\times \\ \times\exp\big[\mathrm{i}\,2(m_{ij}-\Delta)(\tilde\varphi_{ij}-\varphi_{ij})\big]\,d\varphi_{ij}\,, \tag{10.4.65}$$

$$S(1,3,i,j,\Delta) = -\mathrm{i}\,2m_{ij}\int [G(1,3,i,j,\Delta)+S(2,3,i,j,\Delta)]\times \\ \times\exp\big[\mathrm{i}\,2(m_{ij}-\Delta)(\tilde\varphi_{ij}-\varphi_{ij})\big]\,d\varphi_{ij}\,, \tag{10.4.66}$$

$$S(3,1,i,j,\Delta) = -\mathrm{i}\,2m_{ij}\int G(3,1,i,j,\Delta)\exp\big[\mathrm{i}\,2m_{ij}(\varphi_{ij}-\tilde\varphi_{ij})\big]\,d\varphi_{ij}\,, \tag{10.4.67}$$

$$S(3,2,i,j,\Delta) = -\mathrm{i}\,2m_{ij}\int \Big[G(3,2,i,j,\Delta)-\Big(1-\frac{\Delta}{m_{ij}}\Big)S(3,1,i,j,\Delta)\Big]\times \\ \times\exp\big[\mathrm{i}\,2m_{ij}(\varphi_{ij}-\tilde\varphi_{ij})\big]\,d\varphi_{ij}\,. \tag{10.4.68}$$

The mean values of these functions are

$$\langle S(2,3,i,j,\Delta)\rangle = \frac{m_{ij}}{m_{ij}-\Delta}\langle G(2,3,i,j,\Delta)\rangle = 0\,, \tag{10.4.69}$$

$$\langle S(1,3,i,j,\Delta)\rangle = \frac{m_{ij}}{m_{ij}-\Delta}\langle G(1,3,i,j,\Delta)\rangle = \\ = -\tfrac{1}{2}(-1)^{\Delta}\frac{m_{ij}}{m_{ij}-\Delta}\kappa_{ij}a_i^2a_jc_3^{(1)}(a_i,a_j)\,, \tag{10.4.70}$$

$$\langle S(3,1,i,j,\Delta)\rangle = -\langle G(3,1,i,j,\Delta)\rangle = 0\,, \tag{10.4.71}$$

$$\langle S(3,2,i,j,\Delta)\rangle = -\langle G(3,2,i,j,\Delta)\rangle = -\tfrac{1}{3}(-1)^{\Delta}\kappa_{ij}a_i^2 a_j c_3^{(1)}(a_i,a_j)\,. \tag{10.4.72}$$

In taking these quadratures one has to deal with functions (10.3.32) and their generalizations where the integrand δ^{-n} is replaced by some more complicated functions. It is evident that a closed-form representation is possible only by introducing such new types of function.

Finally, equations (10.4.26)–(10.4.29) involve for $\kappa = 3$ and $\kappa = 5$

$$S(\kappa,\kappa,i,j,\Delta) = -\mathrm{i}\,2m_{ij}\int G^{+}(\kappa,\kappa,i,j,\Delta)\exp\left[\mathrm{i}\,2\Delta(\varphi_{ij}-\tilde{\varphi}_{ij})\right]d\varphi_{ij}\,, \tag{10.4.73}$$

and

$$S(\kappa,\kappa+1,i,j,\Delta) = -\mathrm{i}\,2m_{ij}\int G(\kappa,\kappa+1,i,j,\Delta)\times \times\exp\left[\mathrm{i}\,2(2m_{ij}-\Delta)(\varphi_{ij}-\tilde{\varphi}_{ij})\right]d\varphi_{ij}\,. \tag{10.4.74}$$

Critical values (10.4.30), which are excluded in (10.4.73), are

$$G^{*}(\kappa,\kappa,i,j,\Delta) = \langle G(\kappa,\kappa,i,j,\Delta)\exp\mathrm{i}\,2\Delta\varphi_{ij}\rangle\exp(-\mathrm{i}\,2\Delta\varphi_{ij}) \tag{10.4.75}$$
$$\kappa = 3,\ 5,\quad \Delta = 0,\,1$$

with the mean values

$$\langle G(3,3,i,j,0)\rangle = -\langle G(5,5,i,j,0)\rangle = -\tfrac{1}{4}\kappa_{ij}a_i^2 a_j c_3^{(1)}(a_i,a_j)\,, \tag{10.4.76}$$

$$\langle G(3,3,i,j,1)\exp\mathrm{i}\,2\varphi_{ij}\rangle = -\tfrac{1}{4}\kappa_{ij}a_i^2 a_j c_3^{(2)}(a_i,a_j) \tag{10.4.77}$$

and

$$\langle G(5,5,i,j,1)\exp\mathrm{i}\,2\varphi_{ij}\rangle = \tfrac{1}{4}\kappa_{ij}a_i^2 a_j c_3^{(1)}(a_i,a_j)\,. \tag{10.4.78}$$

These values lead directly to the well-known first-order expressions of the matrices (10.4.33) of secular system (10.1.8)

$$A(i,j,0) = \langle G(3,3,i,j,0)\rangle = -\tfrac{1}{4}\kappa_{ij}a_i^2 a_j c_3^{(1)}(a_i,a_j)\,, \tag{10.4.79}$$

$$A(i,j,1) = -\langle G(3,3,i,j,1)\exp\mathrm{i}\,2\varphi_{ij}\rangle = \tfrac{1}{4}\kappa_{ij}a_i^2 a_j c_3^{(2)}(a_i,a_j)\,, \tag{10.4.80}$$

$$B(i,j,0) = \langle G(5,5,i,j,0)\rangle = \tfrac{1}{4}\kappa_{ij}a_i^2 a_j c_3^{(1)}(a_i,a_j) \tag{10.4.81}$$

and

$$B(i,j,1) = -\langle G(5,5,i,j,1)\exp\mathrm{i}\,2\varphi_{ij}\rangle = -\tfrac{1}{4}\kappa_{ij}a_i^2 a_j c_3^{(1)}(a_i,a_j)\,. \tag{10.4.82}$$

The only peculiarity in dealing with quadratures (10.4.73) and (10.4.74) is that for $\Delta = 1$ in (10.4.73) one meets a function of type (10.3.32) with $s = 0$ and integer value $\alpha = 2$. But in this case the integrand kernel does not contain the corresponding harmonic resulting again in the periodic function of φ_{ij}. The mean values of periodic functions (10.4.73) and (10.4.74) are as follows:

$$\langle S(3,3,i,j,\Delta)\rangle = -\Delta\, m_{ij}\langle G(3,3,i,j,1)\rangle = -\tfrac{1}{4}\Delta\, m_{ij}\kappa_{ij}a_i^2 a_j c_3^{(1)}(a_i,a_j)\,, \tag{10.4.83}$$

$$\begin{aligned}\langle S(3,4,i,j,\Delta)\rangle &= \frac{m_{ij}}{\Delta - 2m_{ij}}\langle G(3,4,i,j,\Delta)\rangle = \\ &= \tfrac{1}{4}(-1)^{\Delta}\frac{m_{ij}}{2m_{ij}-\Delta}\kappa_{ij}a_i^2 a_j c_3^{(1)}(a_i,a_j)\,,\end{aligned} \tag{10.4.84}$$

$$\begin{aligned}\langle S(5,5,i,j,\Delta)\rangle &= -\Delta\, m_{ij}\langle G(5,5,i,j,1)\rangle = \\ &= -\frac{1}{2}\Delta\, m_{ij}\kappa_{ij}\left[\left(\frac{a_i}{a_j}\right)^2 - \frac{1}{2}a_i^2 a_j c_3^{(0)}(a_i,a_j)\right],\end{aligned} \tag{10.4.85}$$

$$\langle S(5,6,i,j,0)\rangle = -\tfrac{1}{2}\langle G(5,6,i,j,0)\rangle = -\tfrac{1}{8}\kappa_{ij}a_i^2 a_j c_3^{(1)}(a_i,a_j)\,, \tag{10.4.86}$$

and

$$\begin{aligned}\langle S(5,6,i,j,1)\rangle &= \frac{m_{ij}}{1-2m_{ij}}\langle G(5,5,i,j,1)\rangle = \\ &= \frac{1}{2}\frac{m_{ij}}{1-2m_{ij}}\kappa_{ij}\left[\left(\frac{a_i}{a_j}\right)^2 - \frac{1}{2}a_i^2 a_j c_3^{(0)}(a_i,a_j)\right].\end{aligned} \tag{10.4.87}$$

In version (c) all the elements of the matrices K, L, M, Q, G and S are presented by τ-series. For the first-order theory one may obtain the elements of S by means of quadratures (10.4.61)–(10.4.63), (10.4.65)–(10.4.68), (10.4.73) and (10.4.74) expanding the integrands into τ-series or by means of formulae analogous to (10.4.42)–(10.4.51). For higher orders one should use the technique given in relation to (10.3.49).

In any version one completes the construction of the linear theory by computing matrices (10.4.36)–(10.4.38). Their first-order mean values result from (10.4.69)–(10.4.72) and (10.4.83)–(10.4.87):

$$\langle c(i,j,0)\rangle = -\tfrac{5}{16}\kappa_{ij}a_i^2 a_j c_3^{(1)}(a_i,a_j)\,, \tag{10.4.88}$$

$$\langle c(i,j,1)\rangle = \tfrac{1}{8}m_{ij}\left(1+\frac{4}{m_{ij}-1}-\frac{3}{2m_{ij}-1}\right)\kappa_{ij}a_i^2 a_j c_3^{(1)}(a_i,a_j)\,, \tag{10.4.89}$$

$$\langle d(i,j,0)\rangle = \tfrac{7}{16}\kappa_{ij}a_i^2 a_j c_3^{(1)}(a_i,a_j)\,, \tag{10.4.90}$$

$$\langle d(i,j,1)\rangle = \tfrac{1}{8}m_{ij}\left(-3-\frac{4}{m_{ij}-1}+\frac{1}{2m_{ij}-1}\right)\kappa_{ij}a_i^2 a_j c_3^{(1)}(a_i,a_j)\,, \tag{10.4.91}$$

$$\langle f(i,j,0)\rangle = -\tfrac{1}{8}\kappa_{ij}a_i^2 a_j c_3^{(1)}(a_i,a_j) \tag{10.4.92}$$

and

$$\langle f(i,j,1)\rangle = \frac{m_{ij}^2}{1-2m_{ij}}\kappa_{ij}\left[\left(\frac{a_i}{a_j}\right)^2 - \frac{1}{2}a_i^2 a_j c_3^{(0)}(a_i,a_j)\right]. \tag{10.4.93}$$

In constructing a theory of motion for the Moon or some minor planet there is no necessity to have a very accurate theory of motion for the major planets and it is sufficient to use only Keplerian terms, a first-order intermediary and first-order linear theory. Therefore, the coordinates p_i and w_i of the major planets ($i = 1, 2, \ldots, N$) necessary for the theory of motion for the Moon or a minor planet may be presented in the form

$$p_i = \delta \underset{0}{p_i} + \underset{1}{p_i^{(0)}} + \delta \underset{1}{p_i^{(1)}}, \qquad w_i = \underset{0}{w_i} + \underset{1}{w_i^{(1)}}. \tag{10.4.94}$$

$\delta \underset{0}{p_i}$ and $\underset{0}{w_i}$ stand here for the Keplerian terms determined by (7.3.8)–(7.3.17). The terms $\underset{1}{p_i^{(0)}}$ represent intermediary (10.1.10) with (10.3.29) and (10.3.30). The linear terms $\delta \underset{1}{p_i^{(1)}}$ and $\underset{1}{w_i^{(1)}}$ have already been determined in this section. In explicit form they are as follows:

$$\delta \underset{1}{p_i^{(1)}} = \sum_{j=1}^{N}{}^{(i)} \delta \underset{1}{p_{ij}^{(1)}}, \qquad \underset{1}{w_i^{(1)}} = \sum_{j=1}^{N}{}^{(i)} \underset{1}{w_{ij}^{(1)}}, \tag{10.4.95}$$

$$\delta \underset{1}{p_{ij}^{(1)}} = c(i,j,0)a_i + d(i,j,0)\bar{a}_i + c(i,j,1)a_j + d(i,j,1)\bar{a}_j \tag{10.4.96}$$

and

$$\underset{1}{w_{ij}^{(1)}} = f(i,j,0)b_i + \bar{f}(i,j,0)\bar{b}_i + f(i,j,1)b_j + \bar{f}(i,j,1)\bar{b}_j \,. \tag{10.4.97}$$

Let us recall once again that a_i and b_i are related to the slowly changing elements α_i and β_i by (10.1.7).

10.5 Non-linear Terms

Returning to transformation (10.4.12) from system (10.4.4) to new system (10.4.13) one may determine Γ by iterations on μ in increasing order with respect to the total degree of the polynomial variables ($m = 2, 3, \ldots$) based on the equations

$$U = R + Q\Gamma - \mathcal{N}^{-1}\Gamma_Y \mathcal{N} G^* Y - \mathcal{N}^{-1}(S + \Gamma_Y)\mathcal{N} U^*, \tag{10.5.1}$$

$$U = U^* + U^+ \tag{10.5.2}$$

and

$$\Gamma_t + \mathrm{i}\,(\Gamma_Y\, \mathcal{N}\mathcal{P}Y - \mathcal{N}\mathcal{P}\Gamma) = \mathrm{i}\,\mathcal{N}U^+ \tag{10.5.3}$$

with

$$\Gamma_t = \frac{\partial \Gamma}{\partial t}, \qquad \Gamma_Y = \left\| \frac{\partial \Gamma_\kappa}{\partial Y_\nu} \right\|, \quad (\kappa, \nu = 1, 2, \ldots, 6).$$

The iteration process starts with $U = R$. The separation of (10.5.2) into "critical" (*) and "non-critical" ($^+$) parts has the aim of integrating (10.5.3) without secular terms. Equations (10.5.1)–(10.5.3) generalize equations (9.3.5)–(9.3.7) to the N-dimension case by treating the linear terms separately. In block decomposition, equations (10.5.3) take the form

$$\frac{\partial \Gamma_1}{\partial t} + \mathrm{i}\left(\frac{\partial \Gamma_1}{\partial a}\mathcal{N}a - \frac{\partial \Gamma_1}{\partial \bar{a}}\mathcal{N}\bar{a} + \frac{\partial \Gamma_1}{\partial b}\mathcal{N}b - \frac{\partial \Gamma_1}{\partial \bar{b}}\mathcal{N}\bar{b}\right) - \mathrm{i}\mathcal{N}\Gamma_2 = \mathrm{i}\mathcal{N}U_1\,, \tag{10.5.4}$$

$$\frac{\partial \Gamma_2}{\partial t} + \mathrm{i}\left(\frac{\partial \Gamma_2}{\partial a}\mathcal{N}a - \frac{\partial \Gamma_2}{\partial \bar{a}}\mathcal{N}\bar{a} + \frac{\partial \Gamma_2}{\partial b}\mathcal{N}b - \frac{\partial \Gamma_2}{\partial \bar{b}}\mathcal{N}\bar{b}\right) = \mathrm{i}\mathcal{N}U_2 \tag{10.5.5}$$

and

$$\frac{\partial \Gamma_\kappa}{\partial t} + \mathrm{i}\left(\frac{\partial \Gamma_\kappa}{\partial a}\mathcal{N}a - \frac{\partial \Gamma_\kappa}{\partial \bar{a}}\mathcal{N}\bar{a} + \frac{\partial \Gamma_\kappa}{\partial b}\mathcal{N}b - \frac{\partial \Gamma_\kappa}{\partial \bar{b}}\mathcal{N}\bar{b}\right) - \mathrm{i}\mathcal{N}\Gamma_\kappa = \mathrm{i}\mathcal{N}U_\kappa^+\,, \quad (\kappa = 3,\,5). \tag{10.5.6}$$

The blocks with $k = 4$ and $k = 6$ are not needed since the vectors Y and Γ satisfy the same relations as (10.4.7). The vectors U and F satisfy the same relations as (10.4.8). Evidently, the dimension of the diagonal matrix $\mathcal{N}$ of the mean motions is $6N \times 6N$ in (10.5.1) and (10.5.3) and $N \times N$ in (10.5.4)–(10.5.6).

Rewriting (10.5.1) in scalar form one has

$$\begin{aligned} U_\kappa[i] = R_\kappa[i] + \sum_{j=1}^{N}\sum_{\nu=1}^{6} Q_{\kappa\nu}[i,j]\Gamma_\nu[j] - \\ - \sum_{j=1}^{N}\sum_{\nu=3}^{6}\frac{n_j}{n_i}\left\{\left(S_{\kappa\nu}[i,j] + \frac{\partial \Gamma_\kappa[i]}{\partial Y_\nu[j]}\right)U_\nu^*[j] + \right. \\ \left. + \frac{\partial \Gamma_\kappa[i]}{\partial Y_\nu[j]}\sum_{k=1}^{N} G_{\nu\nu}^*[j,k]Y_\nu[k]\right\}. \end{aligned} \tag{10.5.7}$$

The linear terms found in the previous section serve as initial values in determining the second-degree terms. In general, if expansions (10.1.5) and (10.1.6) are known up to degree $m-1$ inclusive ($m = 2, 3, \ldots$) then by the various methods indicated in Section 10.2 it is possible to find terms of degree m in the components $R_\kappa[i]$ of right-hand members (10.4.6), say

$$R_\kappa^{(m)}[i] = \sum_{p+q+r+s=m} R_{pqrs}^{(\kappa i)}(t)\prod_{j=1}^{N} a_j^{p_j}\bar{a}_j^{q_j}b_j^{r_j}\bar{b}_j^{s_j}\,. \tag{10.5.8}$$

In version (a) the coefficients $R_{pqrs}^{(\kappa i)}$ are expanded again in series like (10.3.1) with condition (10.3.3)

$$R^{(\kappa i)}_{pqrs}(t) = \sum R^{(\kappa i)}_{pqrsk} \exp \mathrm{i}(k\lambda) . \qquad (10.5.9)$$

In a first approximation the functions $U^{(m)}_\kappa[i]$ just coincide with $R^{(m)}_\kappa[i]$. The critical terms $U^*_\kappa[i]$ ($\kappa = 3,\ 5$) occur for odd values of m ($m = 3, 5, \dots$) and correspond to the following relations for the indices:

$$\sum_{j=1}^{N} (p_j - q_j + r_j - s_j) = 1 , \qquad k_j = \delta_{ij} - p_j + q_j - r_j + s_j . \qquad (10.5.10)$$

For even values of m ($m = 2, 4, \dots$) the critical terms satisfying the relations

$$\sum_{j=1}^{N} (p_j - q_j + r_j - s_j) = 0 , \qquad k_j = -p_j + q_j - r_j + s_j \qquad (10.5.11)$$

might occur in subvectors U_1 and U_2. But the relation $U^*_2 = 0$ holds true by itself and serves to check the computation. Critical terms in U_1 are not of importance because they may be cancelled out in (10.5.4) by an appropriate choice of arbitrary constants when integrating (10.5.5). Hence, the coefficients $\Gamma^{(\kappa i)}_{pqrsk}$ are determined by

$$\Gamma^{(2i)}_{pqrsk} = \frac{n_i}{((p-q+r-s+k)n)} U^{(2i)}_{pqrsk} , \qquad (10.5.12)$$

$$\Gamma^{(1i)}_{pqrsk} = \frac{n_i}{((p-q+r-s+k)n)} \left(U^{(1i)}_{pqrsk} + \Gamma^{(2i)}_{pqrsk} \right) , \qquad (10.5.13)$$

and

$$\Gamma^{(\kappa i)}_{pqrsk} = \frac{n_i}{((p-q+r-s+k)n) - n_i} U^{(\kappa i)}_{pqrsk} \quad (\kappa = 3,\ 5) . \qquad (10.5.14)$$

For critical values (10.5.10) and (10.5.11), which lead to zero divisors, the coefficients are chosen as follows:

$$\Gamma^{(2i)}_{pqrsk} = -U^{(1i)}_{pqrsk} , \qquad \Gamma^{(\kappa i)}_{pqrsk} = 0 \quad (\kappa = 1, 3, 5) . \qquad (10.5.15)$$

Then the improved values (with respect to μ) of $U^{(m)}_\kappa[i]$ result from (10.5.7), or in more detail

$$U_\kappa^{(m)}[i] = R_\kappa^{(m)}[i] + \sum_{j=1}^{N}\sum_{\nu=1}^{6} Q_{\kappa\nu}[i,i]\Gamma_\nu^{(m)}[j] -$$

$$-\sum_{j=1}^{N}\frac{n_j}{n_i}\left\{\sum_{\nu=3}^{6} S_{\kappa\nu}[i,j]U_\nu^{*(m)}[j] +\right.$$

$$+\sum_{k=1}^{N}\left(G_{33}^*[j,k]a_k\frac{\partial}{\partial a_j} - \bar{G}_{33}^*[j,k]\bar{a}_k\frac{\partial}{\partial \bar{a}_j} +\right.$$

$$\left.+G_{55}^*[j,k]b_k\frac{\partial}{\partial b_j} - \bar{G}_{55}^*[j,k]\bar{b}_k\frac{\partial}{\partial \bar{b}_j}\right)\Gamma_\kappa^{(m)}[i] +$$

$$+\sum_{l=1}^{E(\frac{m-2}{2})}\left(U_3^{*(2l+1)}[j]\frac{\partial}{\partial a_j} - \bar{U}_3^{*(2l+1)}[j]\frac{\partial}{\partial \bar{a}_j} +\right.$$

$$\left.\left.+U_5^{*(2l+1)}[j]\frac{\partial}{\partial b_j} - \bar{U}_5^{*(2l+1)}[j]\frac{\partial}{\partial \bar{b}_j}\right)\Gamma_\kappa^{(m-2l)}[i]\right\}. \qquad (10.5.16)$$

The process of determining U and Γ is repeated until the required accuracy is achieved.

In version (b) we are interested only in the first-order terms with respect to μ and can write the right-hand members in the form

$$\underset{1}{R_\kappa}[i] = \sum_{j=1}^{N}{}^{(i)}\, \underset{1}{R}(\kappa,i,j)\,. \qquad (10.5.17)$$

Introducing similar designations for U and Γ we have for the terms of degree m

$$\underset{1}{U}^{(m)}(\kappa,i,j) = \underset{1}{R}^{(m)}(\kappa,i,j) + \sum_{\nu=1}^{6}\left\{Q(\kappa,\nu,i,j,0)\,\underset{0}{\Gamma}^{(m)}(\nu,i) +\right.$$

$$\left.+Q(\kappa,\nu,i,j,1)\,\underset{0}{\Gamma}^{(m)}(\nu,j)\right\} - \left\{G^*(3,3,i,j,0)\left(a_i\frac{\partial}{\partial a_i} - \bar{a}_i\frac{\partial}{\partial \bar{a}_i}\right) +\right.$$

$$+G^*(5,5,i,j,0)\left(b_i\frac{\partial}{\partial b_i} - \bar{b}_i\frac{\partial}{\partial \bar{b}_i}\right) + G^*(3,3,i,j,1)\left(a_j\frac{\partial}{\partial a_i} - \bar{a}_j\frac{\partial}{\partial \bar{a}_i}\right) +$$

$$\left.+G^*(5,5,i,j,1)\left(b_j\frac{\partial}{\partial b_i} - \bar{b}_j\frac{\partial}{\partial \bar{b}_i}\right)\right\}\underset{0}{\Gamma}^{(m)}(\kappa,i) -$$

$$-\sum_{l=1}^{E(\frac{m-2}{2})}\left\{\underset{1}{U}^{*(2l+1)}(3,i,j)\frac{\partial}{\partial a_i} - \underset{1}{\bar{U}}^{*(2l+1)}(3,i,j)\frac{\partial}{\partial \bar{a}_i} +\right.$$

$$\left.+\underset{1}{U}^{*(2l+1)}(5,i,j)\frac{\partial}{\partial b_i} - \underset{1}{\bar{U}}^{*(2l+1)}(5,i,j)\frac{\partial}{\partial \bar{b}_i}\right\}\underset{0}{\Gamma}^{(m-2l)}(\kappa,i)\,, \qquad (10.5.18)$$

where for the zeroth order we simply put

$$\underset{0}{\Gamma}_{\kappa}[i] = \underset{0}{\Gamma}(\kappa, i)\,. \tag{10.5.19}$$

Having obtained the terms of some degree in the expansion

$$\underset{1}{U}(\kappa, i, j) = \sum U^{(\kappa ij)}_{pqrsp'q'r's'} a_i^p \bar{a}_i^q b_i^r \bar{b}_i^s a_j^{p'} \bar{a}_j^{q'} b_j^{r'} \bar{b}_j^{s'} \tag{10.5.20}$$

one gets from (10.5.4)–(10.5.6) the corresponding coefficients for $\underset{1}{\Gamma}(\kappa, i, j)$:

$$\Gamma^{(2ij)}_{pqrsp'q'r's'} = -2\mathrm{i}\, m_{ij} \int U^{(2ij)}_{pqrsp'q'r's'} \exp\Big\{ 2\mathrm{i}\, m_{ij} \Big[p - q + r - s + \\ + \Big(1 - \frac{1}{m_{ij}}\Big)(p' - q' + r' - s')\Big](\tilde{\varphi}_{ij} - \varphi_{ij})\Big\}\, d\varphi_{ij}\,, \tag{10.5.21}$$

$$\Gamma^{(1ij)}_{pqrsp'q'r's'} = -2\mathrm{i}\, m_{ij} \int \Big(U^{(1ij)}_{pqrsp'q'r's'} + \Gamma^{(2ij)}_{pqrsp'q'r's'} \Big) \times \\ \times \exp\Big\{ 2\mathrm{i}\, m_{ij} \Big[p - q + r - s + \\ + \Big(1 - \frac{1}{m_{ij}}\Big)(p' - q' + r' - s')\Big](\tilde{\varphi}_{ij} - \varphi_{ij})\Big\}\, d\varphi_{ij} \tag{10.5.22}$$

and for $\kappa = 3,\ 5$

$$\Gamma^{(\kappa ij)}_{pqrsp'q'r's'} = -2\mathrm{i}\, m_{ij} \int U^{+(\kappa ij)}_{pqrsp'q'r's'} \exp\Big\{ 2\mathrm{i}\, m_{ij} \Big[p - q + r - s - 1 + \\ + \Big(1 - \frac{1}{m_{ij}}\Big)(p' - q' + r' - s')\Big](\tilde{\varphi}_{ij} - \varphi_{ij})\Big\}\, d\varphi_{ij}\,. \tag{10.5.23}$$

Critical values U^*_κ may occur only for the indices satisfying the relations

$$\left.\begin{aligned} p + q + r + s + p' + q' + r' + s' &= m\,, \\ p - q + r - s + p' - q' + r' - s' &= \nabla \end{aligned}\right\} \tag{10.5.24}$$

with

$$\nabla = \begin{cases} 0, & \kappa = 1,\ 2 \\ 1, & \kappa = 3,\ 5 \end{cases} \tag{10.5.25}$$

and $m = 2k + \nabla$ $(k = 1, 2, \ldots)$. These relations imply the following ones:

$$\left.\begin{aligned} p + r + p' + r' &= k + \nabla\,, \\ q + s + q' + s' &= k\,. \end{aligned}\right\} \tag{10.5.26}$$

From the structure of (10.5.21)–(10.5.23) it can be seen that the critical values U^*_κ are determined by

$$U^{*(\kappa ij)}_{pqrsp'q'r's'} = \Big\langle U^{(\kappa ij)}_{pqrsp'q'r's'} \exp 2\mathrm{i}(-p + q - r + s + \nabla)\varphi_{ij} \Big\rangle \times \\ \times \exp 2\mathrm{i}(p - q + r - s - \nabla)\varphi_{ij} \tag{10.5.27}$$

with relations (10.5.24) or (10.5.26) for the power indices. Again one should take into account these critical values only for $\kappa = 3$ and $\kappa = 5$. For $\kappa = 2$ this expression turns out to be identically zero whereas for $\kappa = 1$ it is always possible to add to Γ_2 the term

$$\Gamma^{(2ij)}_{pqrsp'q'r's'} = -U^{*(1ij)}_{pqrsp'q'r's'} \tag{10.5.28}$$

for the critical values of the indices. In taking quadratures (10.5.21)–(10.5.23) in a purely periodic form one has to deal again with functions (10.3.32) and their generalizations.

In version (c) it is possible to take quadratures (10.5.21)–(10.5.23) by means of τ-series. In dealing with the second-order terms with respect to μ one should again use the technique of (10.3.49)–(10.3.55). In any version after determination of U and Γ one gets basic GPT series (10.4.34) and (10.4.35) as well as secular system (10.1.8). Indeed, the right-hand members of (10.4.13) result from

$$F = U^* \tag{10.5.29}$$

and transformation (10.1.7) brings system (10.4.13) to (10.1.8) with the right-hand members

$$\Phi = D^{-1}U_3^* , \qquad \Psi = D^{-1}U_5^* . \tag{10.5.30}$$

10.6 The Secular System

The secular system is given by equations (10.1.8) and their conjugates. The matrices A and B are determined by (10.4.33) and to the first order have values (10.4.79)–(10.4.82). From (10.4.81)–(10.4.82) it is seen that any diagonal element of B is equal to the sum of the non-diagonal elements of the same line leading to the zero value of one eigennumber of B. The numerical second-order values of A and B are given by Brumberg and Chapront (1973). The functions Φ and Ψ are determined by (10.5.30). In version (a) these functions may be written as

$$\Phi_i = \sum{}^* U^{(3i)}_{pqrsk} \prod_{j=1}^{N} \alpha_j^{p_j} \bar{\alpha}_j^{q_j} \beta_j^{r_j} \bar{\beta}_j^{s_j} , \tag{10.6.1}$$

$$\Psi_i = \sum{}^* U^{(5i)}_{pqrsk} \prod_{j=1}^{N} \alpha_j^{p_j} \bar{\alpha}_j^{q_j} \beta_j^{r_j} \bar{\beta}_j^{s_j} , \tag{10.6.2}$$

where the asterisk denotes summation over critical values (10.5.10). In the first-order theory of version (b) one can write these functions in the form

$$\underset{1}{\Phi_i} = \sum_{j=1}^{N} {}^{(i)}\Phi(i,j)\,, \qquad \underset{1}{\Psi_i} = \sum_{j=1}^{N} {}^{(i)}\Psi(i,j)\,, \tag{10.6.3}$$

$$\Phi(i,j) = \sum \Phi^{(ij)}_{pqrsp'q'r's'} \alpha_i^p \bar{\alpha}_i^q \beta_i^r \bar{\beta}_i^s \alpha_j^{p'} \bar{\alpha}_j^{q'} \beta_j^{r'} \bar{\beta}_j^{s'} \tag{10.6.4}$$

and

$$\Psi(i,j) = \sum \Psi^{(ij)}_{pqrsp'q'r's'} \alpha_i^p \bar{\alpha}_i^q \beta_i^r \bar{\beta}_i^s \alpha_j^{p'} \bar{\alpha}_j^{q'} \beta_j^{r'} \bar{\beta}_j^{s'} \tag{10.6.5}$$

with numerical coefficients

$$\begin{aligned} \Phi^{(ij)}_{pqrsp'q'r's'} &= (-1)^{p-q+r-s-1} \times \\ &\times \left\langle U^{(3ij)}_{pqrsp'q'r's'} \exp 2\mathrm{i}(-p+q-r+s+1)\varphi_{ij} \right\rangle \end{aligned} \tag{10.6.6}$$

and

$$\begin{aligned} \Psi^{(ij)}_{pqrsp'q'r's'} &= (-1)^{p-q+r-s-1} \times \\ &\times \left\langle U^{(5ij)}_{pqrsp'q'r's'} \exp 2\mathrm{i}(-p+q-r+s+1)\varphi_{ij} \right\rangle . \end{aligned} \tag{10.6.7}$$

The summation in (10.6.4) and (10.6.5) is performed over values p, q, r, s, p', q', r' and s' related by (10.5.26) with $\nabla = 1$. It is easy to calculate the number of terms in these finite sums. Consider again equation (1.3.1) with $m = 4$. The total number of solutions of this equation with fixed k is $L(4,k)$ as determined by (1.3.2). We can write

$$L(4,k) = L_1(4,k) + L_2(4,k)\,, \tag{10.6.8}$$

where $L_1(4,k)$is the number of solutions with an odd value of $i_3 + i_4$ and $L_2(4,k)$ is the number of solutions with an even value of $i_3 + i_4$. One easily finds that

$$L_1(4, 2k-1) = L_2(4, 2k-1) = \tfrac{1}{3}k(k+1)(2k+1)\,, \tag{10.6.9}$$

$$L_1(4, 2k) = \tfrac{2}{3}k(k+1)(k+2) \tag{10.6.10}$$

and

$$L_2(4, 2k) = \tfrac{1}{3}(1+k)(2k^2+4k+3)\,. \tag{10.6.11}$$

The number $L_3^*(2k+1)$ of terms of degree $2k+1$ in (10.6.4) will be

$$L_3^*(2k+1) = L_1(4,k+1)L_1(4,k) + L_2(4,k+1)L_2(4,k)\,. \tag{10.6.12}$$

The number $L_5^*(2k+1)$ of terms of degree $2k+1$ in (10.6.5) will be

$$L_5^*(2k+1) = L_1(4,k+1)L_2(4,k) + L_2(4,k+1)L_1(4,k)\,. \tag{10.6.13}$$

These two numbers are the same. For degrees 3, 5 and 7 (k = 1, 2, 3) they give 20, 100 and 350, respectively. It is possible to give explicit formulae to calculate the third-degree terms in Φ_i and Ψ_i but it is more convenient to derive them by computer.

The techniques for solving secular system (10.1.8) have been considered in Chapter 4.

At present, the GPT has not been completed to such an extent that it can be used in practice. But the techniques presented in this chapter enable one to hope that it will be actually constructed in the not too distant future.

11 GPT Techniques in Minor-Planet and Lunar Problems

11.1 Right-Hand Members

This chapter is devoted to the study of the construction of theories of motion for minor planets (Sections 11.1–11.4) and the Moon (Section 11.5) in the framework of the GPT. This framework is designed for small eccentricities and inclinations. For minor planets with large values of eccentricities and inclinations it is reasonable to use the expansion of the disturbing function of type (6.2.13), which involves Jacobi nomes and actual values of inclinations. Currently, significant progress in analytical studies of minor-planet motion has been achieved by Milani and Knežević (1990, 1992), in particular, in the theory of secular resonances caused by the commensurabilities between the secular motions of perihelia and nodes of major planets and asteroids. Therefore, consideration of the motion of asteroids by the same technique as used in the major-planet theory may bring new insight to this problem.

The motion of a minor planet is considered in this chapter to be the motion of a material point of negligible mass under the attraction of the Sun, of mass m_0, and N major planets of masses m_i $(i = 1, 2, \dots, N)$. In the heliocentric coordinates $\mathbf{r} = (x,\ y,\ z)$ this motion is described by the equations

$$\ddot{\mathbf{r}} = \frac{\partial U}{\partial \mathbf{r}} \tag{11.1.1}$$

with the force function

$$U = \frac{Gm_0}{r} + \sum_{i=1}^{N} Gm_i \left(\frac{1}{\Delta_i} - \frac{\mathbf{r}\mathbf{r}_i}{r_i^3} \right) \tag{11.1.2}$$

with

$$r^2 = x^2 + y^2 + z^2 \,, \qquad r_i^2 = x_i^2 + y_i^2 + z_i^2$$

and

$$\Delta_i^2 = (x - x_i)^2 + (y - y_i)^2 + (z - z_i)^2 \,.$$

The heliocentric coordinates of the major planets are replaced by (7.2.2) and (7.2.3) and similarly one has for the minor-planet coordinates the relations already met in (6.2.43), i.e.

$$x + \mathrm{i}y = a(1-p)\exp \mathrm{i}\lambda\,, \tag{11.1.3}$$

$$z = aw \tag{11.1.4}$$

with

$$\lambda = nt + \varepsilon \tag{11.1.5}$$

and

$$n^2 a^3 = Gm_0\,. \tag{11.1.6}$$

In new variables equations (11.1.1) take the form

$$\ddot{p} + 2\,\mathrm{i}\,n\dot{p} - \tfrac{3}{2}n^2(p+q) = n^2 P \tag{11.1.7}$$

and

$$\ddot{w} + n^2 w = n^2 W \tag{11.1.8}$$

with $q = \bar{p}$ and the right-hand members

$$\begin{aligned} P &= -1 - \frac{1}{2}p - \frac{3}{2}q + \frac{2}{n^2a^2}\frac{\partial U}{\partial q}\,, \\ W &= w + \frac{1}{n^2a^2}\frac{\partial U}{\partial w} \end{aligned} \tag{11.1.9}$$

and

$$\begin{aligned} U = n^2a^2\Bigg\{\frac{a}{r} + \mu\sum_{i=1}^{N}\kappa_i\Bigg[\frac{a}{\Delta_i} - \\ - \frac{1}{2}a^2a_i\frac{(1-p)(1-q_i)\zeta_i + (1-q)(1-p_i)\zeta_i^{-1} + 2ww_i}{r_i^3}\Bigg]\Bigg\}\,. \end{aligned} \tag{11.1.10}$$

Here

$$\zeta_i = \exp \mathrm{i}(\lambda - \lambda_i) \tag{11.1.11}$$

and

$$\mu\kappa_i = \frac{m_i}{m_0}\,, \tag{11.1.12}$$

μ being a small parameter taken in the GPT as $\mu = 10^{-3}$. In addition,

$$r^2 = a^2\left[(1-p)(1-q) + w^2\right],$$

$$r_i^2 = a_i^2\left[(1-p_i)(1-q_i) + w_i^2\right]$$

and

$$\Delta_i^2 = \left[a(1-p) - a_i(1-p_i)\zeta_i^{-1}\right]\left[a(1-q) - a_i(1-q_i)\zeta_i\right] + (aw - a_iw_i)^2\,.$$

Therefore,

$$P = -1 - \frac{1}{2}p - \frac{3}{2}q + (1-p)\left(\frac{a}{r}\right)^3 + \mu \sum_{i=1}^{N} P_i \tag{11.1.13}$$

and

$$W = w - w\left(\frac{a}{r}\right)^3 + \mu \sum_{i=1}^{N} W_i \tag{11.1.14}$$

with

$$P_i = \kappa_i \left\{ \left[1 - p - \frac{a_i}{a}(1-p_i)\zeta_i^{-1}\right]\left(\frac{a}{\Delta_i}\right)^3 + \zeta_i^{-1}\left(\frac{a}{a_i}\right)^2 (1-p_i)\left(\frac{a_i}{r_i}\right)^3 \right\} \tag{11.1.15}$$

and

$$W_i = -\kappa_i \left[\left(w - \frac{a_i}{a} w_i\right)\left(\frac{a}{\Delta_i}\right)^3 + \left(\frac{a}{a_i}\right)^2 w_i \left(\frac{a_i}{r_i}\right)^3 \right]. \tag{11.1.16}$$

These expansions already appeared in (7.2.24)–(7.2.27) in the N-planet problem. By expanding these functions in powers of p, q, w, p_i, q_i and w_i one obtains the expansions similar to (10.2.1) and (10.2.2)

$$P = \sum_{k=0}^{\infty}\sum_{l=0}^{\infty}\sum_{m=0}^{\infty}\Bigg[\varphi_{klm} p^k q^l w^{2m} + \\ + \mu \left(\sum_{i=1}^{N} \kappa_i \sum_{r=0}^{\infty}\sum_{s=0}^{\infty} \sum_{t=\mathrm{E}(\frac{m+1}{2})}^{\infty} \psi^{(i)}_{k,l,r,s,t,t-m} p_i^r q_i^s w_i^{2t-m} \right) p^k q^l w^m \Bigg] \tag{11.1.17}$$

and

$$W = \sum_{k=0}^{\infty}\sum_{l=0}^{\infty}\sum_{m=0}^{\infty}\Bigg[\rho_{klm} p^k q^l w^{2m+1} + \\ + \mu \left(\sum_{i=1}^{N} \kappa_i \sum_{r=0}^{\infty}\sum_{s=0}^{\infty} \sum_{t=\mathrm{E}(\frac{m}{2})}^{\infty} \theta^{(i)}_{k,l,r,s,t,t-m} p_i^r q_i^s w_i^{2t-m+1} \right) p^k q^l w^m \Bigg]. \tag{11.1.18}$$

The constant coefficients φ_{klm} and ρ_{klm} are determined again by (10.2.3) and (10.2.5), respectively. The coefficients $\psi^{(i)}_{klrsmt}$ and $\theta^{(i)}_{klrsmt}$ are quasi-periodic functions of time that depend on the differences $\lambda - \lambda_i$. By analogy with (10.2.4), (10.2.6), (6.2.60) and (6.2.51) they are determined by the explicit expressions

$$\psi^{(i)}_{klrsmt} = (-1)^t \frac{(\frac{3}{2})_m(\frac{1}{2}+m)_{k+r}(\frac{3}{2}+m)_{l+s}}{(1)_m(1)_{k+r}(1)_{l+s}} \left[\left(\frac{a}{a_i}\right)^2 \delta_{k0}\delta_{l0}\delta_{tm}\zeta_i^{-1} + \right.$$
$$+ \frac{(1+r)_k(1+s)_l(1+m-t)_{m+t}}{(1)_k(1)_l(1)_{m+t}} \times$$
$$\left. \times \gamma\left(r+s+m+t, -m-k-r-\tfrac{1}{2}, -m-l-s-\tfrac{3}{2}, -r+s, \frac{a_i}{a}, \zeta_i\right)\right] \tag{11.1.19}$$

and

$$\theta^{(i)}_{klrsmt} = (-1)^t \frac{(\frac{3}{2})_m(\frac{3}{2}+m)_{k+r}(\frac{3}{2}+m)_{l+s}}{(1)_m(1)_{k+r}(1)_{l+s}} \left[-\left(\frac{a}{a_i}\right)^2 \delta_{k0}\delta_{l0}\delta_{tm} + \right.$$
$$+ \frac{(1+r)_k(1+s)_l(1+m-t)_{m+t+1}}{(1)_k(1)_l(1)_{m+t+1}} \times$$
$$\times \gamma\Big(r+s+m+t+1, -m-k-r-\tfrac{3}{2}, -m-l-s-\tfrac{3}{2},$$
$$\left. -r+s, \frac{a_i}{a}, \zeta_i\Big)\right]. \tag{11.1.20}$$

In accordance with the GPT technique one puts

$$p = p^{(0)} + \delta p\,, \tag{11.1.21}$$

$p = p^{(0)}$ being a particular planar quasi-periodic solution of (11.1.7) with $\delta p_i = w_i = 0$ in the right-hand member (implying that the motions of the major planets are in planar quasi-periodic orbits). Right-hand members (11.1.17) and (11.1.18) can now be transformed using (10.2.8) and (11.1.21). One has

$$P = \sum_{k=0}^{\infty}\sum_{l=0}^{\infty}\sum_{m=0}^{\infty} (\delta p)^k (\delta q)^l w^m \Big\{ P_{klm} w^m +$$
$$+ \mu \sum_{i=1}^{N}\sum_{r=0}^{\infty}\sum_{s=0}^{\infty}\sum_{t=\mathrm{E}(\frac{m+1}{2})}^{\infty} P^{(i)}_{k,l,r,s,t,t-m} (\delta p_i)^r (\delta q_i)^s w_i^{2t-m} \Big\} \tag{11.1.22}$$

and

$$W = \sum_{k=0}^{\infty}\sum_{l=0}^{\infty}\sum_{m=0}^{\infty} (\delta p)^k (\delta q)^l w^m \Big\{ W_{klm} w^{m+1} +$$
$$+ \mu \sum_{i=1}^{N}\sum_{r=0}^{\infty}\sum_{s=0}^{\infty}\sum_{t=\mathrm{E}(\frac{m}{2})}^{\infty} W^{(i)}_{k,l,r,s,t,t-m} (\delta p_i)^r (\delta q_i)^s w_i^{2t-m+1} \Big\}, \tag{11.1.23}$$

the coefficients being the functions of the quasi-periodic solutions $p^{(0)}$ and $p_i^{(0)}$:

$$P_{\alpha\beta m} = \sum_{k=0}^{\infty}\sum_{l=0}^{\infty} \frac{(1+k)_\alpha(1+l)_\beta}{(1)_\alpha(1)_\beta} \varphi_{\alpha+k,\beta+l,m}\, p^k q^l\,, \tag{11.1.24}$$

$$P^{(i)}_{\alpha\beta\sigma\nu mt} = \kappa_i \sum_{k=0}^{\infty}\sum_{l=0}^{\infty}\sum_{r=0}^{\infty}\sum_{s=0}^{\infty} \frac{(1+k)_\alpha(1+l)_\beta(1+r)_\sigma(1+s)_\nu}{(1)_\alpha(1)_\beta(1)_\sigma(1)_\nu} \times$$
$$\times\, \psi^{(i)}_{\alpha+k,\beta+l,\sigma+r,\nu+s,m,t}\, p^k q^l p_i^r q_i^s\,, \tag{11.1.25}$$

$$W_{\alpha\beta m} = \sum_{k=0}^{\infty}\sum_{l=0}^{\infty} \frac{(1+k)_\alpha(1+l)_\beta}{(1)_\alpha(1)_\beta} \rho_{\alpha+k,\beta+l,m}\, p^k q^l \tag{11.1.26}$$

and

$$W^{(i)}_{\alpha\beta\sigma\nu mt} = \kappa_i \sum_{k=0}^{\infty}\sum_{l=0}^{\infty}\sum_{r=0}^{\infty}\sum_{s=0}^{\infty} \frac{(1+k)_\alpha(1+l)_\beta(1+r)_\sigma(1+s)_\nu}{(1)_\alpha(1)_\beta(1)_\sigma(1)_\nu} \times$$
$$\times\, \theta^{(i)}_{\alpha+k,\beta+l,\sigma+r,\nu+s,m,t}\, p^k q^l p_i^r q_i^s\,. \tag{11.1.27}$$

These expressions are similar to (10.2.39)–(10.2.42) and they should be evaluated for $p = p^{(0)}$, $p_i = p_i^{(0)}$. Separating linear terms one has

$$P = P^{(0)} - K\,\delta p - L\,\delta q + P' \tag{11.1.28}$$

and

$$W = -Mw + W' \tag{11.1.29}$$

with

$$P^{(0)} = P_{000} + \mu \sum_{i=1}^{N} P^{(i)}_{000000}\,, \tag{11.1.30}$$

$$P' = -\sum_{i=1}^{N} (K_i \delta p_i + L_i \delta q_i) + P^*\,, \tag{11.1.31}$$

and

$$W' = -\sum_{i=1}^{N} M_i w_i + W^*\,. \tag{11.1.32}$$

Again, by analogy with (10.2.11), (10.2.15), (10.2.16) and (10.2.35)–(10.2.37) one gets

$$K = -P_{100} - \mu \sum_{i=1}^{N} P^{(i)}_{100000} = \frac{1}{2} - \frac{1}{2}\left(\frac{a}{r}\right)^3 - \frac{1}{2}\mu \sum_{i=1}^{N} \kappa_i \left(\frac{a}{\Delta_i}\right)^3, \tag{11.1.33}$$

$$L = -P_{010} - \mu \sum_{i=1}^{N} P^{(i)}_{010000} =$$

$$= \frac{3}{2} - \frac{3}{2}\left(\frac{a}{r}\right)^5 (1-p)^2 - \frac{3}{2}\mu \sum_{i=1}^{N} \left(\frac{a}{\Delta_i}\right)^5 \left[1 - p - \frac{a_i}{a}(1-p_i)\zeta_i^{-1}\right]^2 , \tag{11.1.34}$$

$$M = -W_{000} - \mu \sum_{i=1}^{N} W^{(i)}_{0,0,0,0,0,-1} = -1 + \left(\frac{a}{r}\right)^3 + \mu \sum_{i=1}^{N} \kappa_i \left(\frac{a}{\Delta_i}\right)^5 , \tag{11.1.35}$$

$$K_i = -\mu P^{(i)}_{001000} = \frac{1}{2}\mu \kappa_i a^2 a_i \left(\frac{1}{\Delta_i^3} - \frac{1}{r_i^3}\right) \zeta_i^{-1} , \tag{11.1.36}$$

$$L_i = -\,\mu P^{(i)}_{000100} =$$

$$= \tfrac{3}{2}\mu \kappa_i a^2 a_i \zeta_i^{-1} \left\{ \frac{1}{\Delta_i^5}\left[1 - p_i - \frac{a}{a_i}(1-p)\zeta_i\right]^2 - \frac{1}{r_i^5}(1-p_i)^2 \right\} \tag{11.1.37}$$

and

$$M_i = -\mu W^{(i)}_{000000} = \mu \kappa_i a^2 a_i \left(\frac{1}{r_i^3} - \frac{1}{\Delta_i^3}\right) . \tag{11.1.38}$$

It is evident that

$$M = -2K , \qquad M_i = -2K_i\zeta_i . \tag{11.1.39}$$

Expressions (11.1.33)–(11.1.38) should be evaluated again with $p = p^{(0)}$, $p_i = p_i^{(0)}$, $w = w_i = 0$. As far as P^* and W^* are concerned, they should be computed by (11.1.22) and (11.1.23) under the conditions

$$k + l + 2m > \Delta , \qquad k + l + r + s + 2t > \Delta \tag{11.1.40}$$

with $\Delta = 1$ for (11.1.22) and $\Delta = 0$ for (11.1.23). P^* and W^* may also be computed by the closed-form expressions resulting from (10.2.22) and (10.2.23). The substitution of (11.1.21) into (11.1.7) and (11.1.8) transforms the equations of motion of a minor planet as follows:

$$\delta\ddot{p} + 2\mathrm{i}\, n\, \delta\dot{p} + n^2\left[(-\tfrac{3}{2} + K)\delta p + (-\tfrac{3}{2} + L)\delta q\right] = n^2 P' \tag{11.1.41}$$

and

$$\ddot{w} + n^2(1 + M)w = n^2 W' . \tag{11.1.42}$$

The right-hand members P' and W' take a particularly simple form if one can confine them by considering the motion of the major planets in intermediate orbits, i.e.

$$P' = \sum_{\substack{k=0\\k+l+2m>1}}^{\infty} \sum_{l=0}^{\infty} \sum_{m=0}^{\infty} \left(P_{klm} + \mu \sum_{i=1}^{N} P^{(i)}_{k,l,0,0,m,-m} \right) (\delta p)^k (\delta q)^l w^{2m} \tag{11.1.43}$$

and

$$W' = \sum_{\substack{k=0 \\ k+l+2m>0}}^{\infty} \sum_{l=0}^{\infty} \sum_{m=0}^{\infty} \left(W_{klm} + \mu \sum_{i=1}^{N} W^{(i)}_{k,l,0,0,m,-m-1} \right) (\delta p)^k (\delta q)^l w^{2m+1} . \tag{11.1.44}$$

In the general case P' and W' are computed by means of (11.1.31) and (11.1.32).

11.2 The Intermediary

The function $p^{(0)}$ satisfies the equation

$$\ddot{p}^{(0)} + 2\mathrm{i}\, n\dot{p}^{(0)} - \tfrac{3}{2}n^2 \left(p^{(0)} + q^{(0)} \right) = n^2 P^{(0)} \tag{11.2.1}$$

with the right-hand member determined by (11.1.30), or

$$P^{(0)} = \sum_{k=0}^{\infty} \sum_{l=0}^{\infty} \left[\varphi_{kl0} + \mu \sum_{j=1}^{N} \kappa_i \sum_{r=0}^{\infty} \sum_{s=0}^{\infty} \psi^{(i)}_{klrs00} \left(p_i^{(0)} \right)^r \left(q_i^{(0)} \right)^s \right] \times \\ \times \left(p^{(0)} \right)^k \left(q^{(0)} \right)^l . \tag{11.2.2}$$

As in the GPT one can solve equation (11.2.1) in three different forms corresponding to versions (a), (b) or (c) of the GPT. Here we restrict ourselves by considering only version (a). In this version the right-hand member $P^{(0)}$ at each step of approximation with respect to μ has the form

$$P^{(0)} = \sum P^{(0)}_{kk'} \exp \mathrm{i}(k\lambda + k'\lambda') , \tag{11.2.3}$$

$\lambda' = (\lambda_1, \ldots, \lambda_N)$ being N-vector of the mean longitudes of the major planets and $k' = (k_1, \ldots, k_N)$ being N-multi-index. One thus has

$$k + k_1 + \ldots + k_N = 0 . \tag{11.2.4}$$

The solution $p^{(0)}$ is of the same form

$$p^{(0)} = \sum p^{(0)}_{kk'} \exp \mathrm{i}(k\lambda + k'\lambda') \tag{11.2.5}$$

with the coefficients

$$p^{(0)}_{00} = -\tfrac{1}{3} P^{(0)}_{00} \tag{11.2.6}$$

and

$$p_{kk'}^{(0)} = \frac{n^2}{(kn+k'n')^2\left[n^2-(kn+k'n')^2\right]}\Big\{-\tfrac{3}{2}n^2P_{-k,-k'}^{(0)}+ \\ +\left[(kn+k'n')^2-2n(kn+k'n')+\tfrac{3}{2}n^2\right]P_{kk'}^{(0)}\Big\}. \qquad (11.2.7)$$

For the first order in μ these formulae may be replaced by the simpler expressions

$$P^{(0)} = \mu\sum_{i=1}^{N}\kappa_i\sum_{\sigma=-\infty}^{\infty}\psi_\sigma^{(i)}(0,0,0,0,0,0)\zeta_i^\sigma \qquad (11.2.8)$$

and

$$p^{(0)} = \mu\sum_{i=1}^{N}\kappa_i\sum_{\sigma=-\infty}^{\infty}p_\sigma^{(i)}\zeta_i^\sigma \qquad (11.2.9)$$

with the coefficients

$$p_0^{(i)} = -\tfrac{1}{3}\psi_0^{(i)}(0,0,0,0,0,0) \qquad (11.2.10)$$

and

$$p_\sigma^{(i)} = \frac{m_i^2}{\sigma^2(m_i^2-\sigma^2)}\Big[-\tfrac{3}{2}m_i^2\psi_{-\sigma}^{(i)}(0,0,0,0,0,0)+ \\ +\left(\sigma^2-2\sigma m_i+\tfrac{3}{2}m_i^2\right)\psi_\sigma^{(i)}(0,0,0,0,0,0)\Big]. \qquad (11.2.11)$$

Here $\psi_\sigma^{(i)}(k,l,r,s,m,t)$ denotes the coefficient in ζ_i^σ in the expansion of function (11.1.19) in powers of ζ_i, and by analogy with (10.3.31)

$$m_i = \frac{n}{n-n_i}. \qquad (11.2.12)$$

For the first-order intermediary coefficients (11.2.10) and (11.2.11) may be expressed very simply in terms of the symmetrical Laplace coefficients

$$\psi_\sigma^{(i)}(0,0,0,0,0,0) = \left(\frac{a}{a_i}\right)^2\delta_{\sigma,-1}+\frac{1}{2}a^3c_3^{(\sigma)}(a,\,a_i)-\frac{1}{2}a^2a_ic_3^{(\sigma+1)}(a,a_i). \qquad (11.2.13)$$

The construction of the intermediary in versions (b) and (c) may be performed just by repeating the operations of the GPT (Section 10.3).

11.3 Solution Techniques

The system of equations (11.1.41) and (11.1.42) may be solved by the technique of Chapters 9 and 10. The transformation

$$\begin{aligned} \delta p &= \xi - \tfrac{2}{3}h - \tfrac{1}{2}u + \tfrac{3}{2}\bar{u}\,, & w &= v + \bar{v}\,, \\ \delta\dot{p} &= \mathrm{i}\,n(h - \tfrac{1}{2}u - \tfrac{3}{2}\bar{u})\,, & \dot{w} &= \mathrm{i}\,n(v - \bar{v}) \end{aligned} \tag{11.3.1}$$

(imaginary ξ, real h, complex u and v) brings this system to form (10.4.4), where one may simply put $\mathcal{N} = n$. The Jordan matrix $\mathcal{P}$ is now a 6×6 matrix with the non-zero elements

$$\mathcal{P}_{12} = \mathcal{P}_{33} = \mathcal{P}_{55} = 1\,, \qquad \mathcal{P}_{44} = \mathcal{P}_{66} = -1\,. \tag{11.3.2}$$

For the matrix Q expressions (10.4.10) and (10.4.11) remain true but all the quantities that occur there are now just scalars. The 6-vector R of the right-hand member has the components

$$\begin{aligned} R_1 &= -(P' + \bar{P}')\,, & R_2 &= \tfrac{3}{2}(P' - \bar{P}')\,, \\ R_3 &= \tfrac{1}{2}(P' - 3\bar{P}')\,, & R_5 &= -\tfrac{1}{2}W' \end{aligned} \tag{11.3.3}$$

with relations (10.4.8). System (10.4.4) is transformed again by means of (10.4.12) to form (10.4.13), always keeping in mind that one now has $\mathcal{N} = n$ and the elements of the block vectors and matrices are simply scalars.

Determination of those terms not depending on the eccentricities and inclinations of the major planets and proportional to the first-degree of the eccentricity and inclination of a minor planet involves the 6×6 matrices S and H. S may be determined by solving equations (10.4.15)–(10.4.29) by iterations with respect to μ. Restricting ourselves again to version (a) of the GPT one has at each step of the approximation the general representation

$$Q_{\kappa\nu} = \sum Q_{kk'}^{(\kappa\nu)} \exp \mathrm{i}(k\lambda + k'\lambda') \tag{11.3.4}$$

with condition (11.2.4). The matrices G and S are expressed in the same form. From equations (10.4.18)–(10.4.29) we have

$$S_{kk'}^{(21)} = \frac{n}{kn + k'n'} G_{kk'}^{(21)}\,, \tag{11.3.5}$$

$$S_{kk'}^{(11)} = \frac{n}{kn + k'n'} \left(G_{kk'}^{(11)} + S_{kk'}^{(21)} \right)\,, \tag{11.3.6}$$

$$S_{kk'}^{(22)} = \frac{n}{kn + k'n'} \left(G_{kk'}^{(22)} - S_{kk'}^{(21)} \right)\,, \tag{11.3.7}$$

$$S_{kk'}^{(12)} = \frac{n}{kn + k'n'} \left(G_{kk'}^{(12)} - S_{kk'}^{(11)} + S_{kk'}^{(22)} \right)\,, \tag{11.3.8}$$

$$S_{kk'}^{(23)} = \frac{n}{(k+1)n + k'n'} G_{kk'}^{(23)}\,, \tag{11.3.9}$$

$$S_{kk'}^{(13)} = \frac{n}{(k+1)n + k'n'} \left(G_{kk'}^{(13)} + S_{kk'}^{(23)} \right) , \tag{11.3.10}$$

$$S_{kk'}^{(31)} = \frac{n}{(k-1)n + k'n'} G_{kk'}^{(31)} , \tag{11.3.11}$$

$$S_{kk'}^{(32)} = \frac{n}{(k-1)n + k'n'} \left(G_{kk'}^{(32)} - S_{kk'}^{(31)} \right) , \tag{11.3.12}$$

$$S_{kk'}^{(\kappa\nu)} = \frac{n}{kn + k'n'} G_{kk'}^{(\kappa\nu)} , \quad (\kappa, \nu) = (3,3), \ (5,5) \tag{11.3.13}$$

and

$$S_{kk'}^{(\kappa\nu)} = \frac{n}{(k-2)n + k'n'} G_{kk'}^{(\kappa\nu)} , \quad (\kappa, \nu) = (3,4), \ (5,6) . \tag{11.3.14}$$

$n' = (n_1, \ldots, n_N)$ denotes here the N-vector of the mean motions of the major planets and

$$k'n' = k_1 n_1 + \ldots + k_N n_N . \tag{11.3.15}$$

The remaining 8 non-zero elements of S are determined again by the relations of conjugation resulting from (10.4.11) by a reversal of sign. The critical values corresponding to the values $k = k' = 0$ in equations (11.3.5)–(11.3.8) and (11.3.13) should be referred to G^*. Then H is determined by (10.4.33).

To the first order with respect to μ these formulae are significantly simplified. The coefficients K, L and M are expanded into series of form (11.2.9) with the coefficients

$$K_\sigma^{(i)} = -\psi_\sigma^{(i)}(1,0,0,0,0,0) - \tfrac{3}{4} \left(p_\sigma^{(i)} + p_{-\sigma}^{(i)} \right) , \tag{11.3.16}$$

$$L_\sigma^{(i)} = -\psi_\sigma^{(i)}(0,1,0,0,0,0) - \tfrac{3}{4} \left(p_\sigma^{(i)} + 5p_{-\sigma}^{(i)} \right) \tag{11.3.17}$$

and

$$M_\sigma^{(i)} = -\theta_\sigma^{(i)}(0,0,0,0,0,-1) + \tfrac{3}{2} \left(p_\sigma^{(i)} + p_{-\sigma}^{(i)} \right) . \tag{11.3.18}$$

These coefficients may also be expressed in terms of the symmetrical Laplace coefficients by means of

$$\psi_\sigma^{(i)}(1,0,0,0,0,0) = -\tfrac{1}{2}\theta_\sigma^{(i)}(0,0,0,0,0,-1) = \tfrac{1}{4} a^3 c_3^{(\sigma)}(a, \, a_i), \tag{11.3.19}$$

and

$$\psi_\sigma^{(i)}(0,1,0,0,0,0) = \tfrac{1}{4}(3+2\sigma) a^3 c_3^{(\sigma)}(a, \, a_i) - \tfrac{1}{2}(1+\sigma) a^2 a_i c_3^{(\sigma+1)}(a, \, a_i) . \tag{11.3.20}$$

Using (10.4.10) one now computes the coefficients of the expansions

$$Q_{\kappa\nu} = \mu \sum_{i=1}^{N} \kappa_i \sum_{\sigma=-\infty}^{\infty} Q_\sigma^{(i\kappa\nu)} \zeta_i^\sigma .$$

Then the corresponding coefficients of the expansions of the elements $S_{\kappa\nu}$ will be determined from (11.3.5)–(11.3.14) as follows:

$$S_\sigma^{(i21)} = \frac{m_i}{\sigma} Q_\sigma^{(i21)}, \tag{11.3.21}$$

$$S_\sigma^{(i11)} = \frac{m_i}{\sigma} \left(Q_\sigma^{(i11)} + S_\sigma^{(i21)} \right), \tag{11.3.22}$$

$$S_\sigma^{(i22)} = \frac{m_i}{\sigma} \left(Q_\sigma^{(i22)} - S_\sigma^{(i21)} \right), \tag{11.3.23}$$

$$S_\sigma^{(i12)} = \frac{m_i}{\sigma} \left(Q_\sigma^{(i12)} - S_\sigma^{(i11)} + S_\sigma^{(i22)} \right), \tag{11.3.24}$$

$$S_\sigma^{(i23)} = \frac{m_i}{\sigma + m_i} Q_\sigma^{(i23)}, \tag{11.3.25}$$

$$S_\sigma^{(i13)} = \frac{m_i}{\sigma + m_i} \left(Q_\sigma^{(i13)} + S_\sigma^{(i23)} \right), \tag{11.3.26}$$

$$S_\sigma^{(i31)} = \frac{m_i}{\sigma - m_i} Q_\sigma^{(i31)}, \tag{11.3.27}$$

$$S_\sigma^{(i32)} = \frac{m_i}{\sigma - m_i} \left(Q_\sigma^{(i32)} - S_\sigma^{(i31)} \right), \tag{11.3.28}$$

$$S_\sigma^{(i\kappa\nu)} = \frac{m_i}{\sigma} Q_\sigma^{(i\kappa\nu)}, \quad (\kappa, \nu) = (3,3),\ (5,5) \tag{11.3.29}$$

and

$$S_\sigma^{(i\kappa\nu)} = \frac{m_i}{\sigma - 2m_i} Q_\sigma^{(i\kappa\nu)}, \quad (\kappa, \nu) = (3,4),\ (5,6)\,. \tag{11.3.30}$$

As far as the critical terms are concerned, one easily finds from (10.4.10), (11.2.10), (11.2.13) and (11.3.16)–(11.3.20) for $\sigma = 0$ that

$$G_{33}^* = -G_{44}^* = -G_{55}^* = G_{66}^* = -\tfrac{3}{8} G_{12}^* == -\tfrac{1}{4}\mu \sum_{i=1}^{N} \kappa_i a^2 a_i c_3^{(1)}(a,\, a_i) \tag{11.3.31}$$

and

$$G_{11}^* = G_{22}^* = G_{21}^* = 0\,. \tag{11.3.32}$$

Relations (11.3.32) remain true for higher-order approximations as well.

Having obtained the matrices S and H one may proceed to the determination of the 6-vectors Γ and F. The order of smallness of these vectors is caused by the terms of second degree with respect to the eccentricity and inclination of a small planet and by the first-degree terms with respect to major-planet orbital eccentricities and inclinations. Γ is determined by iterations on μ in increasing order of the degree of the components of Y from equations (10.5.1)–(10.5.6), taking into account that now $\mathcal{N} = n$ and the elements of block vectors and matrices are scalars. The components of the vector U can be represented by the power series

$$U_\kappa = \sum U^{(\kappa)}_{pqrs}(t) a^p \bar{a}^q b^r \bar{b}^s \,. \tag{11.3.33}$$

The corresponding coefficients $\Gamma^{(\kappa)}_{pqrs}$ of the components of Γ are then determined from equations (9.3.15)–(9.3.18). Again, the critical terms in the component U_2 are zero due to the structure of equations of celestial mechanics. The critical terms in the component U_1 do not in fact give contribution into U^* because they may be cancelled by an appropriate choice of the arbitrary constant in Γ_2. The critical terms in Γ_3 and Γ_5 form in accordance with (10.5.29) and (10.5.30) the right-hand members of the secular system (10.1.8) where again $\mathcal{N} = n$ and α and β are simply scalars. The diagonal matrix D occurring in (10.5.30) is reduced in the present case just to $\exp \mathrm{i}\,\lambda$.

11.4 The Secular System

Equations (10.1.8) describe the evolution of the slowly changing variables α and β. The functions Φ and Ψ represent series in powers of α, $\bar{\alpha}$, β and $\bar{\beta}$. The specific form of the coefficients of these series depends on the representation of right-hand members (11.1.31) and (11.1.32).

Let, for example, the motion of the major planets be given in a purely trigonometric form based on solution (4.2.4), (4.2.20) and (4.2.31)–(4.2.33) of the secular system for the major planets. Then the coefficients of series (11.3.33) may be written in the form

$$U^{(\kappa)}_{pqrs}(t) = \sum U^{(\kappa pqrs)}_{kk'm} \exp \mathrm{i}(k\lambda + k'\lambda' + m\omega) \,. \tag{11.4.1}$$

In this expression k' is an N-index, m is a $2N$-index and ω is the vector of the slowly changing variables (4.2.32) describing the secular motion of the planetary eccentricities and nodes. One of the component of ω is zero. The corresponding coefficients for the components of Γ will be

$$\Gamma^{(2pqrs)}_{kk'm} = \frac{n}{(p-q+r-s+k)n + k'n' + m\dot{\omega}} U^{(2pqrs)}_{kk'm} \,, \tag{11.4.2}$$

$$\Gamma^{(1pqrs)}_{kk'm} = \frac{n}{(p-q+r-s+k)n + k'n' + m\dot{\omega}} \left(U^{(1pqrs)}_{kk'm} + \Gamma^{(2pqrs)}_{kk'm} \right) \tag{11.4.3}$$

and

$$\Gamma^{(\kappa pqrs)}_{kk'm} = \frac{n}{(p-q+r-s+k-1)n + k'n' + m\dot{\omega}} U^{(\kappa pqrs)}_{kk'm} \,, \quad \kappa = 3,\ 5 \,. \tag{11.4.4}$$

The aim of transformation (10.4.12) is to eliminate the fast variables. Therefore, the critical terms for functions (11.4.1) will satisfy the conditions

$$p - q + r - s + k = \nabla \,, \qquad n' = 0 \tag{11.4.5}$$

with ∇ determined by (10.5.25). In virtue of these conditions the right-hand members Φ and Ψ of secular system (10.1.8) are power series in α, $\bar{\alpha}$, β and $\bar{\beta}$ with coefficients dependent on time only by means of the trigonometric functions of the components of ω.

The terms of the original right-hand members that are linear with respect to the major-planet orbital eccentricities and inclinations have for the first order in μ the form

$$P' = \mu \sum_{i=1}^{N} \kappa_i \left(\psi^{(i)}_{001000}\, \delta \underset{0}{p_i} + \psi^{(i)}_{000100}\, \delta \underset{0}{q_i} \right) , \tag{11.4.6}$$

and

$$W' = \mu \sum_{i=1}^{N} \kappa_i \theta^{(i)}_{000000}\, \underset{0}{w_i} \,. \tag{11.4.7}$$

Substituting by (7.3.8) and (7.3.9)

$$\delta \underset{0}{p_i} = -\tfrac{1}{2} a_i + \tfrac{3}{2} \bar{a}_i \,, \qquad \underset{0}{w_i} = b_i + \bar{b}_i$$

one finds by (11.3.3) that

$$R_1 = -\tfrac{1}{2} \mu \sum_{i=1}^{N} \kappa_i \Big[\big(-\psi^{(i)}_{001000} - \bar{\psi}^{(i)}_{000100} + 3\psi^{(i)}_{000100} + 3\bar{\psi}^{(i)}_{001000} \big) a_i + \\ + \big(3\psi^{(i)}_{001000} + 3\bar{\psi}^{(i)}_{000100} - \psi^{(i)}_{000100} - \bar{\psi}^{(i)}_{001000} \big) \bar{a}_i \Big] , \tag{11.4.8}$$

$$R_2 = \tfrac{3}{4} \mu \sum_{i=1}^{N} \kappa_i \Big[\big(-\psi^{(i)}_{001000} + \bar{\psi}^{(i)}_{000100} + 3\psi^{(i)}_{000100} - 3\bar{\psi}^{(i)}_{001000} \big) a_i + \\ + \big(3\psi^{(i)}_{001000} - 3\bar{\psi}^{(i)}_{000100} - \psi^{(i)}_{000100} + \bar{\psi}^{(i)}_{001000} \big) \bar{a}_i \Big] , \tag{11.4.9}$$

$$R_3 = \tfrac{1}{4} \mu \sum_{i=1}^{N} \kappa_i \Big[\big(-\psi^{(i)}_{001000} + 3\bar{\psi}^{(i)}_{000100} + 3\psi^{(i)}_{000100} - 9\bar{\psi}^{(i)}_{001000} \big) a_i + \\ + \big(3\psi^{(i)}_{001000} - 9\psi^{(i)}_{000100} - \psi^{(i)}_{000100} + 3\bar{\psi}^{(i)}_{001000} \big) \bar{a}_i \Big] , \tag{11.4.10}$$

and

$$R_5 = -\tfrac{1}{2} \mu \sum_{i=1}^{N} \kappa_i \theta^{(i)}_{000000} \left(b_i + \bar{b}_i \right) . \tag{11.4.11}$$

Together with (11.3.19) and (11.3.20) one should have the coefficients

$$\psi_\sigma^{(i)}(0,0,1,0,0,0) = \frac{1}{2}\left(\frac{a}{a_i}\right)^2 \delta_{\sigma,-1} - \frac{1}{4}a^2 a_i c_3^{(\sigma+1)}(a,\, a_i)\,, \tag{11.4.12}$$

$$\begin{aligned}\psi_\sigma^{(i)}(0,0,0,1,0,0) = \frac{3}{2}\left(\frac{a}{a_i}\right)^2 \delta_{\sigma,-1} - \frac{1}{2}\sigma a^3 c_3^{(\sigma)}(a,\, a_i) - \\ - \frac{1}{4}(1-2\sigma)a^2 a_i c_3^{(\sigma+1)}(a,\, a_i)\end{aligned} \tag{11.4.13}$$

and

$$\theta_\sigma^{(i)}(0,0,0,0,0,0) = -\left(\frac{a}{a_i}\right)^2 \delta_{\sigma,0} + \frac{1}{2}a^2 a_i c_3^{(\sigma)}(a,\, a_i)\,. \tag{11.4.14}$$

From (11.4.8)–(11.4.11) one finds the critical values

$$\begin{aligned}U_3^* = \tfrac{1}{4}\mu \sum_{i=1}^{N} \kappa_i \big[-\psi_1^{(i)}(0,0,1,0,0,0) + 3\psi_{-1}^{(i)}(0,0,0,1,0,0) + \\ + 3\psi_1^{(i)}(0,0,0,1,0,0) - 9\psi_{-1}^{(i)}(0,0,1,0,0,0) \big] a_i \zeta_i\end{aligned} \tag{11.4.15}$$

and

$$U_5^* = -\tfrac{1}{2}\mu \sum_{i=1}^{N} \kappa_i \theta_1^{(i)}(0,0,0,0,0,0) b_i \zeta_i\,. \tag{11.4.16}$$

Using values (11.4.12)–(11.4.14) one obtains the classic expressions

$$\Phi = \tfrac{1}{4}\mu \sum_{i=1}^{N} \kappa_i a^2 a_i c_3^{(2)}(a,\, a_i)\alpha_i \tag{11.4.17}$$

and

$$\Psi = -\tfrac{1}{4}\mu \sum_{i=1}^{N} \kappa_i a^2 a_i c_3^{(1)}(a,\, a_i)\beta_i\,. \tag{11.4.18}$$

As noted earlier, the most advanced results including order 2 with respect to μ and degree 3 with respect to the eccentricity and inclination of a minor planet have been achieved by Milani and Knežević (1990, 1992). They also investigated some very interesting cases of secular resonance. The aim of this section was only to show that the GPT technique may be applied to minor-planet theories, enabling one to use the same computational procedures for major- and minor-planet orbits.

11.5 The Lunar Problem

This section deals with the basic equations of the theory of motion of the Moon, demonstrating that this theory may be also constructed within the GPT framework. In contrast to the minor-planet theory the ratios of the semi-major axes of the disturbed and disturbing bodies are small quantities in the lunar theory and they may be used as the expansion parameters. On the other hand, in this case one cannot neglect the ratio of the mass of the Moon to that of the Earth. Below we reproduce the derivation of the basic equations of the lunar theory based on the work of Ash (1965). Let the letters s, e, m and c refer to the Sun, Earth, Moon and Earth–Moon barycentre, respectively. Let i denote some disturbing major planet $(i = 1, 2, 4, \ldots, N)$ from Mercury $(N = 1)$ to Neptune $(N = 8)$. The quantities with two letters denote relative coordinates. For example, x_i is the position vector of the planet i in some inertial coordinate system, $x_{is} = x_i - x_s$ is the heliocentric position vector of the planet i and r_i and r_{is} are the lengths of these vectors, respectively. The original equations of motion of the Moon, Earth, Sun and major planets are as follows:

$$\ddot{x}_m = GM_e \frac{x_{em}}{r_{em}^3} + GM_s \frac{x_{sm}}{r_{sm}^3} + G \sum_{j=1}^{N}{}^{(3)} M_j \frac{x_{jm}}{r_{jm}^3} + \frac{1}{M_m} F_m \,, \tag{11.5.1}$$

$$\ddot{x}_e = GM_m \frac{x_{me}}{r_{me}^3} + GM_s \frac{x_{se}}{r_{se}^3} + G \sum_{j=1}^{N}{}^{(3)} M_j \frac{x_{je}}{r_{je}^3} + \frac{1}{M_e} F_e \,, \tag{11.5.2}$$

$$\ddot{x}_s = GM_e \frac{x_{es}}{r_{es}^3} + GM_m \frac{x_{ms}}{r_{ms}^3} + G \sum_{j=1}^{N}{}^{(3)} M_j \frac{x_{js}}{r_{js}^3} \tag{11.5.3}$$

and

$$\ddot{x}_i = GM_s \frac{x_{si}}{r_{si}^3} + GM_e \frac{x_{ei}}{r_{ei}^3} + GM_m \frac{x_{mi}}{r_{mi}^3} + G \sum_{j=1}^{N}{}^{(3,i)} M_j \frac{x_{ji}}{r_{ji}^3} \tag{11.5.4}$$

$$(i = 1, 2, 4, \ldots, N) \,.$$

F_m and F_e in (11.5.1) and (11.5.2) stand for the additive forces acting on the Moon and the Earth, respectively, caused mainly by the asphericities of the Earth's and Moon's figures. With the mass of the Earth–Moon barycentre $M_c = M_e + M_m$ one has

$$x_c = \frac{1}{M_c} (M_e x_e + M_m x_m) \tag{11.5.5}$$

and

$$x_{es} = x_{cs} - \frac{M_m}{M_c} x_{me} \,, \tag{11.5.6}$$

$$x_{ms} = x_{cs} + \frac{M_e}{M_c} x_{me} \,. \tag{11.5.7}$$

Instead of the equations (11.5.1)–(11.5.4) of absolute motion it is sufficient to consider the equations of relative motion in terms of x_{me}, x_{cs} and x_{is}:

$$\begin{aligned}\ddot{x}_{me} = &-GM_c\frac{x_{me}}{r_{me}^3} + GM_s\left(\frac{x_{es}}{r_{es}^3} - \frac{x_{ms}}{r_{ms}^3}\right) + \\ &+ G\sum_{j=1}^{N}{}^{(3)} M_j\left(\frac{x_{jm}}{r_{jm}^3} - \frac{x_{je}}{r_{je}^3}\right) + \frac{1}{M_m}F_m - \frac{1}{M_e}F_e \,,\end{aligned} \tag{11.5.8}$$

$$\begin{aligned}\ddot{x}_{cs} = &-G(M_s + M_c)\frac{x_{cs}}{r_{cs}^3} + G\sum_{j=1}^{N}{}^{(3)} M_j\left(\frac{x_{jc}}{r_{jc}^3} - \frac{x_{js}}{r_{js}^3}\right) + \\ &+ A + \sum_{j=1}^{N}{}^{(3)} A_j + \frac{1}{M_c}(F_e + F_m)\end{aligned} \tag{11.5.9}$$

and

$$\begin{aligned}\ddot{x}_{is} = &-G(M_s + M_i)\frac{x_{is}}{r_{is}^3} + G\sum_{j=1}^{N}{}^{(3,i)} M_j\left(\frac{x_{ji}}{r_{ji}^3} - \frac{x_{js}}{r_{js}^3}\right) + \\ &+ GM_c\left(\frac{x_{ci}}{r_{ci}^3} - \frac{x_{cs}}{r_{cs}^3}\right) + T_i\end{aligned} \tag{11.5.10}$$

with the vectors

$$A = G(M_s + M_c)\left(\frac{x_{cs}}{r_{cs}^3} - \frac{M_e}{M_c}\frac{x_{es}}{r_{es}^3} - \frac{M_m}{M_c}\frac{x_{ms}}{r_{ms}^3}\right) , \tag{11.5.11}$$

$$A_j = GM_j\left(\frac{x_{cj}}{r_{cj}^3} - \frac{M_e}{M_c}\frac{x_{ej}}{r_{ej}^3} - \frac{M_m}{M_c}\frac{x_{mj}}{r_{mj}^3}\right) \tag{11.5.12}$$

and

$$T_i = GM_e\left(\frac{x_{ei}}{r_{ei}^3} - \frac{x_{es}}{r_{es}^3}\right) + GM_m\left(\frac{x_{mi}}{r_{mi}^3} - \frac{x_{ms}}{r_{ms}^3}\right) - GM_c\left(\frac{x_{ci}}{r_{ci}^3} - \frac{x_{cs}}{r_{cs}^3}\right) . \tag{11.5.13}$$

The terms A, A_j and T_i result from explicitly introducing the Earth–Moon barycentre instead of the Earth and the Moon being considered separately. Putting

$$\omega_s = \frac{x_{me}x_{cs}}{r_{me}r_{cs}} , \qquad \omega_j = \frac{x_{me}x_{cj}}{r_{me}r_{cj}} \tag{11.5.14}$$

from (11.5.6) and (11.5.7) one has for any natural n

$$\frac{1}{r_{es}^n} = \frac{1}{r_{cs}^n}\sum_{k=0}^{\infty}\left(\frac{M_m}{M_c}\right)^k\left(\frac{r_{me}}{r_{cs}}\right)^k C_k^{n/2}(\omega_s) \tag{11.5.15}$$

and

$$\frac{1}{r_{ms}^n} = \frac{1}{r_{cs}^n}\sum_{k=0}^{\infty}(-1)^k\left(\frac{M_e}{M_c}\right)^k\left(\frac{r_{me}}{r_{cs}}\right)^k C_k^{n/2}(\omega_s) \tag{11.5.16}$$

and similarly for r_{ej} and r_{mj}. $C_k^{n/2}(\omega)$ stand here for Gegenbauer polynomials. Restricting ourselves to the second-degree terms with respect to r_{me}/r_{cs} and r_{me}/r_{cj} and remembering that

$$C_0^{n/2}(\omega)=1\,,\quad C_1^{n/2}(\omega)=n\omega\,,\quad C_2^{n/2}(\omega)=\tfrac{1}{2}n\left[(n+2)\omega^2-1\right] \tag{11.5.17}$$

one obtains the approximate values

$$A = 3G(M_s+M_c)\frac{M_eM_m}{M_c^2}\left(\frac{r_{me}}{r_{cs}}\right)^2\frac{1}{r_{cs}^2}\left[\frac{x_{me}}{r_{me}}\omega_s-\frac{x_{cs}}{r_{cs}}\left(\frac{5}{2}\omega_s^2-\frac{1}{2}\right)\right], \tag{11.5.18}$$

$$A_j = 3GM_j\frac{M_eM_m}{M_c^2}\left(\frac{r_{me}}{r_{cj}}\right)^2\frac{1}{r_{cj}^2}\left[\frac{x_{me}}{r_{me}}\omega_j-\frac{x_{cj}}{r_{cj}}\left(\frac{5}{2}\omega_j^2-\frac{1}{2}\right)\right] \tag{11.5.19}$$

and

$$T_i = 3GM_c\frac{M_eM_m}{M_c^2}\left\{\left(\frac{r_{me}}{r_{cs}}\right)^2\frac{1}{r_{cs}^2}\left[\frac{x_{me}}{r_{me}}\omega_s-\frac{x_{cs}}{r_{cs}}\left(\frac{5}{2}\omega_s^2-\frac{1}{2}\right)\right]-\right.$$
$$\left.-\left(\frac{r_{me}}{r_{ci}}\right)^2\frac{1}{r_{ci}^2}\left[\frac{x_{me}}{r_{me}}\omega_i-\frac{x_{ci}}{r_{ci}}\left(\frac{5}{2}\omega_i^2-\frac{1}{2}\right)\right]\right\}. \tag{11.5.20}$$

As can be seen from numerical estimations, the effects of A_j and T_i are negligibly small. On the contrary, the term A in equation (11.5.9) should be taken into account. In practice one first deals with equations (11.5.9) and (11.5.10) with zero values of A, A_j, T_i, F_e and F_m (the theory of motion of N major planets), then one treats equation (11.5.8) (the theory of motion of the Moon) and afterwards one takes into account the terms neglected earlier.

Neglecting the terms A, A_j, T_i, F_e and F_m, replacing the index c by the number $j=3$ for the Earth–Moon barycentre and designating the heliocentric position vectors x_{is} and distances r_{ij} by $\mathbf{r}_i$ and Δ_{ij}, respectively, one can represent (11.5.9) and (11.5.10) in usual form (3.1.1) or (7.2.1) with force function (6.2.53). In the same manner, neglecting in (11.5.8) the terms F_e and F_m, designating x_{me} by $\mathbf{r}$ and using expansions like (11.5.15) and (11.5.16) in powers of r/r_3 and r/Δ_{3j} one obtains the equations of the geocentric motion of the Moon in the form

$$\ddot{\mathbf{r}} = -GM_c\frac{\mathbf{r}}{r^3} + \frac{GM_s}{r_3^3}\sum_{k=0}^{\infty}\epsilon_k\left(\frac{r}{r_3}\right)^k\left[\mathbf{r}_3\frac{r}{r_3}C_{k+1}^{3/2}(\omega_s) - \mathbf{r}C_k^{3/2}(\omega_s)\right] +$$
$$+G\sum_{j=1}^{N}{}^{(3)}\frac{M_j}{\Delta_{3j}^3}\sum_{k=0}^{\infty}\epsilon_k\left(\frac{r}{\Delta_{3j}}\right)^k\left[(\mathbf{r}_3-\mathbf{r}_j)\frac{r}{\Delta_{3j}}C_{k+1}^{3/2}(\omega_j) - \mathbf{r}C_k^{3/2}(\omega_j)\right] \tag{11.5.21}$$

with

$$\epsilon_k = \left(\frac{M_m}{M_c}\right)^{k+1} + (-1)^k\left(\frac{M_e}{M_c}\right)^{k+1} \tag{11.5.22}$$

and ω_s and ω_j ($j = 1, 2, 4, \ldots, N$) determined by (11.5.14), or in new designations

$$\omega_s = \frac{\mathbf{r}\mathbf{r}_3}{rr_3}, \tag{11.5.23}$$

and

$$\omega_j = \frac{\mathbf{r}(\mathbf{r}_3 - \mathbf{r}_j)}{r\Delta_{3j}}. \tag{11.5.24}$$

With the aid of the force function U the equations of the geocentric lunar motion may be written as (11.1.1) with

$$U = \frac{GM_c}{r} + GM_s\left(\frac{M_c}{M_e}\frac{1}{r_{ms}} + \frac{M_c}{M_m}\frac{1}{r_{es}}\right) +$$
$$+G\sum_{j=1}^{N}{}^{(3)}M_j\left(\frac{M_c}{M_e}\frac{1}{r_{mj}} + \frac{M_c}{M_m}\frac{1}{r_{ej}}\right) \tag{11.5.25}$$

or after expansion in powers of r/r_3 and r/Δ_{3j}

$$U = \frac{GM_c}{r} + \frac{GM_s}{r_3}\sum_{k=2}^{\infty}\epsilon_{k-2}\left(\frac{r}{r_3}\right)^k P_k(\omega_s) +$$
$$+G\sum_{j=1}^{N}{}^{(3)}\frac{M_j}{\Delta_{3j}}\sum_{k=2}^{\infty}\epsilon_{k-2}\left(\frac{r}{\Delta_{3j}}\right)^k P_k(\omega_j), \tag{11.5.26}$$

$P_k(\omega)$ being Legendre polynomials. Introducing now the mean motions n and n_j of the Moon and major planets, respectively,

$$n^2a^3 = GM_c, \tag{11.5.27}$$

$$n_j^2a_j^3 = G(M_s + M_j) \quad (j = 1, 2, \ldots, N) \tag{11.5.28}$$

one has

$$U = n^2a^2\left(\frac{a}{r} + U_s + U_p\right) \tag{11.5.29}$$

with the solar part U_s,

$$U_s = \left(\frac{n_3}{n}\right)^2 \frac{M_s}{M_s+M_c} \sum_{k=2}^{\infty} \epsilon_{k-2}\left(\frac{a}{a_3}\right)^{k-2} \left(\frac{r}{a}\right)^k \left(\frac{a_3}{r_3}\right)^{k+1} P_k(\omega_s) , \quad (11.5.30)$$

and the planetary part U_p,

$$U_p = \sum_{j=1}^{N} {}^{(3)} \left(\frac{n_j}{n}\right)^2 \frac{M_j}{M_s+M_j} \sum_{k=2}^{\infty} \epsilon_{k-2}\left(\frac{a}{a_j}\right)^{k-2} \left(\frac{r}{a}\right)^k \left(\frac{a_j}{\Delta_{3j}}\right)^{k+1} P_k(\omega_j) . \quad (11.5.31)$$

It can be seen that the main small parameters are $(M_j/M_s)(n_j/n)^2$ for U_p and $(n_3/n)^2$ for U_s.

The factors ϵ_k may be eliminated by introducing an additive power parameter. Indeed, one may put in U_s (Bec et al., 1973)

$$\sigma = \frac{a}{a_3}\frac{M_e - M_m}{M_c} , \qquad \rho = \frac{a}{a_3}\frac{M_e M_m}{M_e^2 - M_m^2} , \quad (11.5.32)$$

enabling one to represent $\epsilon_{k-2}(a/a_3)^{k-2}$ by a polynomial in σ and ρ of degree $k-2$. For example,

$$\begin{aligned} \epsilon_0\left(\frac{a}{a_3}\right)^0 &= 1 , & \epsilon_1\left(\frac{a}{a_3}\right)^1 &= -\sigma , \\ \epsilon_2\left(\frac{a}{a_3}\right)^2 &= \sigma^2 + \sigma\rho , & \epsilon_3\left(\frac{a}{a_3}\right)^3 &= -\sigma^3 - 2\sigma^2\rho \end{aligned} \quad (11.5.33)$$

and so on. Needless to say, this modification is not needed if one uses numerical values of the masses. In U_p one may simply put $\epsilon_k = 1$ for any k.

Under transformation (7.2.2), (7.2.3) and (11.1.3), (11.1.4) the equations of motion of the Moon take form (11.1.7) and (11.1.8) with right-hand members (11.1.9). Force function (11.5.29) is expanded without any difficulty in powers of p, q, w, p_j, q_j and w_j by means of the PS and Keplerian processors. Indeed, from (11.5.23) and (11.5.24) it follows that

$$\omega_s = \frac{1}{2}\frac{a}{r}\frac{a_3}{r_3}\Big[(1-p)(1-q_3)\zeta_3 + (1-q)(1-p_3)\zeta_3^{-1} + 2ww_3\Big] \quad (11.5.34)$$

and

$$\begin{aligned} \omega_j = \frac{1}{2}\frac{a}{r}\frac{a_j}{\Delta_{3j}}\Big\{&(1-p)\Big[\frac{a_3}{a_j}(1-q_3)\zeta_3 - (1-q_j)\zeta_j\Big] + \\ &+ (1-q)\Big[\frac{a_3}{a_j}(1-p_3)\zeta_3^{-1} - (1-p_j)\zeta_j^{-1}\Big] + 2w\Big(\frac{a_3}{a_j}w_3 - w_j\Big)\Big\} . \end{aligned} \quad (11.5.35)$$

In version (a) of the GPT ω_s may be represented by a Poisson series with polynomial variables p, q, w, p_3, q_3, w_3 and trigonometric variable $\lambda - \lambda_3$. ω_j

is a Poisson series with polynomial variables p, q, w, p_i, q_i, w_i and trigonometric variables $\lambda - \lambda_i$ ($i = 1, 2, \ldots, N$) provided that the ratios a_3/a_j in expanding ω_j are taken numerically. The Legendre polynomials $P_k(\omega_s)$ and $P_k(\omega_j)$ may be represented in the same form. As a result, the right-hand members will be represented by series like (11.1.17) and (11.1.18) with additive power parameters σ, ρ and a/a_j ($j = 1, 2, \ldots, N$). All further steps repeat the technique given above. The mean motions n and n_j as well as the masses M_j ($j = 1, 2, \ldots, N$) may be taken numerically. One takes as an intermediary the quasi-periodic solution $p = p^{(0)}$, $w = 0$, which involves zero values of the lunar and planetary orbital eccentricities and inclinations. However, in contrast to Hill's variational curve this solution will contain parallactic terms as well as indirect (caused by the terms $p_3^{(0)}$ in U_s) and direct planetary inequalities. This solution has the form of a multiple Fourier series with numerical coefficients. Linear (with respect to the lunar eccentricity and inclination) and nonlinear inequalities are determined just as in the GPT or in a minor-planet theory. The secular system (10.1.8) for the Moon will contain terms due to the secular evolution of the lunar perigee and node as well as of that of the major planets.

The technique outlined here for the representation of the motion of the Moon has been partly realized by Brumberg and Ivanova (1985).

The application of versions (b) and (c) of the GPT to the lunar theory is quite evident. In principle, it may be possible to construct the solution of the main problem of the lunar theory ($p_3 = w_3 = 0$, $M_j = 0$, $j = 1, 2, 4, \ldots, N$) in a closed form using version (b) of the GPT, but because of the slow convergence with respect to n_3/n such a form might not be very efficient in practice.

Anyway, this section demonstrates that the theory of motion of the Moon can be developed in the same form and under the same principles as the theory of motion of the major planets.

References

Abad A. and Deprit A. (1991): Elliptic Function Processor. IAU 21th General Assembly, Buenos-Aires

Abramowitz M. and Stegun I.A. (1965): Handbook of Mathematical Functions. Dover Publs., New York

Abu-El-Ata N. and Chapront J. (1975): Développements analytiques de l'inverse de la distance en Mécanique Céléste. Astron. Astrophys. **38**, 57–66

Aksnes K. (1972): On the Use of Hill Variables in Artificial Satellite Theory: Brouwer's Theory. Astron. Astrophys. **17**, 70–75

Alfriend K.T., Dasenbrock R. and Kaufman B. (1977a): Luni–Solar Perturbations in the Extended Phase Space Representation of the Vinti Problem. IAF 28 Congress, Paper 77–209

Alfriend K.T., Dasenbrock R., Pickard H. and Deprit A. (1977b): The Extended Phase Space Formulation of the Vinti Problem. Celes. Mech. **16**, 441–458

Ash M.E. (1965): Generation of Planetary Ephemerides on an Electronic Computer. Lincoln Lab. Techn. Rep. No. 391, 1–82, M.I.T.

Babaev I.O. and Krasinsky G.A. (1978): Development of the Force Function of the Nonspherical Body in Arbitrary Coordinate System. Bull. ITA **14**, 334–341 (in Russian)

Barton D., Bourne S.R. and Fitch J.P. (1970a): An Algebra System. Computer J. **13**, 32–39

Barton D., Bourne S.R. and Horton J.R. (1970b): The Structure of the Cambridge Algebra System. Computer J. **13**, 243–247

Bec A., Kovalevsky J. and Meyer C. (1973): Contribution to an Analytical Lunar Theory. Moon **8**, 434–442

Bond V.R. and Janin G. (1981): Canonical Orbital Elements in Terms of an Arbitrary Independent Variable. Celes. Mech. **23**, 159–172

Bretagnon P. and Francou G. (1988): Planetary Theories in Rectangular and Spherical Variables. VSOP87 Solutions. Astron. Astrophys. **202**, 309–315

Broucke R. (1969): Perturbations in Rectangular Coordinates by Iteration. Celes. Mech. **1**, 110–126

Broucke R. (1970): How to Assemble a Keplerian Processor. Celes. Mech. **2**, 9–20

Broucke R.A. (1971a): Construction of Rational and Negative Powers of a Series. Comm. Assoc. Comp. Mach. **14**, 32–35

Broucke R. (1971b): Solution of the N-Body Problem with Recurrent Power Series. Celes. Mech. **4**, 110–115

Broucke R. (1973): Ten Subroutines for the Manipulation of Chebyshev Series. Comm. Assoc. Comp. Mach. **16**, 254–256

Broucke R. (1974a): Computation of Solar Perturbations with Poisson Series. Lect. Notes Math. **362**, 237–259

Broucke R. (1974b): A Note on Velocity-Related Series Expansions in the Two-Body Problem. Celes. Mech. **10**, 469–474

Broucke R. (1978): On Pfaff's Equations of Motion in Dynamics: Applications to Satellite Theory. Celes. Mech. **18**, 207–222

Broucke R.A. (1989): A Fortran-Based Poisson Series Processor and Its Applications in Celestial Mechanics. Celes. Mech. **45**, 255–265

Broucke R. and Carthwaite K. (1969): A Programming System for Analytical Series Expansion on a Computer. Celes. Mech. **1**, 271–284

Broucke R. and Cefola P. (1973): A Note on the Relations Between True and Eccentric Anomalies in the Two-Body Problem. Celes. Mech. **7**, 388–389

Broucke R. and Smith G. (1971): Expansion of the Planetary Disturbing Function. Celes. Mech. **4**, 490–499

Brouwer D. and Clemence G.M. (1961): Methods of Celestial Mechanics. Academic Press, New York and London

Brown E.W. and Shook C.A. (1933): Planetary Theory. Cambridge Univ. Press

Brumberg E. (1992): Perturbed Two-Body Motion with Elliptic Functions. Proc. 25th Symposium on Celestial Mechanics (eds. H. Kinoshita and N. Nakai), 139–155, Tokyo

Brumberg E. and Fukushima T. (1994): Expansions of Elliptic Motion Based on Elliptic Function Theory. Celes. Mech. **60**, 69–89

Brumberg V.A. (1963): The Series of Polynomials in the Problem of Three Bodies. Bull. ITA **9**, 234–256 (in Russian)

Brumberg V.A. (1966): Representation of the Coordinates of the Planets by Trigonometric Series. Trans. ITA **11**, 3–88 (in Russian)

Brumberg V.A. (1967): Development of the Perturbative Function in Satellite Problems. Bull. ITA **11**, 73–83 (in Russian)

Brumberg V.A. (1970): Application of Hill's Lunar Method in General Planetary Theory. In: Periodic Orbits, Stability and Resonances (ed. G.E.O. Giacaglia), 410–450, Reidel, Dordrecht

Brumberg V.A. (1974): An Iterative Method of General Planetary Theory. In: The Stability of the Solar System and of Small Stellar Systems (ed. Y. Kozai), 139–155, Reidel, Dordrecht

Brumberg V.A. (1978): Perturbation Theory in Rectangular Coordinates. Celes. Mech. **18**, 319–336

Brumberg V.A. (1991): Essential Relativistic Celestial Mechanics. Hilger, Bristol

Brumberg V.A. (1992): General Planetary Theory Revisited with the Aid of Elliptic Functions. Proc. 25th Symposium on Celestial Mechanics (eds. H. Kinoshita and N. Nakai), 156–171, Tokyo

Brumberg V.A. (1994): General Planetary Theory in Elliptic Functions. Celes. Mech. **59**, 1–36

Brumberg V.A. and Chapront J. (1973): Construction of a General Planetary Theory of the First Order. Celes. Mech. **8**, 335–356

Brumberg V.A. and Ivanova T.V. (1985): On the Solution of the Secular System of the Equations of Motion of the Moon in Trigonometric Form. Bull. ITA **15**, 424–439 (in Russian)

Brumberg V.A., Evdokimova L.S. and Kochina N.G. (1971): Analytical Methods for the Orbits of Artificial Satellites of the Moon. Celes. Mech. **3**, 197–221

Brumberg V.A., Evdokimova L.S. and Skripnichenko V.I. (1975): Quasiperiodic Intermediate Orbits of the Major Planets and Zero-Order Resonances. Astron. Zh. **52**, 420–431 (in Russian) (English translation: Soviet Astronomy **19**, 255-260)

Brumberg V.A., Evdokimova L.S. and Skripnichenko V.I. (1978): Mathematical Results of the General Planetary Theory in Rectangular Coordinates. In: Dynamics of Planets and Satellites and Theories of their Motion (ed. V. Szebechely), 33–48, Reidel, Dordrecht

Brumberg V.A., Tarasevich S.V. and Vasiliev N.N. (1989): Specialized Celestial Mechanics Systems for Symbolic Manipulation. Celes. Mech. **45**, 149–162

Byrd P.F. and Friedman M.D. (1971): Handbook of Elliptic Integrals for Engineers and Scientists. Springer, Berlin, Heidelberg and New York

Campbell J.A. (1972): An Exercise in Symbolic Programming: Computation of General Normalized Inclination Functions. Celes. Mech. **6**, 187–197

Chapront J. (1970): Construction d'une théorie littérale planétaire jusqu'au second ordre des masses. Astron. Astrophys. **7**, 175–203

Chapront J. (1977): Le calcul direct d'une ephéméride de planète. Astron. Astrophys. **61**, 7–11

Chapront J. (1982): The Fourier–Chebyshev Approximation for Time Series with a Great Many Terms. Celes. Mech. **28**, 415–430

Chapront J. (1984): Approximation Methods in Celestial Mechanics. Application to Pluto's Motion. Celes. Mech. **34**, 165–184

Chapront-Touzé M. and Chapront J. (1988): ELP2000-85: A Semi-Analytical Lunar Ephemeris Adequate for Historical Times. Astron. Astrophys. **190**, 342–352

Charlier C.L. (1902, 1907): Die Mechanik des Himmels. **1**, **2**, Verl. von Veit und Co., Leipzig

Cherniack J.R. (1973): A More General System for Poisson Series Manipulation. Celes. Mech. **7**, 107–121

Cherry T.M. (1924): Note on the Employment of Angular Variables in Celestial Mechanics. Monthly Notices RAS **84**, 729

Coffey S. and Deprit A. (1980): Fast Evaluation of Fourier Series. Astron. Astrophys. **81**, 310–315

Cunningham L.E. (1970): On the Computation of the Spherical Harmonic Terms Needed During the Numerical Integration of the Orbital Motion of an Artificial Satellite. Celes. Mech. **2**, 207–216

Dasenbrock R.R. (1973): Algebraic Manipulation by Computer. Naval Research Lab. Report No. 7564, 1–69

Deprit A. (1969): Canonical Transformations Depending on a Small Parameter. Celes. Mech. **1**, 12–30

Deprit A. (1976): Ideal Frames for Perturbed Keplerian Motion. Celes. Mech. **13**, 253–263

Deprit A. (1979): Note on Lagrange's Inversion Formula. Celes. Mech. **20**, 325–327

Deprit A. (1981): The Elimination of the Parallax in Satellite Theory. Celes. Mech. **24**, 111–153

Deprit A. and Elipe A. (1993): Complete Reduction of the Euler–Poinsot Problem. J. Astronaut. Sci. **41**, 603–628

Deprit A. and Ferrer S. (1989): Simplifications in the Theory of Artificial Satellites. J. Astronaut. Sci. **37**, 451–463

Deprit A. and Miller B. (1989): Simplify or Perish. Celes. Mech. **45**, 189–200

Deprit A., Elipe A. and Ferrer S. (1994): Linearization: Laplace vs. Stiefel. Celes. Mech. **58**, 151–201

Deprit A., Pickard H.M. and Poplarchek W. (1978): Compression of Ephemerides by Discrete Chebyshev Approximations. AIAA/AAS Astrodynamics Conf. Paper No. 78–1403, 1–9

Deprit A., Poplarchek W. and Deprit–Bartholomé A. (1975): Compression of Ephemerides. Celes. Mech. **11**, 53–58

Deprit A., Henrard J., Price J.F. and Rom A. (1969): Birkhoff's Normalization. Celes. Mech. **1**, 222–251

Drożyner A. (1977a): Recurrent Calculation of Gravitational Acceleration of a Satellite. Acta Astronomica **27**, 15–22

Drożyner A. (1977b): An Algorithm for Recurrent Calculation of Gravitational Acceleration. Artificial Satellites **12**, 33–39

Duriez L. (1977): Théorie générale planétaire en variables elliptiques. Astron. Astrophys. **54**, 93–112

Duriez L. (1978): Correspondance entre une théorie générale planétaire en variables elliptiques et la théorie classique de Le Verrier. In: Dynamics of Planets and Satellites and Theories of their Motion (ed. V. Szebechely), 15–32, Reidel, Dordrecht

Dziobek D. (1888): Die mathematischen Theorien der Planetenbewegungen. Verl. Joh. Barth, Leipzig

Egorova A.V. (1960): Perturbations in the Motion of an Earth's Artificial Satellite Caused by the Moon, the Sun and the Non-Sphericity of the Earth. Bull. ITA **7**, 815–821 (in Russian)

Erdélyi A. (ed.) (1953): Higher Transcendental Functions. **1**, McGraw Hill, New York

Erdélyi A. (ed.) (1954): Higher Transcendental Functions. **2**, McGraw Hill, New York

Fateman R.J. (1974): On the Multiplication of Poisson Series. Celes. Mech. **10**, 243–247

Feagin T. and Gottlieb R.G. (1971): Generalization of Lagrange's Implication Function Theorem to N-Dimensions. Celes. Mech. **3**, 227–231

Ferrandiz J.M., Ferrer S. and Sein-Echaluce M.L. (1987): Generalized Elliptic Anomalies. Celes. Mech. **40**, 315–328

Fisher D. (1972): Analytic Short-Period Lunar and Solar Perturbations of Artificial Satellites. Celes. Mech. **6**, 447–467

Fukushima T. (1991): Numerical Computation of Elliptic Integrals and Functions. Proc. 24th Symposium on Celestial Mechanics (eds. H. Kinoshita and H. Yoshida), 158-178, Tokyo

Giacaglia G.E.O. (1972): Perturbation Methods in Non-Linear Systems. Springer, New York, Heidelberg, Berlin

Giacaglia G.E.O. (1975): Evaluation of Geopotential and Luni-Solar Perturbations by a Recursive Algorithm. Appl. Mech. Res. Lab. Paper No. 1073, Univ. of Texas at Austin

Giacaglia G.E.O. (1980): Transformations of Spherical Harmonics and Applications to Geodesy and Satellite Theory. Studia geoph. et geod. **24**, 1–11

Gradshteyn I.S. and Ryzhik I.M. (1980): Table of Integrals, Series and Products. Academic Press, New York

Gyldén H. (1893, 1908): Traité analytique des orbites absolues des huit planètes principales. **1**, **2**, Stockholm

Henrard J. (1989): A Survey of Poisson Series Processors. Celes. Mech. **45**, 245–253

Hill G.W. (1905): Integrals of Planetary Motion Suitable for an Indefinite Length of Time. Astron. J. **25**, 1 = Coll. Works **4**, 320, 1907

Hill G.W. (1907): Application of the Delaunay Transformation in the Planetary Theories. Coll. Works **4**, 345

Hill G.W. (1913): Motion of a System of Material Points under the Action of Gravitation. Astron. J. **27**, 171–182

Hori G. (1966): Theory of General Perturbations with Unspecified Canonical Variables. Publ. Astron. Soc. Japan **18**, 287–296

Howland R.A., Jr (1977): An Accelerated Elimination Technique for the Solution of Perturbed Hamiltonian Systems. Celes. Mech. **15**, 327–352

Howland R.A., Jr (1979): An Improved Transformation–Elimination Technique for the Solution of Perturbed Hamiltonian Systems. Celes. Mech. **19**, 95–110

Howland R.A. (1986): Quadratic Analytical Solution of General Systems through Formal Hamiltonization. Celes. Mech. **39**, 329–340

Hughes S. (1981): The Computation of Tables of Hansen Coefficients. Celes. Mech. **25**, 101–107

Izsak I.G. (1963): A Note on Perturbation Theory. Astron. J. **68**, 559–561

Jarnagin M.P. (1965): Expansions in Elliptic Motion. Astron. Papers Am. Ephemeris **18**, 1–659

Jefferys W.H. (1970): A FORTRAN-Based List Processor for Poisson Series. Celes. Mech. **2**, 474–480

Jefferys W.H. (1971): Automated Closed Form Integration of Formulas in Elliptic Motion. Celes. Mech. **3**, 390–394

Jefferys W.H. (1972): A Precompiler for the Formula Manipulation System TRIGMAN. Celes. Mech. **6**, 117–124

Jefferys W.H. and Ries L.M. (1975): Theory of Enceladus and Dione. Astron. J. **80**, 876–884

Jupp A.H. (1975): On Broucke's Velocity-Related Series Expansions in the Two-Body Problem. Celes. Mech. **12**, 513–518

Kaula W.M. (1961): Analysis of Gravitational and Geometric Aspects of Geodetic Utilization of Satellites. Geophys. J. **5**, 104–133

Kaula W.M. (1962): Development of the Lunar and Solar Disturbing Functions for a Close Satellite. Astron. J. **67**, 300

Klioner S.A. (1992): Some Typical Algorithms of the Perturbation Theory within *Mathematica* and Their Analysis. Proc. 25th Symposium on Celestial Mechanics (eds. H. Kinoshita and H. Nakai), 172–182, Tokyo

Knuth D.E. (1973): The Art of Computer Programming. **1–3**, Addison-Wesley

Krasinsky G.A. (1973a): Basic Equations of the Planetary Theory. In: Minor Planets (ed. N.S.Samoilova-Yakhontova), 81–107, Nauka, Moscow (in Russian)

Krasinsky G.A. (1973b): Method for Calculating the Perturbations of Close Satellites Caused by the Asphericity of the Earth. Astron. Zh. **50**, 1076–1084 (in Russian) (English translation: Sov. Astronomy **17**, 680–685, 1974)

Kuzmin A.V. (1980): LRFOR: A System for Operations on Infinite-Precision Integers and Rational Numbers. Algorithms of Celestial Mechanics No. 33, ITA, Leningrad (in Russian)

Lanczos C. (1956): Applied Analysis. Prentice Hall, Tokyo

Laskar J. (1990): Manipulation des séries. In: Modern Methods in Celestial Mechanics (eds. D. Benest et Cl. Froeschlé), 89–107, Editions Frontiéres, Gift-sur-Yvette

LeSchack A.R. and Sconzo P. (1968): FORMAC Language and its Application to Celestial Mechanics. Astron. J. **73**, 217–220

Lieske J.H. (1977): Theory of Motion of Jupiter's Galilean Satellites. Astron. Astrophys. **56**, 333–352

Magnus W. and Oberhettinger F. (1948): Formulas and Theorems for the Special Functions of Mathematical Physics. Chelsea, New York

Markeev A.P. and Sokolsky A.G. (1976): Some Computation Algorithms to Normalize the Hamiltonian Systems. Inst. Appl. Math. Preprint No. 31, Moscow (in Russian)

Markeev A.P. and Sokolsky A.G. (1978): Method for Investigation of Lyapunov's Periodic Motions in Hamiltonian Systems and Its Realization on a Computer. Trans. ITA **17**, 62–68 (in Russian)

Metris G. (1991): Mean Values of Particular Functions in the Elliptic Motion. Celes. Mech. **52**, 79–84

Milani A. and Knežević Z. (1990): Secular Perturbation Theory and Computation of Asteroid Proper Elements. Celes. Mech. **49**, 347–411

Milani A. and Knežević Z. (1992): Asteroid Proper Elements and Secular Resonances. Icarus **98**, 211–232

Musen P. (1963): On a Modification of Hansen's Lunar Theory. J. Geophys. Res. **68**, 1439–1456

Musen P. (1965a): On the High Order Effects in the Methods of Krylov–Bogoliubov and Poincaré. J. Astronaut. Sci. **12**, 129–134

Musen P. (1965b): On the General Perturbations of the Position Vectors of a Planetary System. J. des Observateurs **48**, 11–27

Musen P. and Carpenter L. (1963): On the General Planetary Perturbations in Rectangular Coordinates. J. Geophys. Res. **68**, 2727–2734

Nacozy P. (1977): The Intermediate Anomaly. Celes. Mech. **16**, 309–313

Newcomb S. (1876): General Integrals of Planetary Motion. Smiths. Contrib. to Knowledge **21**, No. 281

Osácar C. and Palacián J. (1994): Decomposition of Functions for Elliptic Orbits. Celes. Mech. **60**, 207–223

Palacián J.F. (1992): Teoría del satélite artificial: Armónicos teserales y su relegación mediante simplificaciones algebraicas. Tesis Doctoral, Universidad de Zaragoza

Palacián J.F. (1994): An Analytical Solution for Artificial Satellites at Low Altitudes. In: Dynamics and Astrometry of Natural and Artificial Celestial Bodies (eds. K. Kurzyńska, F. Barlier, P.K. Seidelmann and I. Wytrzyszczak), 365–370, Astron. Obs. A. Mickiewicz University, Poznań, Poland

Papadakos D.N. (1983): Generalized *F* and *G* Series and Convergence of the Power Series Solution of the *N*-Body Problem. Celes. Mech. **30**, 275–282

Petrovskaya M.S. (1970): Expansions of the Negative Powers of Mutual Distance Between Bodies. Celes. Mech. **3**, 121–128

Poincaré H. (1905): Leçons de mécanique céleste. **1**, Gauthier Villars, Paris

Press W.H., Teukolsky S.A., Vetterling W.T. and Flannery B.P. (1992): Numerical Recipes in FORTRAN: the Art of Scientific Computing. Cambridge Univ. Press, Cambridge

Richardson D.L. (1982): A Third-Order Intermediate Orbit for Planetary Theory. Celes. Mech. **26**, 187–195

Richardson D.L. (1989): PARSEC: An Interactive Poisson Series Processor for Personal Computing Systems. Celes. Mech. **45**, 267–274

Rom A. (1970): Mechanized Algebraic Operations (MAO). Celes. Mech. **1**, 301–319

Rom A. (1971): Echeloned Series Processor (ESP). Celes. Mech. **3**, 331–345

Sack R.A. (1964): Generalization of Laplace's Expansion to Arbitrary Powers and Functions of the Distance between Two Points. J. Math. Phys. **5**, 245–251

Schneider M. (1993): Himmelsmechanik. **2**, Systemmodelle. BI–Wiss., Mannheim, Leipzig, Wien, Zürich

Sconzo P., LeSchack A.R. and Tobey R. (1965): Symbolic Computation of *f* and *g* Series by Computer. Astron. J. **70**, 269–271

Shelus P.J. and Jefferys W.H. (1975): A Note on an Attempt at More Efficient Poisson Series Evaluation. Celes. Mech. **11**, 75–78

Stiefel E. and Scheifele G. (1971): Linear and Regular Celestial Mechanics. Springer, Heidelberg

Stiefel E., Rössler M., Waldvogel J. and Burdet C.A. (1966): Report to NASA on Methods of Regularization for Computing Orbits in Celestial Mechanics. Juris Deuck, Zürich

Stumpff K. (1943–1947): Über die Reihenentwicklung der rechtwinkligen Bahnkoordinaten im Zweikörperproblem. Astron. Nachr. **274**, 49–68 (1943); **275**, 203–222 (1947)

Stumpff K. (1959): Himmelsmechanik. **1**, Das Zweikörperproblem, DVW, Berlin

Stumpff K. (1974): Himmelsmechanik. **3**, Allgemeine Störungen, DVW, Berlin

Whittaker E.T. and Watson G.N. (1935): A Course of Modern Analysis. Cambridge Univ. Press

Williams C.A., Van Flandern T. and Wright E.A. (1987): First Order Planetary Perturbations with Elliptic Functions. Celes. Mech. **40**, 367–391

Yuasa M. and Hori G. (1979): New Approach to the Planetary Theory. In: Dynamics of the Solar System (ed. R.L. Duncombe), 69–72, Reidel, Dordrecht

Yuasa M. and Hori G. (1984): Some Devices Concerning the Development of the Disturbing Function. J. Fac. Sci. Technol., Kinki Univ., No. 20, 29–32

Subject Index